Springer Collected Works in Mathematics

For further volumes:
http://www.springer.com/series/11104

Photo by Barry Evans

Irving Kaplansky

Selected Papers
and Other Writings

Reprint of the 1995 Edition

Irving Kaplansky (1917 – 2006)
Mathematical Sciences Research Institute
Berkeley, CA
USA

ISSN 2194–9875
ISBN 978-1-4614-9452-2   (Softcover)
        978-1-4612-5354-9   (Hardcover)
DOI 10.1007/978-1-4612-5352-5
Springer New York Heidelberg Dordrecht London

Library of Congress Control Number: 2012954381

Mathematical Subject Classification (1991): 01A75, 13-02-16-02

Printed on acid-free paper

Springer is part of Springer Science+Business Media (www.springer.com)

Preface

It is not often that one gets to write a preface to a collection of one's own papers. The most urgent task is to thank the people who made this book possible. That means first of all Hy Bass who, on behalf of Springer-Verlag, approached me about the idea. The late Walter Kaufmann-Bühler was very encouraging; Paulo Ribenboim helped in an important way; and Ina Lindemann saw the project through with tact and skill that I deeply appreciate.

My wishes have been indulged in two ways. First, I was allowed to follow up each selected paper with an afterthought. Back in my student days I became aware of the *Gesammelte Mathematische Werke* of Dedekind, edited by Fricke, Noether, and Ore. I was impressed by the editors' notes that followed most of the papers and found them very useful. A more direct model was furnished by the collected papers of Lars Ahlfors, in which the author himself supplied afterthoughts for each paper or group of papers. These were tough acts to follow, but I hope that some readers will find at least some of my afterthoughts interesting.

Second, I was permitted to add eight previously unpublished items. My model here, to a certain extent, was the charming little book, *A Mathematician's Miscellany* by J. E. Littlewood. In picking these eight I had quite a selection to make —from fourteen loose-leaf notebooks of such writings. Here again I hope that at least some will be found to be of interest.

Ina Lindemann suggested that I should date these additional pieces. Here is my effort. I refer to these writings by the letters A to H, as in the Table of Contents.

A. This arose while I was teaching Chicago's undergraduate course in modern algebra in the fall of 1981.
B. This was noted after I pondered Nagata's splendid theorem for the nth time— some time during the 1970s.
C., E., and F. were written shortly after the papers and book to which they refer.
D. I can only offer a guess: perhaps around 1980.
G. was stimulated by the paper of André Weil to which reference is made. I believe I first read this carefully around 1985.
H. arose in connection with the last time I taught Chicago's graduate course in modern algebra in the spring of 1982. I decided to insert a short interlude on

commutativity theorems. This inspired me to look again at the subject. By the time I was done I had added something new. It was presented at a 1987 Chicago conference honoring Yitz Herstein.

Some biographical details are customary in a collection such as this. There are fragments of automathography (if I may purloin from Halmos a word he coined) in the afterthoughts. If there exists a reader who would like more, I venture to mention the interview in *More Mathematical People*. More autobiography is to appear in an article entitled "Reminiscences" in the proceedings of a 1993 conference celebrating 150 years of mathematics at the University of Toronto.

I have left for the end some important thanks. Each afterthought and additional writing was circulated in preprint form. I was fortunate to receive many helpful comments. I hereby express my gratitude, but not by name: The reviewers are just too numerous.

Irving Kaplansky
Mathematical Sciences Research Institute
March 1994

Contents

---

*Numbers in brackets refer to the Bibliography on pp. ix–xiii.

Contents

Bibliography of the Publications of Irving Kaplansky

1. On a generalization of the "problème des rencontres," *Amer. Math. Monthly* 46(1939), 159–61.
2. Maximal fields with valuations, *Duke Math. J.* 9(1942), 303–21; 3-264.*
3. Some remarks on relatively complete fields, *Bull. Amer. Math. Soc.* 48(1942), 744–47; 4-71 (with O. F. G. Schilling).
4. A characterization of the normal distribution, *Ann. Math. Statist.* 14(1943), 197–98; 4-279.
5. Solution of the "problème des ménages," *Bull. Amer. Math. Soc.* 49(1943), 784–85; 5-86.
6. Symbolic solution of certain problems in permutations, *Bull. Amer. Math. Soc.* 50(1944), 906–14; 6-159.
7. Lucas's tests for Mersenne numbers, *Amer. Math. Monthly* 52(1945), 188–90; 6-254.
8. A contribution to von Neumann's theory of games, Ann. of Math. (2) 46(1945), 474–79; 7-214.
9. The asympototic distribution of runs of consecutive elements, *Ann. Math. Statist.* 16(1945), 200–03; 7-208.
10. A note on groups without isomorphic subgroups, *Bull. Amer. Math. Soc.* 51(1945), 529–30; 7-2.
11. Maximal fields with valuations, II, *Duke Math. J.* 12(1945), 243–48; 7-3.
12. A common error concerning kurtosis, *J. Amer. Statist. Assoc.* 40(1945), 259; 7-20.
13. Note on the preceding paper, *Bull. Amer. Math. Soc.* 51(1945), 437–38; 7-12 (with Harry Pollard).
14. Multiple matching and runs by the symbolic method, *Ann. Math. Statist.* 16(1945), 272–77; 7-309 (with John Riordan).
15. On a problem of Kurosch and Jacobson, *Bull. Amer. Math. Soc.* 52(1946), 496–500; 8-63, 708.
16. Rings with a finite number of primes, I. *Trans. Amer. Math. Soc.* 60(1946), 468–77; 8-434 (with I. S. Cohen).
17. The asymptotic number of Latin rectangles, *Amer. J. Math.* 68(1946), 230–36; 7-407 (with Paul Erdös).
18. Sequences of plus and minus, *Scripta Math.* 12(1946), 73–5; 8-126 (with Paul Erdös).

---

*This indicates vol. 3, p. 264 of *Math. Reviews*. Starting with vol. 20 of *MR*, the review number is used.

19. The problème des ménages, *Scripta Math.* 12(1946), 113–24; 8-365 (with John Riordan).
20. The problem of the rooks and its applications, *Duke Math. J.* 13(1946), 259–68; 7-508 (with John Riordan).
21. Topological rings, *Amer. J. Math.* 69(1947), 153–83; 8-434.
22. Lattices of continuous functions, *Bull. Amer. Math. Soc.* 53(1947) 617–23; 8-587.
23. Semi-automorphisms of rings, *Duke Math. J.* 14(1947), 521–25; 9-172.
24. Topological methods in valuation theory, *Duke Math. J.* 14(1947), 527–41; 9-172.
25. Locally compact rings, *Amer. J. Math.* 70(1948), 447–59; 9-562.
26. Lattices of continuous functions, II, *Amer. J. Math.* 70(1948), 626–34; 10-127.
27. Dual rings, *Ann. of Math.* (2) 49(1948), 689–701; 10-7.
28. Rings with a polynomial identity, *Bull. Amer. Math. Soc.* 54(1948), 575–80; 10-7.
29. Topological rings, *Bull. Amer. Math. Soc.* 54(1948), 809–26; 10-179.
30. Polynomials in topological fields, *Bull. Amer. Math. Soc.* 54(1948), 909–16; 10-280.
31. Regular Banach algebras, *J. Indian Math. Soc.* (N.S.) 12(1948) 57–62; 10-549.
32. Topological representation of algebras, *Trans. Amer. Math. Soc.* 63(1948), 457–81; 10-7 (with Richard F. Arens).
33. Groups with representations of bounded degree, *Canadian J. Math.* 1(1949), 105–12; 10-428.
34. Normed algebras, *Duke Math. J.* 16(1949), 399–418; 11-115.
35. Primary ideals in group algebras, *Proc. Nat. Acad. Sci. U.S.A.* 35(1949), 133–36; 10-428.
36. Elementary divisors and modules, *Trans. Amer. Math. Soc.* 66(1949), 464–91; 11-155.
37. Forms in infinite-dimensional spaces, *Anais Acad. Brasil. Ci.* 22(1950), 1–17; 12-238.
38. Quelques résultats sur les anneaux d'opérateurs, *C. R. Acad Sci. Paris* 231(1950), 485–86; 12-186.
39. The Weierstrass theorem in fields with valuations, *Proc. Amer. Math. Soc.* 1(1950), 356–57; 12-6.
40. Topological representation of algebras, II, *Trans. Amer. Math. Soc.* 68(1950), 62–75; 11-317.
41. Locally compact rings, II, *Amer. J. Math.* 73(1951), 20–4; 12-584.
42. Projections in Banach algebras, *Ann. of Math.* (2) 53(1951), 235–49; 13-48.
43. A theorem on division rings, *Canadian J. Math.* 3(1951), 290–92; 13-101.
44. A theorem on rings of operators, *Pacific J. Math.* 1(1951), 227–32; 14-291.
45. Semi-simple alternative rings, *Portugaliae Math.* 10(1951), 37–50; 13-8.
46. Group algebras in the large, *Tohoku Math. J.* (2) 3(1951), 249–56; 14-58.
47. The structure of certain operator algebras. *Trans. Amer. Math. Soc.* 70(1951), 219–55; 13-48.
48. Rings for which every module is a direct sum of cyclic modules, *Math. Z.* 54(1951), 97–101; 13-202 (with I. S. Cohen).
49. A generalization of Ulm's theorem, *Summa Brasil. Math.* 2(1951), 195–202; 14-128 (with George W. Mackey).
50. Locally compact rings, III, *Amer. J. Math.* 74(1952), 929–35, 14-348.
51. Algebras of type I, *Ann. of Math.* (2) 56(1952), 460–72; 14-291.
52. Representations of separable algebras, *Duke Math. J.* 19(1952), 219–22; 13-619.
53. Orthogonal similarity in infinite-dimensional spaces, *Proc. Amer. Math. Soc.* 3(1952), 16–25; 13-754.
54. Symmetry of Banach algebras, *Proc. Amer. Math. Soc.* 3(1952), 396–99; 14-58.
55. Some results on Abelian groups, *Proc. Nat. Acad. Sci. U.S.A.* 38(1952), 538–40; 14-133.

56. Modules over Dedekind rings and valuation rings, *Trans. Amer. Math. Soc.* 72(1952), 327–40; 13-719.

57. Modules over operator algebras, *Amer. J. Math.* 75(1953), 839–58; 15-327.

58. Products of normal operators, *Duke Math. J.* 20(1953), 257–60; 14-883.

59. Quadratic forms, *J. Math. Soc. Japan* 5(1953), 200–7; 15-500.

60. Completely continuous normal operators with property *L*, *Pacific J. Math.* 3(1953), 721–24; 15-440.

61. Dual modules over a valuation ring, I, *Proc. Amer. Math. Soc.* 4(1953), 213–19; 14-720.

62. Infinite-dimensional quadratic forms admitting composition, *Proc. Amer. Math. Soc.* 4(1953), 956–60; 15-596.

63. *Infinite Abelian Groups*, University of Michigan Press, Ann Arbor, 1954, v + 91 pp.; 16-444.

64. Ring isomorphisms of Banach algebras, *Canadian J. Math.* 6(1954), 374–81; 16-49.

65. Any orthocomplemented complete modular lattice is a continuous geometry, *Ann. of Math.* (2) 61(1955), 524–41; 19-524.

66. *An Introduction to Differential Algebra*, Actualités Sci. Ind., No. 1251 Publ. Inst. Math. Univ. Nancago, No. 5, Hermann, Paris, 1957. 63 pp. 20, no. 177. Second edition, 1976, 57, no. 297.

67. *Problems in the Theory of Rings*. Report of a conference on linear algebras, June 1956, pp. 1–3. National Academy of Sciences, National Research Council, Washington, Publ. 502, v + 60 pp. (1957); 20, no. 3179.

68. Projective modules, *Ann. of Math.* (2) 68(1958), 372–77; 20, no. 6453.

69. On the dimension of modules and algebras, X. A right hereditary ring which is not left hereditary. *Nogoya Math. J.* 13(1958), 85–88; 20, no. 7049.

70. Lie algebras of characteristic *p*. *Trans. Amer. Math. Soc.* 89(1958), 149–83; 20, no. 5799.

71. Functional analysis, Some aspects of analysis and probability, pp. 1–34. *Surveys in Applied Mathematics*, vol. 4. John Wiley and Sons, Inc., New York; Chapman and Hall, Ltd., London; 1958, xi + 243 pp.; 21, 56.

72. *An Introduction to Differential Algebra*, trans. into Russian by G. I. Kleinerman; M. M. Postnikov, ed., Moscow, 1959; 22, no. 2606 (translation of no. 66).

73. *Topological Algebra*, Notas Mat. No. 16 (1959); 21, no. 5907.

74. A characterization of Prüfer rings, *J. Indian Math. Soc.* 24(1961), 279–81; 23, no. A2443.

75. The splitting of modules over integral domains, *Arch. Math.* 13(1962), 341–43; 26, no. 2479.

76. Decomposability of modules, *Proc. Amer. Math. Soc.* 13(1962), 532–35; 25, no. 1187.

77. *R*-sequences and homological dimension, *Nagoya Math. J.* 20(1962), 195–99; 31, no. 251.

78. Lie algebras, *Lectures on Modern Mathematics*, vol. I, pp. 115–32, Wiley, New York, 1963; 31, no. 2355.

79. The homological dimension of a quotient field, *Nagoya Math. J.* 27(1966), 139–42; 33, no. 2664.

80. Composition of binary quadratic forms, *Studia Math.* 31(1968), 523–30; 58, no. 3328.

81. Submodules of quaternion algebras, *Proc. Lon. Math. Soc.* 19(1969), 219–32; 39, no. 1496.

82. Abstract quadratic forms, *Canadian J. of Math.* 21(1969), 1218–33 (with R. J. Shaker); 40, no. 2701.

83. Commutative rings, *Canadian Math. Congress*, 1969. Reprinted as pages 1–12 of Jeffrey-Williams Lectures 1968–1972, *Can. Math. Congress*, 1972.

84. *Infinite Abelian Groups*, 2d ed., Univ. of Michigan Press, Ann Arbor, 1969; 38, no. 2208 (2d ed. of #63).

85. *Rings of Operators*, Benjamin Cummings, New York 1968; 39, no. 6092.

86. *Fields and Rings*, Univ. of Chicago Press, Chicago 1969; 2d ed. 1972; 42, no. 4345 and 50, no. 2139.

87. *Linear Algebra and Geometry, a Second Course*, Allyn and Bacon, 1969; 40, no. 2689. Corrected reprinting, Chelsea, 1974.

88. Fröhlich's local quadratic forms, *J. für reine und angew. Math.* 239–240(1970), 74–77; 41, no. 1644.

89. *Commutative Rings*, Allyn and Bacon, Boston 1970; 40, no. 7234. Corrected reprinting with added notes, Univ. of Chicago Press, Chicago, 1974; 49, no. 10674.

90. "Problems in the theory of rings" revisited, *Amer. Math. Monthly* 77(1970), 445–54; 41, no. 3510.

91. *Lie Algebras and Locally Compact Groups*, Univ. of Chicago Press, Chicago, 1971; 43, no. 2145. Reprinted with corrections and additions, 1974. Russian translation by E. A. Gorina, Moscow, 1974.

92. Algebraic and analytic aspects of operator algebras, *Conference Board of the Mathematical Sciences*, 1970; 47, no. 845.

93. Adjacent prime ideals, *J. of Algebra* 20(1972), 94–97; 47, no. 6668.

94. *Set Theory and Metric Spaces*, Allyn and Bacon, Boston 1972; 46, no. 11795. Corrected reprinting, Chelsea, 1977.

95. A curious property of three-by-three matrices and an application to orders, *Linear Alg. and Its Appl.* 6(1973), 269–72; 47, no. 3417 (with William Butler).

96. Infinite-dimensional Lie algebras, *Scripta Math.* 29(1973), 237–41; 53, no. 8170.

97. Equal sums of sixth powers, *J. of Number Theory* 6(1974), 401–3 (with S. Brudno); 51, no. 8026.

98. *Matters Mathematical*, Harper and Row, New York, 1974 (with I. N. Herstein). Corrected reprintiing, Chelsea, 1978. 58, no. 15763.

99. Infiinite-dimensional Lie algebras II, *Linear and Multilinear Algebra* 3(1975), 61–65 (with R. Kibler); 53, no. 8171.

100. Simple supersymmetries, *J. of Math. Physics* 17(1976), 228–31 (with Peter Freund); 53, no. 7249.

101. Three-dimensional division algebras, *J. of Algebra*, 40(1976), 384–91.

102. Three-dimensional division algebras II, *Houston J. of Math.* 1(1975), 63–79; 55, 5689a, b reviews both this and the preceding paper [101].

103. The Engel-Kolchin theorem revisited, pp. 233–37 in *Contributions to Algebra* (a collection of papers dedicated to Ellis Kolchin), Academic Press, New York 1977; 57, no. 3156.

104. Infinite-dimensional Lie algebras III, *Bull. Inst. Math. Acad. Sinica* 6(1978), 368–77; 58, no. 22192.

105. Composition of quadratic and alternate forms, *C. R. Math. Rep. Acad. Sci. Canada* 1(1979), 87–90; 80e, no. 10024.

106. Five theorems on Abelian groups, pp. 47–51 in *Topics in Algebra*, Proc. of the 18th Summer Research Institute of the Australian Mathematical Society, Springer Lecture Notes no. 697, 1978; 80d, no. 20048.

107. The early work of Saunders Mac Lane on valuations and fields, pp. 519–524 in *Saunders Mac Lane: Selected Papers*, Springer, 1979. (I. Kaplansky was also the general editor of this book, 80k, no. 01064.)

108. Jacobson's lemma revisited, *J. of Algebra* 62(1980), 473–76; 81d, no. 17006.

109. Abraham Adrian Albert, pp. 3–22 in vol. 51 of *NAS Biographical Memoirs*, 1980; 91a; no. 01041.
110. Superalgebras, *Pacific J. of Math.* 86(1980), 93–98; 81j, no. 17006.
111. The Virasoro algebra, *Comm. Math. Phys.* 86(1982), 49–54; 84c, no. 17012.
112. Normal antilinear operators on a Hilbert space, pp. 379–85 in *Contemporary Mathematics*, vol. 13, American Mathematical Society, 1982; 85e, no. 47031.
113. Some simple Lie algebras of characteristic 2, pp. 127–29 in *Lie Algebras and Related Topics*, Lecture Notes in Math., vol. 933, 1982; 83k, no. 17011.
114. Harish-Chandra modules over the Virasoro algebra, pp. 217–31 in *Infinite Dimensional Groups with Applications*, MSRI Publications, vol. 4, Springer-Verlag, 1985 (with L. J. Santharoubane); 87d, no. 17013.
115. A cofinal coloring theorem for partially ordered algebras, *Order* 1(1985), 259–63 (with G. Bergman); 86h, no. 06036.
116. Dual Vector Spaces, pp. 151–161 in *Group Representations, Ergodic Theory, Operator Algebras, and Mathematical Physics*, MSRI Publications, vol. 6, Springer-Verlag, 1987; 88c, no. 01034.
117. Algebras with many derivations, pp. 431–38 in *Aspects of Mathematics and Its Applications*, North-Holland, Amsterdam 1986; 88a, no. 17002.
118. *CCR*-rings, *Pacific J. Math.* 137(1989), 155–57; 90a, no. 16011.
119. Algebraic polar decomposition, *SIAM J. on Matrix Analysis and Applications* 11(1990), 213–17; 91j, no. 15011.
120. Nilpotent elements in Lie algebras, *J. of Algebra* 133(1990), 467–71; 91g, no. 07018.
121. Z_2-graded algebras, *Illinois J. Math* 35(1991), 85–92; 92a, no. 17005.
122. Integers uniquely represented by certain ternary forms, preprint, 11 pp. To appear in a volume honoring Paul Erdös.
123. A quasi-commutative ring that is not neocommutative, preprint, 2 pp. To appear in *Proc. Amer. Math. Soc.*
124. The first nontrivial genus of positive definite ternary forms, preprint, 12 pp. To appear in *Mathematics of Computation*.
125. To Harold Widom on his 60th birthday, preprint, 5 pp. To appear in the proceedings of a conference on Wiener-Hopf and Toeplitz Operators held at UC Santa Cruz, September 20–22, 1992.
126. Ternary positive quadratic forms that represent all odd positive integers, to appear in *Acta Arithmetica*.

Introduction

Irving Kaplansky, "Kap" to those who have had the good fortune to know him personally, has generously responded to the urging of friends and colleagues by offering this selection from his mathematical work. It contains twenty-two published papers, each accompanied by an "Afterthought," plus eight short, previously unpublished pieces.

Viewed through his bibliography, and these selections in particular, Kap's mathematical career conveys the impression of an expedition across a vast mathematical landscape, with frequent revisits to past sojourns. I personally joined this expedition for only a brief passage, as Kap's student during the years 1955–59 at the University of Chicago.

The mathematical atmosphere then in Chicago was exhilirating. Weil, Chern, Zygmund, and Spanier were still there, the legacy of the brilliant leadership of Marshall Stone, himself then an elder statesman. Dieudonné came from Northwestern to attend colloquia. Mac Lane, Kap's own Ph.D. advisor from Harvard days, ran the department like a nurturing, but authoritative, patriarch. Among the instructors were Steve Smale and Dick Swan. The graduate students included Paul Cohen and John Thompson.

Kap is an algebraist, par excellence. He is a master of formulas yet displays few of them, communicating instead mainly with carefully crafted words, terse yet conversational. You will find no pictures or diagrams here. While an accomplished combinatorist, he has a strong taste for the infinite, particularly the threshold between the countable and beyond. (The paper on projective modules [68]—proving that a direct summand of a direct sum of countably generated modules is again such—is a stunning example of this.) His prodigious problem-solving skills, which, at Toronto, achieved first place in the first Putnam Exam—so earning him a fellowship to Harvard—continued throughout his carreer. The paper [2], based on Kap's 1941 thesis, definitively resolves two questions of Krull on maximal valuations rings: conditions for uniqueness of an immediate maximal extension, and its power series structure in the equal characteristic case. And the 1990 paper [119], written while Kap directed the Mathematical Sciences Research Institute, affirms a strengthened formulation of a conjecture of Choudhury and Horn in applied linear algebra, concerning conditions for the existence of an algebraic polar decomposition, $A = QS$, with Q orthogonal and S symmetric.

More than a problem solver, Kap is also a structuralist, exemplifying the axiomatic spirit of Emmy Noether. In several areas (valuation rings, locally compact rings, operator algebras, continuous geometries, rings with polynomial identity, module theory, Lie superalgebras, and so forth) he has developed substantial parts of the general structure theory, often in near-optimum generality.

These qualities, technical virtuosity combined with conceptual elegance and synthesis, are well illustrated by the papers in this volume. But these papers, impressive though they are, do not give the whole picture. Kap has never been a mathematician obsessed with his own work and reputation. As befits the environment in Chicago, he values the broad culture of mathematics and swims eagerly in its intellectual currents (not to mention those of Lake Michigan), often well beyond his own immediate research. He seeks to comprehend major new developments and to transmit them lucidly to students and colleagues.

When I came to Chicago, I first encountered Kap through his little red book on Infinite Abelian Groups [63]. This gem of exposition contains his joint work with Mackey [49], reproduced here, generalizing Ulm's Theorem to mixed modules of rank 1 over a complete discrete valuation ring. One of my first courses with Kap was on Lie algebras and the then-recent solution of Hilbert's fifth problem by Gleason and Montgomery-Zippin. During that period, with the recent publication of *Homological Algebra* by Cartan-Eilenberg and Serre's paper FAC, Grothendieck's monumental refounding of alebraic geometry was getting under way; but it was accessible mainly to insiders in Paris and Cambridge, Massachusetts. This stimulated a major development of commutative homological algebra, pioneered largely by work of Auslander-Buchsbaum and Serre. They had discovered, for example, that smoothness of a point on an algebraic variety is characterized by finiteness of the global homological dimension of its local ring. Moreover, Auslander-Buchsbaum proved, by homological methods, that all such regular local rings are factorial. In his typical audacious manner, Kap undertook offering a course on these new results and methods. This course, and its sequels, introduced a generation of Chicago students to this field. While Kap himself published several important works in this area, including a book [89], and three of the papers [68], [79], and [93] in this volume, these do not convey the full impact of his timely and inspired courses of lectures on students and young colleagues during that period.

I was drawn to Kap by his qualities as a teacher. With Emil Artin, who had directed my freshman calculus course at Princeton, Kap shared the attitude that what matters is not only what you prove, but how you prove it. Proof analysis is a cultivated virtue. Essential ideas, when disarmed of superfluous features, should flow easily, if not with inevitability. And the same proof should not be allowed to masquerade in different guises. Sammy Eilenberg, another of my mentors, is also a strong disciple of this philosophy. It is expressed consistently in all of Kap's writings in this volume.

A modest, but charming, illustration of this attitude can be found in the paper [D] in this volume. In an associative ring R with no nonzero nilpotent ideals, the intersection of the prime ideals is 0. Kap shows how to drop the associativity assumption, and more. Let L be a complete lattice, with least element 0, and equipped with a product, xy, satisfying: $xy \le x \cap y$; $(x \cup y)z = xz \cup yz$; $z(x \cup y) = zx \cup zy$; and, if $u \ne 0$, there is a "finitely generated" $x \ne 0$, $x \le u$. "Finitely generated" here means that if (y_i) is a chain and $x \le \cup y_i$, then $x \le$ some y_i. Call p prime if $xy \le p$ implies x or y is $\le p$. (For the problem at hand, take L to be the lattice of two-sided ideals of R, with xy the ideal generated by the ring-theoretic product.)

THEOREM. *Assume that $x^2 = 0$ only for $x = 0$. Then the intersection of the primes of L is 0.*

Another model for L is te lattice of normal subgroups of a group, with xy the commutator of x and y, whence a bonus application.

At the risk of quickly transcending my technical competence, let me offer some comments on some of the general themes in the papers that follow. A major theme in Kap's early career is topological ring theory. This appears in his thesis on valuations (cf. [2]), continues through [22] ($C(X)$, as a lattice, determines X), [25] and sequels (on the structure of locally compact rings), [39] (a non-Archimedean Weierstrass-Stone theorem), and culminates with an important series of papers, including [42], [44], [47], [58], and [65], on operator algebras, the latter research largely inspired by the work of von Neumann. Kap made some fundamental contributions to the then-emerging theory of C^*-algebras, for example his constantly used Density Theorem [44]. The papers [42] and [47] furnished a starting point for the fundamental later work of Kadison and Glimm on Type I C^*-algebras. The paper [47] contains a number of innovative ideas, including an approximation to sheaf-theoretic constructions on the primitive ideal space, and a stratification of that space using polynomial identities. Moreover, it contains a noncommutative Weierstrass-Stone Theorem for the algebras studied.

The paper [65] is a tour de force of the projection techniques that Kap had developed. Following von Neumann, a continuous geometry is a complete complemented modular lattice L in which the lattice operations are continuous (in a suitable sense). Kap shows that, if one assumes that L is orthocomplemented (there is a canonical involutive complementation operator), then the continuity conditions are automatic.

In a quite different context, von Neumann also inspired the earlier paper [8] in game theory. Kap gives a definitive treatment of an $m \times n$ two-person, zero-sum game, the 2×2 case having been handled by von Neumann himself.

The 1949 groundbreaking paper [28] proves that a primitive ring with a polynomial identity is finite over its center. This is a cornerstone of the theory of PI-rings, which has since blossomed in the hands of Amitsur, Procesi, Artin-Schelter, and others, and has led to the foundations of a type of noncommutative algebraic geometry.

Of the remaining papers that haven't already been mentioned, I note only the following. The paper [93] provides the first counterexample to a question of Krull about preservation of adjacency for contraction of prime ideals in integral extensions. Paper [80] provides an elegant treatment of Gauss' composition of binary quadratic forms, over a Bezout domain. The paper [110], stimulated by the physicist Peter Freund, records Kap's approach to the classification of Lie superalgebras possessing a nondegenerate invariant form. Though the results were superseded by the work of V. Kac, Kap's methods retain some independent interest.

The Other Writings section includes several expository gems. The last of these, "Commutativity Revisited," offers a substantial addition to the large literature on criteria for commutativity of a ring.

To conclude, let me say that it is a great honor for me to be able to introduce this selection of Kap's work, to pay tribute to his impressive mathematical achievements, and to thank him for his inspiration as a teacher and example. And I must add my personal thanks to his wife, Chellie, for her constant support and generosity.

Hyman Bass
New York
September 1994

Reprinted from DUKE MATHEMATICAL JOURNAL
Vol. 9, No. 2, June, 1942

MAXIMAL FIELDS WITH VALUATIONS

BY IRVING KAPLANSKY

1. Introduction. A field with a valuation is said to be *maximal* if it possesses no proper immediate extensions, i.e., if every extension of the field must enlarge either the value group or the residue class field. This definition is due to F. K. Schmidt, but was first published by Krull ([4], p. 191). In the same paper Krull succeeded in proving that any field with a valuation possesses at least one immediate maximal extension and that any field of formal power series is maximal in its natural valuation. These facts led Krull to propound the following two queries.

(1) Is the immediate maximal extension of a field uniquely determined?

(2) If a maximal field K has the same characteristic as its residue class field, is K necessarily a power series field?

These two closely related questions form the central problem of this investigation. The answer to the first is obtained in §3 (Theorem 5), as follows. The immediate maximal extension is always unique if the residue class field has characteristic ∞; but if the latter has characteristic p, then a pair of conditions which we have labelled "hypothesis A" must be satisfied. It is then not difficult to obtain the answer to the second question in §4. In fact, with the same hypothesis, the answer is again affirmative, provided factor sets are admitted in the construction of the power series field (Theorem 6). Granted an additional hypothesis, it is furthermore possible to dispense with factor sets (Theorem 8). In §5, examples are given to show that the conclusions of the preceding theorems may fail if hypothesis A is not fulfilled.

The notion of pseudo-convergence, borrowed from Ostrowski ([9], p. 368), appears to be a natural tool for investigations of maximality, and it is employed consistently throughout the paper. The reason for this is to be found in Theorem 4, which shows that pseudo-convergence provides us with an *intrinsic* characterization of maximality.

2. Pseudo-convergence and maximality. Throughout this section K will always denote a field with a valuation V on an ordered Abelian group Γ, B its valuation ring, and $\Re$ its residue class field.[1]

DEFINITION. A well-ordered set $\{a_\rho\}$ of elements of K, without a last element, is said to be *pseudo-convergent* if

$$(1) \qquad\qquad V(a_\sigma - a_\rho) < V(a_\tau - a_\sigma)$$

Received November 3, 1941; presented to the American Mathematical Society, February 22, 1941. The author wishes to express his thanks to Professor MacLane for his assistance in the preparation of this paper.

[1] For these definitions, cf. [4] and [7].

303

for all $\rho < \sigma < \tau$.[2]

LEMMA 1.[3] *If $\{a_\rho\}$ is pseudo-convergent, then either*
(i) $V(a_\rho) < V(a_\sigma)$ *for all* $\rho < \sigma$, *or*
(ii) $V(a_\rho) = V(a_\sigma)$ *from some point on, i.e., for all* $\rho, \sigma \geqq$ *some ordinal* λ.

Proof. Suppose that (i) does not hold, i.e., that $V(a_\rho) \geqq V(a_\sigma)$ for some $\rho < \sigma$. Then $V(a_\tau)$ must equal $V(a_\sigma)$ for all $\tau > \sigma$. For, if not, we would have

$$V(a_\tau - a_\sigma) = \min [V(a_\sigma), V(a_\tau)] \leqq V(a_\sigma),$$

while $V(a_\sigma - a_\rho) \geqq V(a_\sigma)$, so that the inequality (1) could not possibly hold.

LEMMA 2. *If $\{a_\rho\}$ is pseudo-convergent, then $V(a_\sigma - a_\rho) = V(a_{\rho+1} - a_\rho)$ for all $\rho < \sigma$.*

Proof. We may assume $\sigma > \rho + 1$. From the inequality

$$V(a_{\rho+1} - a_\rho) < V(a_\sigma - a_{\rho+1}),$$

and the identity

$$a_\sigma - a_\rho = (a_\sigma - a_{\rho+1}) + (a_{\rho+1} - a_\rho),$$

we deduce that

$$V(a_\sigma - a_\rho) = \min [V(a_\sigma - a_{\rho+1}), V(a_{\rho+1} - a_\rho)]$$
$$= V(a_{\rho+1} - a_\rho).$$

As a consequence of Lemma 2, we can unambiguously introduce the abbreviation γ_ρ for $V(a_\sigma - a_\rho)$ $(\rho < \sigma)$. We note that $\{\gamma_\rho\}$ is a monotone increasing set of elements of Γ.

DEFINITION.[4] An element x of K is said to be a *limit* of the pseudo-convergent set $\{a_\rho\}$ if $V(x - a_\rho) = \gamma_\rho$ for all ρ.

DEFINITION. The set of all elements y of K such that $Vy > \gamma_\rho$ for all ρ forms an (integral or fractional) ideal in the valuation ring B; this ideal we call the *breadth*[5] of $\{a_\rho\}$.

The limit of a pseudo-convergent set is by no means unique; however, given one limit, it is easy to describe the totality of limits.

LEMMA 3. *Let $\{a_\rho\}$ be pseudo-convergent with breadth $\mathfrak{A}$, and let x be a limit of $\{a_\rho\}$. Then an element is a limit of $\{a_\rho\}$ if and only if it is of the form $x + y$, with $y \in \mathfrak{A}$.*

Proof. If z is any other limit, it follows from

$$x - z = (x - a_\rho) - (z - a_\rho)$$

[2] Cf. [9], p. 368. The inequalities here read in the opposite sense because Ostrowski uses an exponential valuation.
[3] Cf. [9], p. 369.
[4] This definition does not always coincide with Ostrowski's, [9], p. 375.
[5] A translation of "Breite", [9], p. 368.

that $V(x - z) > \gamma_\rho$ for all ρ, whence $x - z$ lies in $\mathfrak{A}$. Conversely, if $y \in \mathfrak{A}$, then

$$Vy > \gamma_\rho = V(x - a_\rho),$$

and so $V(x + y - a_\rho) = \gamma_\rho$, whence $x + y$ is a limit of $\{a_\rho\}$.

Let the field L be an extension of K, with a valuation which is an extension of V. If the value group and residue class field of L coincide with Γ and $\mathfrak{R}$, respectively, we say that L is an *immediate* extension of K. If K admits no proper immediate extensions, K is said to be *maximal*. It will now be our object to prove that maximality is equivalent to the possession of a limit by every pseudo-convergent set; half of this equivalence is obtained in the following theorem.

THEOREM 1. *Let L be an immediate extension of K. Then any element in L but not in K is a limit of a pseudo-convergent set of elements of K, without a limit in K.*

Proof.[6] Let z be an element in L but not in K, and let S denote the totality of values $V(z - a)$, with a in K. Certainly S does not include the symbol ∞. Further, S cannot have a greatest member γ. For, suppose $V(z - g) = \gamma$, $g \in K$; let $c \in K$ have value γ, and let $d \in K$ be a representative of the residue class of $(z - g)/c$. Then $V(z - g - cd) > \gamma$, where $g + cd \in K$, a contradiction.

From the set S select a well-ordered cofinal subset[7] $\{\alpha_\rho\}$; since S has no greatest member, $\{\alpha_\rho\}$ cannot have a last term. Choose elements $a_\rho \in K$ with $V(z - a_\rho) = \alpha_\rho$. The identity

$$a_\sigma - a_\rho = (z - a_\rho) - (z - a_\sigma),$$

together with the inequality

$$V(z - a_\rho) < V(z - a_\sigma) \qquad\qquad (\rho < \sigma),$$

then imply

$$(2) \qquad\qquad V(a_\sigma - a_\rho) = V(z - a_\rho) \qquad\qquad (\rho < \sigma),$$

whence $\{a_\rho\}$ is pseudo-convergent with z as limit.

Suppose that $\{a_\rho\}$ had the further limit z_1 in K. Then, by Lemma 3,

$$V(z - z_1) > V(a_\sigma - a_\rho) \qquad\qquad (\rho < \sigma).$$

Combining this with (2) and using the fact that $\{a_\rho\}$ has no last member, we obtain

$$V(z - z_1) > V(z - a_\rho) = \alpha_\rho$$

for all ρ; and this is a contradiction, since $\{\alpha_\rho\}$ is cofinal in S.

Next, we must show that if some pseudo-convergent set $\{a_\rho\}$ in K lacks a limit, then K is not maximal. This will be accomplished by adjoining to K a

---

[6] It is perhaps worth remarking that Theorem 1 and the preceding lemmas do not depend on the commutativity of either $\mathfrak{R}$ or Γ.

[7] [3], p. 129.

limit of $\{a_\rho\}$ and then proving that the resulting extension is immediate.   Since we shall later be interested in questions of uniqueness, Theorems 2 and 3 will also include some preliminary results on uniqueness.

We borrow from Ostrowski the following two lemmas ([9], p. 371, IV and III). His proofs are readily adapted for the more general case under consideration here.

LEMMA 4.   *Let β_1, $\cdots$, β_m be any elements of an ordered Abelian group Γ, and further let $\{\gamma_\rho\}$ be a well-ordered, monotone increasing set of elements of Γ, without a last element.   Let t_1, $\cdots$, t_m be distinct positive integers.   Then there will exist an ordinal μ and an integer k $(1 \leqq k \leqq m)$ such that*

$$\beta_i + t_i\gamma_\rho > \beta_k + t_k\gamma_\rho$$

for all $i \neq k$ and $\rho > \mu$.

LEMMA 5.   *If $\{a_\rho\}$ is pseudo-convergent in K, and $f(x)$ is a polynomial with coefficients in K, then $\{f(a_\rho)\}$ is ultimately pseudo-convergent.*[8]

By combining Lemmas 1 and 5 we can make a useful deduction concerning the set $\{Vf(a_\rho)\}$, namely that for all sufficiently large ρ and σ either

(3) $$Vf(a_\rho) = Vf(a_\sigma)$$

or

(4) $$Vf(a_\rho) < Vf(a_\sigma) \qquad\qquad (\rho < \sigma)$$

must hold.   The distinction between these two cases will persist throughout the discussion, and for convenience we introduce the following definitions.

DEFINITIONS.   A pseudo-convergent set $\{a_\rho\}$ in K is said to be of *transcendental type* (with respect to K) if (3) holds for every polynomial $f(x)$ with coefficients in K; if, on the other hand, (4) holds for at least one polynomial $f(x)$, we shall say that $\{a_\rho\}$ is of *algebraic type*.

THEOREM 2.   *If there is a pseudo-convergent set $\{a_\rho\}$ of transcendental type in K, without a limit in K, then there exists an immediate transcendental extension $K(z)$ of K.   The valuation of $K(z)$ can be specifically defined as follows: for any polynomial $f(z)$ with coefficients in K we define $Vf(z)$ to be the fixed value which $Vf(a_\rho)$ ultimately assumes.   In the resulting valuation, $K(z)$ is an immediate extension of K, and z is a limit of $\{a_\rho\}$.*

Conversely, if $K(u)$ is any extension of K, with a valuation which is an extension of V such that u is a limit of $\{a_\rho\}$, then $K(u)$ and $K(z)$ are analytically equivalent over K.[9]

---

[8] It is to be noted that Ostrowski's pseudo-convergence need only hold from some point on.

[9] By an analytic equivalence over K we mean a value preserving isomorphism which is the identity on K.

Proof.[10] We must first verify that the above definition actually defines a valuation of $K(z)$, i.e., we must show that

(5) $$V[g(z)h(z)] = Vg(z) + Vh(z)$$

and

(6) $$V[g(z) + h(z)] \geq \min [Vg(z), Vh(z)]$$

for all rational functions $g(z)$ and $h(z)$. But the truth of (5) and (6) follows at once from the truth of the corresponding equations with z replaced by a_ρ.

Next, we wish to show that, with this valuation, $K(z)$ is an immediate extension of K. By definition, $Vf(z) = Vf(a_\rho)$ for large ρ, so there is clearly no extension of the value group. To prove the same for the residue class field, it will suffice to take any polynomial $f(z)$ with $Vf(z) = 0$, and find an element $b \in K$ with $V[f(z) - b] > 0$, for then $f(z)$ and b will lie in the same residue class. Since $\{f(a_\rho)\}$ is ultimately pseudo-convergent, we have

$$V[f(a_\tau) - f(a_\sigma)] > V[f(a_\sigma) - f(a_\rho)] \geq 0$$

for sufficiently large $\rho < \sigma < \tau$. But, by definition,

$$V[f(z) - f(a_\sigma)] = V[f(a_\tau) - f(a_\sigma)]$$

for large τ. Therefore, $V[f(z) - f(a_\sigma)] > 0$ so that $f(z)$ and $f(a_\sigma)$ lie in the same residue class.

To show that z is a limit of $\{a_\rho\}$, we observe that

(7) $$V(z - a_\rho) = V(a_\sigma - a_\rho) \qquad (\rho < \sigma)$$

for large ρ. An application of Lemma 1 to the pseudo-convergent set $\{z - a_\rho\}$ yields that $\{V(z - a_\rho)\}$ is monotone increasing. Then from the identity

$$(z - a_\rho) - (z - a_\sigma) = a_\sigma - a_\rho,$$

we obtain (7) for all ρ.

It remains to prove the final statement of Theorem 2. Regardless of the characteristic of K, it is possible to form a Taylor expansion for a polynomial $f(u)$ of degree m:

(8) $$f(u) - f(a_\rho) = (u - a_\rho)f_1(a_\rho) + \cdots + (u - a_\rho)^m f_m(a_\rho),$$

where $f_k(u)$ may be thought of as replacing the formal expression $f^{(k)}(u)/k!$. (See, for example, [1], p. 165, Ex. 2, or [2].) By hypothesis it is possible to cut into $\{a_\rho\}$ so far that the values of $f\{a_\rho\}, f_1(a_\rho), \cdots, f_m(a_\rho)$ are all independent of ρ. We shall suppose that this has been done, and let us write β_i for $Vf_i(a_\rho)$ $(i = 1, \cdots, m)$. We apply Lemma 4 with $t_i = i$ $(i = 1, \cdots, m)$ and $\gamma_\rho = V(u - a_\rho)$. Since

$$\beta_i + i\gamma_\rho = V[(u - a_\rho)^i f_i(a_\rho)],$$

---

[10] Similar results are proved in [4], p. 194 and [9], p. 374.

it follows that for sufficiently large ρ some one of the terms

$$(u - a_\rho)^i f_i(a_\rho) \qquad\qquad (i = 1, \cdots, m)$$

has smaller value than all the others. This means that the value of the right member of (8) increases monotonically with ρ for large ρ. Since $Vf(a_\rho)$ is fixed, this is possible only if $Vf(u) = Vf(a_\rho)$ for large ρ, which in turn implies that $Vf(u) = Vf(z)$. We have obtained an explicit analytical equivalence over K between the fields $K(u)$ and $K(z)$.

THEOREM 3. *If* $\{a_\rho\}$ *is a pseudo-convergent set of algebraic type in* K, *without a limit in* K, *then there exists an immediate algebraic extension* $K(z)$ *of* K, *which can be explicitly obtained as follows. Among the polynomials* $f(x)$ *for which* (4) *holds, choose one of least degree* n—*say* $q(x)$. *Let* z *be a root of* $q(x) = 0$, *and for any polynomial* $f(z)$ *of degree less than* n, *define* $Vf(z)$ *to be the fixed value which* $Vf(a_\rho)$ *ultimately assumes. In the resulting valuation,* $K(z)$ *is an immediate extension of* K, *and* z *is a limit of* $\{a_\rho\}$.

Conversely, if u *is a root of* $q(x) = 0$, *and if* $K(u)$ *has a valuation which is an extension of* V *such that* u *is a limit of* $\{a_\rho\}$, *then* $K(u)$ *and* $K(z)$ *are analytically equivalent over* K.

Proof. First, it is necessary to remark that the polynomial $q(x)$ is irreducible and of degree $\geqq 2$. For, if $q(x) = b(x - c)$, then $V(c - a_\rho)$ increases monotonically for large ρ. But $\{c - a_\rho\}$ is pseudo-convergent; by Lemma 1, $V(c - a_\rho)$ increases monotonically for all ρ, whence it follows that

$$V(c - a_\rho) = V(a_\sigma - a_\rho) \qquad\qquad (\rho < \sigma),$$

and c is a limit of $\{a_\rho\}$, contrary to hypothesis. Again, if $q(x) = q_1(x)q_2(x)$, where q_1 and q_2 are polynomials of degree less than n, then $V[q_1(a_\rho)q_2(a_\rho)]$ increases monotonically for large ρ; the same must, therefore, hold for either $Vq_1(a_\rho)$ or $Vq_2(a_\rho)$, contradicting the minimal choice of $q(x)$.

The remainder of the proof, with one exception, is a duplication of the proof of Theorem 2, the discussion being, of course, confined to polynomials of degree less than n. The one exceptional point is the proof of the multiplicative character of the valuation of $K(z)$, and this proof we shall now give.

$K(z)$ consists of polynomials in z of degree less than n, with coefficients in K. The product $h(z)$ of two such polynomials $f(z)$ and $g(z)$ is defined by an equation of the form

$$f(z)g(z) = h(z) + k(z)q(z).$$

We have, for all ρ,

(9) $$f(a_\rho)g(a_\rho) - h(a_\rho) = k(a_\rho)q(a_\rho).$$

Now, for large ρ, the value of the right member of (9) increases monotonically, while $V[f(a_\rho)g(a_\rho)]$ and $Vh(a_\rho)$ are fixed for large ρ. This is possible only if

$$Vh(a_\rho) = V[f(a_\rho)g(a_\rho)]$$

for large ρ, whence, by the definition of V on $K(z)$,

$$Vh(z) = Vf(z) + Vg(z),$$

as desired.

Upon combining Theorems 1, 2, and 3, we obtain

THEOREM 4. *A field with a valuation is maximal if and only if it contains a limit for each of its pseudo-convergent sets.*

3. Uniqueness of the maximal extension.

It was proved by Krull ([4], Th. 24, p. 191) that any field with a valuation possesses at least one immediate maximal extension. It is natural to inquire whether this extension is uniquely determined. More precisely, if N and N' are two immediate maximal extensions of K, we ask whether there exists between N and N' an analytical equivalence over K.

It is first of all clear from Theorems 2 and 3 that N or N' can be obtained from K by a transfinite series of adjunctions of limits of pseudo-convergent sets. If we can demonstrate that each of these adjunctions takes place in a unique fashion, we shall have obtained an affirmative answer to the question of uniqueness. In the case of transcendental pseudo-convergent sets, uniqueness is already assured us by Theorem 2. It only remains to examine sets of algebraic type, and here, as will appear, uniqueness can indeed fail.

By the use of Theorem 3, we can reformulate the question as follows. Suppose the pseudo-convergent set $\{a_\rho\}$ is of algebraic type in K, and let $q(x)$ be a polynomial of least degree such that $Vq(a_\rho)$ ultimately increases monotonically. Let N be any immediate maximal extension of K.

(*) Does N contain a limit of $\{a_\rho\}$ which is also a root of $q(x) = 0$?

It is now clear that the answer to the question of the uniqueness of the maximal extension hinges entirely on the answer to the question (*).

We shall adopt the following fixed notation for the discussion. Let the degree of q be n. Let q_i denote the i-th formal derivative of q. Cut into $\{a_\rho\}$ sufficiently far so that $Vq_i(a_\rho)$ $(i = 1, \cdots, n)$ is independent of ρ, equal, say, to β_i.[11] Denote $V(a_\sigma - a_\rho)$ $(\rho < \sigma)$ by γ_ρ. Finally, let $\Re$, the residue class field of K, have characteristic p. We treat explicitly only the case where p is finite, but as a matter of fact the proof can be read equally well for the case $p = \infty$; it is only necessary to replace throughout all powers of p by unity.

First, we prove a simple number-theoretic lemma.

LEMMA 6. *If p is prime, and r is a positive integer prime to p, $r > 1$, then $\binom{p^t r}{p^t}$ is prime to p, for any integer $t \geq 0$.*

Proof.

$$\binom{p^t r}{p^t} = \frac{p^t r(p^t r - 1) \cdots (p^t r - p^t + 1)}{p^t (p^t - 1) \cdots 1}.$$

---

[11] The fact that some of the β's may be infinite does not vitiate any of the arguments.

In the numerator of this fraction, the first factor is divisible by precisely p^t, while the remaining ones are not divisible by p^t. Hence, for every factor m occurring in the numerator, the factor $m - p^t(r - 1)$ which occurs in the denominator will be divisible by p to precisely the same power. This gives the desired result.

LEMMA 7. *If $i = p^t$, $j = p^t r$ with $r > 1$, $(r, p) = 1$, then*

$$\beta_i + i\gamma_\rho < \beta_j + j\gamma_\rho$$

for all sufficiently large ρ.

Proof. We form a Taylor expansion for $q_i(a_\rho)$. In doing so it is necessary to introduce certain binomial coefficients ([2], p. 226).

(10)
$$q_i(a_\sigma) - q_i(a_\rho) = (i + 1)(a_\sigma - a_\rho)q_{i+1}(a_\rho) + \cdots$$
$$+ \binom{j}{i}(a_\sigma - a_\rho)^{j-i}q_j(a_\rho) + \cdots + \binom{n}{i}(a_\sigma - a_\rho)^{n-i}q_n(a_\rho).$$

Consider the right member of (10) with $\rho < \sigma$. By Lemma 4, there will be among these terms precisely one of least value, provided ρ is sufficiently large. The value of this term must then equal the value of the left member of (10), which in turn is not less than β_i. It follows that the term

$$\binom{j}{i}(a_\sigma - a_\rho)^{j-i}q_j(a_\rho),$$

occurring in (10), must also have value not less than β_i. But, by Lemma 6, $\binom{j}{i}$ has value zero. Therefore,

$$\beta_i \leqq (j - i)\gamma_\rho + \beta_j$$

and the result follows at once from the fact that $\{\gamma_\rho\}$ is monotone increasing.

LEMMA 8. *There is an integer h, which is a power of p, such that for all sufficiently large ρ*

(11) $$\beta_i + i\gamma_\rho > \beta_h + h\gamma_\rho \qquad (i \neq h)$$

and

(12) $$Vq(a_\rho) = \beta_h + h\gamma_\rho.$$

Proof. Consider

(13) $$q(a_\sigma) = q(a_\rho) + (a_\sigma - a_\rho)q_1(a_\rho) + \cdots + (a_\sigma - a_\rho)^n q_n(a_\rho)$$

with $\rho < \sigma$. Applying Lemma 4 to the terms on the right of (13) other than $q(a_\rho)$, we find that for large ρ there is precisely one of them, say $(a_\sigma - a_\rho)^h q_h(a_\rho)$, of least value; this proves (11). Now, keeping ρ fixed and varying σ, we observe that $Vq(a_\sigma)$ increases monotonically. This is possible only if

$$Vq(a_\rho) = V[(a_\sigma - a_\rho)^h q_h(a_\rho)]$$
$$= \beta_h + h\gamma_\rho.$$

That h is a power of p is an immediate consequence of Lemma 7.

Throughout the succeeding discussion, we shall reserve the letter h for the integer occurring in Lemma 8.

LEMMA 9. *If y is a limit of $\{a_\rho\}$, then*

$$(14) \qquad Vq(y) > \beta_h + h\gamma_\rho$$

for all ρ, and

$$(15) \qquad Vq_i(y) = \beta_i \qquad\qquad (i = 1, \cdots, n).$$

Proof. We have

$$(16) \qquad q(y) = q(a_\rho) + (y - a_\rho)\, q_1(a_\rho) + \cdots + (y - a_\rho)^n q_n(a_\rho),$$

where, by definition, $V(y - a_\rho) = \gamma_\rho$. By Lemma 8, the terms of least value on the right of (16) are $q(a_\rho)$ and $(y - a_\rho)^h q_h(a_\rho)$. Therefore, $Vq(y) \geqq Vq(a_\rho)$, and, in view of the monotone increasing character of $\{\gamma_\rho\}$, we obtain (14). To prove (15) we form the Taylor expansion

$$(17) \qquad q_i(y) - q_i(a_\rho) = (i + 1)(y - a_\rho)q_{i+1}(a_\rho) + \cdots + \binom{n}{i}(y - a_\rho)^{n-i}q_n(a_\rho).$$

By Lemma 4, the value of the right member of (17) increases for large ρ; hence, $Vq_i(y) = Vq_i(a_\rho)$.

The following result is not required in the present connection, but, for convenience, the proof will be given at this point.

LEMMA 10. *Suppose the value group Γ is Archimedean, and suppose further that the breadth of $\{a_\rho\}$ is not the zero ideal. Let*

$$q^*(x) = \sum_{i=p^u} (x - a_\theta)^i q_i(a_\theta),$$

the summation ranging as indicated only over powers of p. Then for sufficiently large θ, and $\rho > \theta$,

$$(18) \qquad Vq^*(a_\rho) = Vq(a_\rho) = \beta_h + h\gamma_\rho\,.$$

Proof. Suppose i is not a power of p, and let k be the highest power of p dividing i. By Lemma 7, for large ρ,

$$(19) \qquad \beta_i + i\gamma_\rho > \beta_k + k\gamma_\rho\,.$$

The hypothesis that the breadth of $\{a_\rho\}$ is not the zero ideal means that the real numbers γ_ρ approach a finite limit S. From (19), a fortiori,

$$\beta_i + iS > \beta_k + kS.$$

It follows that we can choose a fixed μ so large that

$$(20) \qquad \beta_i + i\gamma_\mu > \beta_k + k\gamma_\rho$$

for all ρ. Thus, for every i not a power of p, there is a corresponding μ such that (20) holds. Let θ be any ordinal exceeding all these μ's. Then

$$(21) \qquad \beta_i + i\gamma_\theta > \beta_k + k\gamma_\rho \geqq \beta_h + h\gamma_\rho$$

holds for every i not a power of p.   Next, we write

$$(22) \qquad q(a_\rho) = \sum_{i=0}^{n} (a_\rho - a_\theta)^i q_i(a_\theta).$$

Since, by Lemma 7, $Vq(a_\rho) = \beta_h + h\gamma_\rho$, it follows from (21) that we can eject from (22) the terms for which i is not a power of p without altering the value of the right hand member; this yields (18) at once.

We are now able to obtain our principal result on the uniqueness of the immediate maximal extension.   In the event that the residue class field $\Re$ has finite characteristic p, the requisite hypothesis is contained in the following two statements:

(1) Any equation of the form

$$x^{p^n} + a_1 x^{p^{n-1}} + \cdots + a_{n-1} x^p + a_n x + a_{n+1} = 0,$$

with coefficients in $\Re$, has a root in $\Re$.[12]

(2) The value group Γ satisfies $\Gamma = p\Gamma$.

For convenience, we shall refer to this pair of conditions as "hypothesis A". If the characteristic of $\Re$ is infinite, we shall further agree that hypothesis A is vacuous.   The proof of the following theorem can then be read for this case in the light of our previous remark that all p-th powers are to be replaced by unity.[13]

THEOREM 5.   *Let the field K have a valuation with value group Γ and residue class field $\Re$, such that $\Re$ and Γ satisfy hypothesis A.   Then the immediate maximal extension of K is uniquely determined up to analytical equivalence over K.*

Proof.   As we remarked above, it follows from Theorems 2 and 3 that we need only prove the following statement: if $\{a_\rho\}$ is pseudo-convergent of algebraic type in K with $q(x)$ for a minimal polynomial, and if N is any immediate maximal extension of K, then N contains a limit of $\{a_\rho\}$ which is also a root of $q(x) = 0$.   This is done by a transfinite approximation, for which purpose it is convenient first to prove the following lemma.   (The symbols γ_ρ, β_i and h are used as defined above.)

LEMMA 11.   *If for some limit $t \in N$ of $\{a_\rho\}$ we have $Vq(t) = \alpha$, we can obtain the better approximation $Vq(t^*) > \alpha$, where $t^* \in N$ is a limit of $\{a_\rho\}$ such that*

$$V(t^* - t) = \max_{i=p^u} (\alpha - \beta_i)/i,$$

i ranging as indicated over the powers of p $(1 \leqq i \leqq n)$.

[12] Concerning this rather unusual hypothesis we wish to remark that it definitely falls short of algebraic closure.   An example is provided by taking the Galois field of p elements $(p > 2)$ and closing it off with respect to extensions of odd degree.

[13] In fact, the whole discussion could be greatly shortened if we were interested in this case only.   In particular, the transfinite approximation in Theorem 5 could be replaced by a simple application of the Hensel-Rychlik theorem.

Proof. Write

(23) $$\delta = \max (\alpha - \beta_i)/i,$$

the range of i being the powers of p, as it will be throughout the proof. Taking $i = h$ in (23) and using Lemma 9, we obtain

(24) $$\delta > (\beta_h + h\gamma_\rho - \beta_h)/h = \gamma_\rho$$

for all ρ. Let $k \epsilon N$ be any element of value δ. (This is possible, as hypothesis A implies $\delta \epsilon \Gamma$.) For any $z \epsilon N$,

(25) $$q(t + kz)/q(t) = \sum_{j=0}^{n} k^j z^j q_j(t)/q(t).$$

In the polynomial (25) the coefficient of z^j has value $j\delta + \beta_j - \alpha$. If j is a power of p, we have

(26) $$j\delta + \beta_j - \alpha \geq 0$$

by (23). If j is not a power of p, and i is the highest power of p dividing j, then

$$j\delta + \beta_j - \alpha > i\delta + \beta_i - \alpha \geq 0$$

by Lemma 7, (24) and (26). Taking these facts together, we observe that if we replace each coefficient in (25) by its residue class, we obtain a polynomial with coefficients in $\Re$, say $\bar{F}(z)$, of precisely the type used in hypothesis A. Hence, $\Re$ contains a root $\bar{z}_1$ of $\bar{F}(z) = 0$. If $z_1 \epsilon N$ is any representative of the residue class $\bar{z}_1$, we then have $V[q(t + kz_1)/q(t)] > 0$ or $Vq(t + kz_1) > \alpha$. Also, by the choice of k, $V(kz_1) = \delta$. By (24), kz_1 lies in the breadth of $\{a_\rho\}$, whence by Lemma 3, $t + kz_1$ is a limit of $\{a_\rho\}$. With the choice of $t^* = t + kz_1$, we have therefore proved Lemma 11.

We now resume the proof of Theorem 5. We are going to select a transfinite set of elements $\{t_\mu\}$ of N such that:

(1) each t_μ is a limit of $\{a_\rho\}$;
(2) if $Vq(t_\mu) = \alpha_\mu$, then $\alpha_\mu < \alpha_\nu$, $(\mu < \nu)$;
(3) $V(t_\nu - t_\mu) = \max (\alpha_\mu - \beta_i)/i$ $(\mu < \nu)$, the range of i again being the powers of p.

Let us first observe that the proof of Theorem 5 can then be easily completed; for, a cardinal number consideration shows that the choice of the t's must terminate with the appearance of an element $t_\zeta \epsilon N$, which is a limit of $\{a_\rho\}$, and for which $Vq(t_\zeta) = \infty$, or $q(t_\zeta) = 0$. For t_1, we choose any limit of $\{a_\rho\}$ in N (there is at least one by Theorem 4), and suppose t_μ has been chosen for all $\mu < \lambda$ so as to satisfy (1), (2), and (3) for $\mu < \nu < \lambda$.

(i) λ a limit number. Then (2) and (3) imply

$$V(t_\nu - t_\mu) < V(t_\theta - t_\nu) \qquad (\mu < \nu < \theta < \lambda),$$

showing that $\{t_\mu\}_{\mu<\lambda}$ is pseudo-convergent. Let $t_\lambda \epsilon N$ be any limit of $\{t_\mu\}_{\mu<\lambda}$.

As an immediate consequence of the definition of limit, we have

$$(27) \qquad\qquad V(t_\lambda - t_\mu) = \max (\alpha_\mu - \beta_i)/i \qquad\qquad (\mu < \lambda).$$

From (27) and Lemma 9, as in (24), we have $V(t_\lambda - t_\mu) > \gamma_\rho$, so that $t_\lambda - t_\mu$ is in the breadth of $\{a_\rho\}$, and t_λ is a limit of $\{a_\rho\}$ by Lemma 3. Finally, we must prove

$$(28) \qquad\qquad\qquad Vq(t_\lambda) > \alpha_\mu \qquad\qquad\qquad (\mu < \lambda).$$

We write

$$(29) \qquad q(t_\lambda) = q(t_\mu) + (t_\lambda - t_\mu)q_1(t_\mu) + \cdots + (t_\lambda - t_\mu)^n q_n(t_\mu).$$

For j a power of p, by Lemma 9,

$$(30) \qquad V[(t_\lambda - t_\mu)^j q_j(t_\mu)] = \beta_j + j \max (\alpha_\mu - \beta_i)/i \geqq \alpha_\mu \, ;$$

while for j not a power of p, (30) follows a fortiori from Lemma 7. Applying these facts to (29), we obtain $Vq(t_\lambda) \geqq \alpha_\mu$, which suffices to prove (28).

(ii) λ not a limit number. Here $t_{\lambda-1}$ is given, and by Lemma 11, we can find a limit t_λ of $\{a_\rho\}$ such that

$$(31) \qquad\qquad\qquad Vq(t_\lambda) > \alpha_{\lambda-1}$$

and

$$(32) \qquad\qquad V(t_\lambda - t_{\lambda-1}) = \max (\alpha_{\lambda-1} - \beta_i)/i.$$

From (31) and (32), respectively, (28) and (27) readily follow. With this the induction is complete.

4. **The structure of maximal fields.** The results obtained in §3 will now enable us to obtain explicit theorems on the structure of maximal fields and their representations as power series fields.

If $\Re$ is any field and Γ is any ordered Abelian group, the set of all formal series

$$\sum a_\rho t^{\alpha_\rho} \qquad\qquad (a_\rho \, \epsilon \, \Re, \, \alpha_\rho \, \epsilon \, \Gamma, \, \{\alpha_\rho\} \text{ well-ordered})$$

form a field, when addition and multiplication are defined in the usual formal fashion. This field we may denote by $\Re(t^\Gamma)$.[14] In $\Re(t^\Gamma)$ we can introduce a valuation V by setting

$$V(\sum a_\rho t^{\alpha_\rho}) = \alpha_1 \qquad\qquad (a_1 \neq 0).$$

Krull has proved that in this valuation $\Re(t^\Gamma)$ is maximal ([4], p. 193).

We now pose the converse query: is a maximal field K, with value group Γ and residue class field $\Re$, analytically isomorphic to $\Re(t^\Gamma)$? As the first step in obtaining such a representation, we must find a subfield M of K which can serve as the coefficient field. Let H denote the homomorphism mapping every

[14] For power series fields cf. [4], [8], [10].

$a \epsilon K$ with $Va \geq 0$ into its residue class, and mapping every $a \epsilon K$ with $Va < 0$ into ∞.[15] Then the property we desire for M is represented by the equation $H(M) = \Re$.

LEMMA 12. *Let K have the same characteristic as its residue class field $\Re$, and suppose K is algebraically perfect, and satisfies the Hensel-Rychlik theorem.[16] Then K possesses a subfield M with $H(M) = \Re$.*

Proof. The proof requires only a slight amplification of Lemma 2, [7]. If P and $\mathfrak{P}$ denote the prime subfields of K and $\Re$, respectively, we necessarily have $H(P) = \mathfrak{P}$. We build up M by successive adjunctions as in MacLane's proof, and the only point that needs further investigation is the case of an inseparable algebraic extension. Suppose then that we have $H(N) = \mathfrak{N}$, and we wish to obtain an extension of N corresponding to $N(\bar{a}^{1/p})$, $\bar{a} \epsilon \mathfrak{N}$. Let $a \epsilon N$ be the representative of $\bar{a}$. By hypothesis $a^{1/p} \epsilon K$, and $N(a^{1/p})$ provides the desired extension.

Next we must obtain a set of elements playing the rôle of the elements $\{t^\alpha\}$ in a power series field. It will, however, in general be necessary to admit a factor set for the multiplication of these elements.

LEMMA 13. *With the same hypothesis as in Lemma 12, and with a fixed choice of the field M, there exists a set $\{t^\alpha\}$ in K, with $Vt^\alpha = \alpha$ for every $\alpha \epsilon \Gamma$, and with*

$$t^\alpha t^\beta = c_{\alpha,\beta} t^{\alpha+\beta} \qquad (c_{\alpha,\beta} \epsilon M),$$

where $c_{\alpha,\beta}$ is a factor set.

Proof. By well-known methods, it is possible to choose a rationally independent basis $\{\zeta_\rho\}$ for Γ, i.e., a set $\{\zeta_\rho\}$ such that every $\alpha \epsilon \Gamma$ has a unique representation as a sum of ζ's with rational coefficients. For t^ζ we choose any element of value ζ; and if α is a sum of ζ's with integral coefficients, we choose for t^α the product of the corresponding elements t^ζ raised to the appropriate powers. Suppose finally that α is of the form

$$\alpha = r_1 \zeta_1 + \cdots + r_m \zeta_m$$

with not all the r_i integral. Let r be the L.C.M. of the denominators of r_1, $\cdots$, r_m; then $t^{r\alpha}$ has already been assigned. We shall now show that it is possible to select an element t^α such that

$$(33) \qquad (t^\alpha)^r = at^{r\alpha} \qquad (a \epsilon M).$$

Since K is perfect, it suffices to take the case $(r, p) = 1$. Let $z \epsilon K$ have value α, and let a be the M-representative of $z^r/t^{r\alpha}$. Then $at^{r\alpha}/z^r$ lies in the same residue class as 1, and, by the Hensel-Rychlik theorem, has an r-th root in K.

---

[15] A detailed statement of the connection between H and V is given in [7].

[16] In [4], p. 178, this theorem is proved on the hypothesis of completeness, a weaker condition than maximality.

Therefore, $at^{r\alpha}$ also has an r-th root, and this is our choice for t^α. For any other $\beta \,\epsilon\, \Gamma$ we will have the similar equations:

$$(t^\beta)^s = bt^{s\beta} \qquad (b \,\epsilon\, M),$$ (34)

$$(t^{\alpha+\beta})^u \, ct^{u(\alpha+\beta)} \qquad (c \,\epsilon\, M).$$ (35)

From (33), (34), and (35) we obtain

$$(t^\alpha t^\beta / t^{\alpha+\beta})^{rsu} = a^{su}b^{ru}/c^{rs} \,\epsilon\, M.$$

Since M is a coefficient field, it follows from this that $t^\alpha t^\beta / t^{\alpha+\beta} \,\epsilon\, M$, as desired.

Before we can obtain our structure theorem, it will be necessary to prove the following result.

LEMMA 14. *Let N be a maximal field of characteristic p, with value group Γ, and residue class field $\Re$, and suppose $\Re$ is perfect and $\Gamma = p\Gamma$. Then N is perfect.*

Proof. Suppose that, on the contrary, $a \,\epsilon\, N$ has no p-th root in N. We construct the extension $M = N(a^{1/p})$. Because N is complete in Krull's sense, the valuation of N extends to M in the following unique manner.[17] Any b in M but not in N will satisfy an irreducible equation of the form

$$x^p + c_1 x^{p-1} + \cdots + c_p = 0 \qquad (c_i \,\epsilon\, N)$$

and to b we assign the value $V(c_p)/p$. Plainly this involves no extension of Γ. Furthermore, since any element in M is a p-th root of some element in N, it follows readily that there is no extension of $\Re$. Hence M is an immediate extension of N, contrary to the hypothesis that N is maximal.

The corresponding lemma for the characteristic unequal case will also be needed, but here a stronger hypothesis must be made.

LEMMA 15. *Let N be maximal and of infinite characteristic, while the residue class field $\Re$ has characteristic p, and suppose Γ and $\Re$ satisfy hypothesis A (cf. Theorem 5). Then every element of N has a p-th root in N.*

Proof. We employ a transfinite approximation. Since this is carried out in virtually the same fashion as in Theorem 5, we shall here merely summarize the method.

First, it suffices to prove $a^{1/p} \,\epsilon\, N$ for $Va = 0$. We make an inductive choice of elements $\{t_\rho\}$, with $V(t_\rho^p - a) = \alpha_\rho$, such that

$$\alpha_\rho < \alpha_\sigma \qquad (\rho < \sigma),$$

$$V(t_\sigma - t_\rho) = \max(\alpha_\rho/p, \, \alpha_\rho - Vp) \qquad (\rho < \sigma).$$

For t_1 we choose any element of N with $V(t_1^p - a) > 0$. (There is such an element, since hypothesis A implies that the residue class of a has a p-th root.) Suppose we have chosen t_ρ for all $\rho < \lambda$. If λ is not a limit number, it follows just as in Lemma 11 that there exists t_λ with $V(t_\lambda^p - a) > \alpha_{\lambda-1}$, and $V(t_\lambda - t_{\lambda-1}) = \max(\alpha_{\lambda-1}/p, \, \alpha_{\lambda-1} - Vp)$. If λ is a limit number, then $\{t_\rho\}_{\rho<\lambda}$ is pseudo-con-

---

[17] [4], p. 180.

vergent; for t_λ we then choose any limit of $\{t_\rho\}_{\rho<\lambda}$. When the approximation terminates, we obtain a p-th root of a in N, as desired.

We are now able to prove our first structure theorem. For brevity let us denote by $\mathfrak{K}(t^\Gamma, c_{\alpha,\beta})$ the power series field in which multiplication takes place according to the rule: $t^\alpha t^\beta = c_{\alpha,\beta} t^{\alpha+\beta}(c_{\alpha,\beta} \in \mathfrak{K})$.

THEOREM 6. *Let the maximal field N, with value group Γ and residue class field $\mathfrak{K}$, have the same characteristic as K; and suppose $\mathfrak{K}$ and Γ satisfy hypothesis A. Then N is analytically isomorphic to a power series field $\mathfrak{K}(t^\Gamma, c_{\alpha,\beta})$.*

Proof. By Lemma 14, N is algebraically perfect. We are then able to apply Lemmas 12 and 13, obtaining a coefficient field M which we may identify with $\mathfrak{K}$ and a set of representatives $\{u^\alpha\}$ with

$$u^\alpha u^\beta = c_{\alpha,\beta} u^{\alpha+\beta} \qquad\qquad (c_{\alpha,\beta} \in \mathfrak{K}).$$

Form the subfield K of N obtained by adjoining to $\mathfrak{K}$ all the elements u^α, and let K' denote the analogous subfield of $\mathfrak{K}(t^\Gamma, c_{\alpha,\beta})$, i.e., the field obtained by adjoining to $\mathfrak{K}$ all elements t^α. Let T be the natural map of K' on K, i.e., T is the identity on $\mathfrak{K}$ and $Tt^\alpha = u^\alpha$. Plainly T is a homomorphism; but, moreover, T preserves values, and so must actually be an analytic isomorphism. Now N and $\mathfrak{K}(t^\Gamma, c_{\alpha,\beta})$ are immediate maximal extensions of K and K', respectively. By Theorem 5 and our hypothesis, N and $\mathfrak{K}(t^\Gamma, c_{\alpha,\beta})$ are analytically isomorphic.

It is natural to inquire in what circumstances the factor set occurring in Theorem 6 can be dispensed with. For this purpose we require an extension theorem of wider scope than Theorem 5. The investigation also yields a uniqueness theorem for the characteristic unequal case, as will appear in Theorem 8.

THEOREM 7.[18] *Let the field K have value group Γ and residue class field $\mathfrak{K}$, and let the two maximal extensions L and L' of K have value group Δ and residue class field $\mathfrak{L}$, which may be proper extensions of Γ and $\mathfrak{K}$. Then if Δ and $\mathfrak{L}$ satisfy hypothesis A, and if every element of $\mathfrak{L}$ has an n-th root in $\mathfrak{L}$ for all n, L and L' are analytically equivalent over K.*

Proof. It will suffice to prove that L and L' contain analytically equivalent subfields N and N' with value group Δ and residue class field $\mathfrak{L}$; for then L and L' are analytically equivalent by Theorem 5. We will build up N and N' through a transfinite succession of fields paralleling adjunctions in $\mathfrak{L}/\mathfrak{K}$ and Δ/Γ, and it suffices to consider the case of a single adjunction.

Residue class field adjunction. A transcendental or separable algebraic extension is handled exactly as in [6], Theorem 3; to treat an inseparable algebraic extension, we observe that, by hypothesis and Lemmas 14 and 15, every element in L has a p-th root in L; so we are again able to use MacLane's method.

Value group adjunction. Consider an extension $\Gamma(\alpha)$ of Γ. If α is rationally

---

[18] This is the non-discrete analogue of [6], Theorem 3, and [5].

independent of Γ, simply let $a \in L$, $a' \in L'$ be any elements of value α. Then for any polynomial $f(x) = \sum c_i x^i$ with coefficients in K we have

$$(36) \qquad\qquad Vf(a) = Vf(a') = \min V(c_i a^i),$$

showing that $K(a)$ and $K(a')$ are analytically equivalent. If α is rationally dependent on Γ, we have $n\alpha = \beta$ for some $\beta \in \Gamma$. Let $b \in K$ have value β; we wish to show that b has an n-th root in L. Using Lemmas 14 and 15 we reduce the consideration to the case $(n, p) = 1$. Let $z \in L$ have value α. By hypothesis, the residue class of b/z^n has an n-th root in L. By the Hensel-Rychlik theorem, b/z^n has an n-th root in L, whence b has an n-th root, say a, in L. Likewise b has an n-th root a' in L'. Then (36) holds for polynomials of degree less than n, showing again that $K(a)$ and $K(a')$ are analytically equivalent.

We now obtain our second structure theorem.

Theorem 8. *Let the maximal field K have value group Γ and residue class field $\Re$, and suppose that Γ and $\Re$ satisfy hypothesis A, and that every element of $\Re$ has an n-th root in $\Re$, for all n. Then K is uniquely determined, up to analytic isomorphism by $\Re$, Γ, its characteristic, and in the characteristic unequal case, Vp.*

Proof. Let P be the prime subfield of K. The valuation of P is uniquely determined by the given data, up to analytic isomorphism. Theorem 8 then follows from Theorem 7.

Corollary. *In the equal characteristic case, every field with a valuation is analytically isomorphic to a subfield of a suitable power series field.*

5. **Counter-examples.** We will show by an example that without hypothesis A the conclusion of Theorem 5 may fail.[19]

Let $\Re$ be a field of characteristic p, and let Γ be the additive group of all rational numbers. Let K be the subfield of the power series field $\Re(t^\Gamma)$ obtained by adjoining to $\Re$ all the elements t^α; K is then the field of all quotients of linear expressions in the t's, with coefficients in $\Re$. Consider the pseudo-convergent sequence $\{a_i\}$:

$$a_i = t^{-1/2} + t^{-1/4} + \cdots + t^{-1/2^i}.$$

We wish first to show that $\{a_i\}$ is of transcendental type in K. Now the breadth of $\{a_i\}$ is in fact precisely the valuation ring of K; at any rate, it is not the zero ideal. If $\{a_i\}$ were of algebraic type in K, then by Lemma 10 there would exist elements $c_j \in K$ ($j = 0, \cdots, n + 1$) such that the value of

$$z_i = c_0 a_i^{p^n} + c_1 a_i^{p^{n-1}} + \cdots + c_{n-1} a_i^p + c_n a_i + c_{n+1}$$

increases monotonically for large i. We can suppose without loss of generality that $Vc_j \geqq 0$ ($0 \leqq j \leqq n$), $Vc_j = 0$ for at least one j in the same range, and

---

[19] The principle upon which this example is constructed is the same as in the first of the counter-examples of [7].

further that all the c's are actually polynomials in the t's. We now imagine z_i multiplied out in full, and consider the portion of z_i consisting of terms t^α with $\alpha < 0$; call this portion w_i. Suppose c_j begins with the term $d_j \in \Re$ (d_j may be zero); then the contribution of $c_j a_i^{p^{n-j}}$ to w_i will consist of $d_j a_i^{p^{n-j}}$ together with some other terms, the latter being fixed for large i. Moreover, for different j's, the contributions $d_j a_i^{p^{n-j}}$ are elementwise distinct, at any rate if $p \neq 2$. It must then be the case that, for large i, w_i will once and for all contain a fixed term of least value, and this statement is incompatible with the previous assertion that Vz_i increases monotonically for large i.

Now suppose that hypothesis A is violated because the equation

$$(37) \qquad g(x) = x^{p^n} + b_1 x^{p^{n-1}} + \cdots + b_n x = b,$$

with coefficients in $\Re$, has no root in $\Re$. The formal series

$$a = t^{-1/2} + t^{-1/4} + t^{-1/8} + \cdots$$

is a limit of $\{a_i\}$, and likewise $g(a)$ is a limit of $\{g(a_i)\}$, which along with $\{a_i\}$ is of transcendental type in K. By Lemma 3, $g(a) + b$ is also a limit of $\{g(a_i)\}$. Hence, by Theorem 2, $g(a)$ and $g(a) + b$ are both transcendental over K and the mapping $g(a) \to g(a) + b$ provides an analytic automorphism of the field $K[g(a)] = L$, say. The adjunction of a to L, which is plainly immediate, can be paralleled by the corresponding adjunction of a root a' of the equation

$$(38) \qquad\qquad g(x) = g(a) + b.$$

Let N and N' be any immediate maximal extensions of $L(a)$ and $L(a')$, respectively. Then there cannot exist any analytic isomorphism between N and N' which leaves L elementwise fixed. For, if there were such an isomorphism, then N', like N, would contain a root of $g(x) = g(a)$. But N' already contains a root of (38). Hence N' would contain a root of (37); any such root would necessarily have value zero, and, taking residue classes, we would obtain a root of (37) in $\Re$, contrary to hypothesis.

We have thus shown that the violation of hypothesis A may entail the existence of inequivalent maximal extensions. But this example still leaves another question unanswered, for it might nevertheless be true that all the immediate maximal extensions of the above field K are analytically isomorphic to $\Re(t^\Gamma)$. (This actually occurs in the discrete finite rank case; MacLane [7] gives examples of non-unique extensions, but Schilling [10] has proved that all such fields are power series fields.) However, in the non-discrete case which we are considering, uniqueness fails even in this broader sense. To show this, a somewhat more complicated example is needed.

We shall use the same notation as in the preceding example and in addition the abbreviations $p^n = q$ and $t^{1/2^n p^n} = w_n$. Suppose that hypothesis A is violated, this time by the fact that the element $b \in \Re$ has no p-th root in $\Re$. We adjoin to the field K the following elements in turn: a, $(a + bt)^{1/p} = u_1$,

$(u_1 + bw_1)^{1/p} = u_2, \cdots, (u_n + bw_n)^{1/p} = u_{n+1}$, etc. To assign a valuation to these extensions, we argue as before. First, we find that

$$(39) \qquad u_n^q = a + bt + b^p t^{1/2} + \cdots + b^{q/p} t^{1/2^{n-1}}.$$

Since a is a limit of $\{a_i\}$, so is u_n^q by Lemma 3. Hence, u_n is a limit of $\{a_i^{1/q}\}$ and, again by Lemma 3, so is $u_n + bw_n$. Also $\{a_i^{1/q}\}$, along with $\{a_i\}$, is of transcendental type in K. By Theorem 2, the mapping $u_n + bw_n \rightarrow a^{1/q}$ provides an analytical equivalence over K between the fields $K(u_n)$ and $K(a^{1/q})$. Then the extension $K(u_{n+1})$ of $K(u_n)$ can be given a valuation paralleling that of $K(a^{1/pq})$, and in this valuation $K(u_{n+1})$ is an immediate extension of $K(u_n)$.

Let N be any immediate maximal extension of the field $K(a, u_1, u_2, \cdots)$. We shall prove that N is not analytically isomorphic to a power series field.

First we need the following elementary observation. If in a power series field M of characteristic p we have a pseudo-convergent set $\{a_\rho\}$ such that each a_ρ has a p^m-th root in M, then M contains some limit of $\{a_\rho\}$ which also has a p^m-th root in M. To prove this it suffices to construct the power series y which agrees with a_ρ for all terms of value less than $V(a_{\rho+1} - a_\rho)$. Then y is a limit of $\{a_\rho\}$, and, since p^m-th root extraction goes termwise, y has a p^m-th root in M.

Now, in our case, a_i has a p^m-th root in N for all m. If N were a power series field, it would, therefore, contain a limit z of $\{a_i\}$, with a p^m-th root in N for all m. By Lemma 3, $V(z - a) \geqq 0$. Write $z = a + c + z_1$, where $c \epsilon \Re$, $Vz_1 > 0$, say, for definiteness, $Vz_1 > 1/2^{n-1}$. Now $a + c + z_1$ is a p^n-th (or q-th) power in N; together with (39) this implies that $c - b^{q/p} t^{1/2^{n-1}} +$ terms of higher value is a q-th power in N. This means that the residue class of c has a q-th root, whence, since $\Re$ is a coefficient field, c has a q-th root in $\Re$. Subtracting c, and repeating the argument, we obtain that $b^{q/p}$ has a q-th root in $\Re$, i.e., b has a p-th root in $\Re$, contrary to our initial assumption.

Remark. This example is easily duplicated if, instead of $\Re$, it is Γ that is imperfect, i.e., if $\Gamma \neq p\Gamma$. But the author has not succeeded in constructing a counter-example on the assumption that the general equation (37) lacks a root. Thus the possibility remains open that a weaker condition than hypothesis A will suffice to ensure that a maximal field in the equal characteristic case is a power series.

BIBLIOGRAPHY

1. A. A. ALBERT, *Modern Higher Algebra*, Chicago, 1937.
2. H. HASSE AND F. K. SCHMIDT, *Noch eine Begründung der Theorie der höheren Differentialquotienten*, Journal für Mathematik, vol. 177(1937), pp. 215–237.
3. F. HAUSDORFF, *Mengenlehre*, first edition, Berlin, 1914.
4. W. KRULL, *Allgemeine Bewertungstheorie*, Journal für Mathematik, vol. 167(1932), pp. 160–196.
5. S. MACLANE, *Note on the relative structure of p-adic fields*, Annals of Mathematics, vol. 41(1940), pp. 751–753.

6. S. MacLane, *Subfields and automorphism groups of p-adic fields*, Annals of Mathematics, vol. 40(1939), pp. 423–442.
7. S. MacLane, *The uniqueness of the power series representation of certain fields with valuations*, Annals of Mathematics, vol. 39(1938), pp. 370–382.
8. S. MacLane, *The universality of formal power series fields*, Bulletin of the American Mathematical Society, vol. 45(1939), pp. 888–890.
9. A. Ostrowski, *Untersuchungen zur arithmetischen Theorie der Körper*, Mathematische Zeitschrift, vol. 39(1935), pp. 269–404.
10. O. F. G. Schilling, *Arithmetic in fields of formal power series in several variables*, Annals of Mathematics, vol. 38(1937), pp. 551–576.

Harvard University.

Afterthought

This paper is a substantial part of my 1941 Harvard thesis, written under Saunders Mac Lane. For some background on how it came about, see [5].

The main result found an application in the work of Ax and Kochen [1], but in later simplified versions of their work it became irrelevant. More recent relevant papers are [2] and [6].

In this afterthought I shall concentrate on Hypothesis A, or more precisely, the portion of Hypothesis A referring to the residue class field. At the time, as I more or less confessed, I found this hypothesis puzzling. Things became a lot clearer when Whaples [7] proved the following theorem. (I adopt Whaples' terminology: Over a field of characteristic p a p-polynomial is one of the form

$$a_0 x^{p^n} + a_1 x^{p^{n-1}} + \cdots + a_{n-1} x^p + a_n x + a_{n+1}.)$$

THEOREM. *Let K be a field of characteristic p. Then the following statements are equivalent*: (a) *Every p-polynomial has a root in K, and* (b) *K has no extensions with degree divisible by p.*

The proof of Whaples used cohomology of groups. It is reported in [4, p. 51] that Delon found a simple proof, which apparently has not been published.

Proof. In the direction (b) → (a) the proof is quite easy. Let f be a p-polynomial with coefficients in K. We are to show that f has a root in K. Write $f = g + c$, where c is the constant term of f. Then g is an additive polynomial. By our hypothesis, the irreducible factors of f have degrees prime to p. Let h, of degree r, be an irreducible factor of f. Fix a root t of h (in a possibly larger field). The roots of f have the form $t + a_i$ where the a_is are the roots of g. Note that the roots of g form an additive group. The sum of the roots of h lies in K. This gives us $rt + b \varepsilon K$, where b is a sum of a subset of the a_is and is therefore a root of g. Likewise, $r^{-1}b$ is a root of g (note that the integer r is prime to p and therefore invertible iin K). Then $t + r^{-1}b$ is a root of f, and it lies in K, as required.

The following proof of (a) → (b) was shown to me by David Leep. Statement (a) implies that K has no inseparable extensions. It will therefore suffice to take a normal extension L of K with degree n, and show that p does not divide n. By the normal basis theorem there is a vector space basis $c = c_1, \ldots, c_n$ of L over K, with the cs the roots of an irreducible polynomial q over K. The elements

$$1, c, c^p, c^{p^2}, \ldots, c^{p^{n-1}}$$

are linearly dependent over K since $[L:K] = n$. Therefore there exist elements $d_0, d_1, \ldots, d_{n-1}, e$ in K such that the p-polynomial

$$s(x) = d_{n-1} x^{p^{n-1}} + \cdots + d_1 x^p + d_0 x + e$$

has c as a root. It follows that all the c_is are roots of s. Therefore the elements $c_2 - c, \ldots, c_n - c$ are roots of $s(x) - e = 0$. Since these $n - 1$ roots are clearly linearly independent over K, they are a fortiori linearly independent over the prime field of integers mod p. This implies that the additive group G generated by the elements $c_2 - c, \ldots, c_n - c$ contains p^{n-1} distinct elements, which are precisely the roots of $s(x) - e$. The roots of $s(x)$ are then given by $G + c$. By hypothesis one of these roots lies in K; call it M. There exist integers $y_2, \ldots, y_n$ such that

$$M = y_2(c_2 - c) + \cdots + y_n(c_n - c) + c.$$

In this equation take the trace from L to K. When we note that $c, c_2, \ldots, c_n$ all have the same trace and that this trace is nonzero, we get $nM \neq 0$. Hence n is not divisible by p.

In conclusion I offer two additional remarks:

1. The following excellent point is made in [4]. Now that Hypothesis A has been recast in a more civilized form, should there not be a more civilized discussion of immediate maximal extensions that bypasses pseudo-convergence? There is, and they do it.

2. There is a very close connection between maximality of fields with valuations and the concept of injectivity. This has been noted repeatedly in the literature. One consequence is that the existence of immediate maximal extensions, proved by Krull [3] by using a "power series" representation with no algebraic properties, is a direct consequence of the existence and uniqueness of injective envelopes. In this connection I note that if one has an a priori argument establishing a cardinal number bound on essential extensions of a module A, then one quickly sees that A has an injective envelope. But I do not know such an argument.

References

1. J. Ax and S. Kochen, Diophantine problems over local fields I, *Amer. J. of Math.* 87(1965), 605–30.
2. R. Brown, Automorphisms and isomorphisms of real Henselian fields, *Trans. Amer. Math. Soc.* 307(1988), 675–703.
3. W. Krull, Allgemeine Bewertungstheorie, *J. reine angew. Math.* 167(1931), 160–96.
4. F.-V. Kuhlmann, M. Pank, and P. Roquette, Immediate and purely wild extensions of valued fields, *Manuscripta Math.* 55(1986), 39–67.
5. S. Mac Lane, *Selected Papers*, pp. 519–24, Springer, New York, 1979.
6. S. Warner, Nonuniqueness of immediate maximal extensions of a valuation, *Math. Scand.* 56(1985), 191–202.
7. G. Whaples, Galois cohomology of additive polynomials and n-th power mappings of fields, *Duke Math. J.* 24(1957), 143–50.

SOLUTION OF THE "PROBLÈME DES MÉNAGES"

IRVING KAPLANSKY

The *problème des ménages* asks for the number of ways of seating n husbands and n wives at a circular table, men alternating with women, so that no husband sits next to his wife. Despite the considerable literature devoted to this problem (cf. the appended bibliography), the following simple solution seems to have been missed.

It is convenient first to solve two preliminary problems, perhaps of some interest in themselves.

LEMMA 1. *The number of ways of selecting k objects, no two consecutive, from n objects arrayed in a row is $_{n-k+1}C_k$.*

Let $f(n, k)$ be the desired number. We split the selections into two subsets: those which include the last of the n objects and those which do not. The former are $f(n-2, k-1)$ in number (since further selection of the second last object is forbidden); the latter are $f(n-1, k)$ in number. Hence

$$f(n, k) = f(n - 1, k) + f(n - 2, k - 1),$$

and, combining this with $f(n, 1) = n$, we readily prove by induction that $f(n, k) = _{n-k+1}C_k$.

LEMMA 2. *The number of ways of selecting k objects, no two consecutive, from n objects arrayed in a circle is $_{n-k}C_k n/(n - k)$.*

This differs from the preceding problem only in the imposition of the further restriction that no selection is to include both the first and last objects; and the number of such selections which are otherwise acceptable is $f(n-4, k-2)$. Hence the desired result is $f(n, k) - f(n-4, k-2) = _{n-k}C_k n/(n-k)$.

Presented to the Society, September 13, 1943; received by the editors May 4, 1943.

We now restate the *problème des ménages* in the usual fashion by observing that the answer is $2n!u_n$, where u_n is the number of permutations of $1, \cdots, n$ which do not satisfy any of the following $2n$ conditions: 1 is 1st or 2nd, 2 is 2nd or 3rd, $\cdots$, n is nth or 1st. Now let us select a subset of k conditions from the above $2n$ and inquire how many permutations of $1, \cdots, n$ there are which satisfy all k; the answer is $(n-k)!$ or 0 according as the k conditions are compatible or not. If we further denote by v_k the number of ways of selecting k compatible conditions from the $2n$, we have, by the familiar argument of inclusion and exclusion, $u_n = \sum(-1)^k v_k (n-k)!$. It remains to evaluate v_k, for which purpose we note that the $2n$ conditions, when arrayed in a circle, have the property that only consecutive ones are not compatible. It follows from Lemma 2 that $v_k = {}_{2n-k}C_k 2n/(2n-k)$, and hence

$$u_n = n! - \frac{2n}{2n-1}\,{}_{2n-1}C_1(n-1)! + \frac{2n}{2n-2}\,{}_{2n-2}C_2(n-2)! - \cdots .$$

From this result it follows without difficulty that $u_n/n! \to e^{-2}$ as $n \to \infty$.

BIBLIOGRAPHY

A. Cayley, *A problem of arrangements*, Proceedings of the Royal Society of Edinburgh vol. 9 (1878) pp. 338–341.

E. Lucas, *Théorie des nombres*, Paris, 1891, pp. 491–495.

P. A. MacMahon, *Combinatory analysis*, vol. 1, Cambridge, 1915, pp. 253–254.

E. Netto, *Lehrbuch der Combinatorik*, Berlin, 1927, pp. 75–80.

J. Touchard, *Sur un problème de permutations*, C. R. Acad. Sci. Paris vol. 198 (1934) pp. 631–633.

HARVARD UNIVERSITY

Afterthought

When I wrote this paper I (of course) did not know that I had been anticipated by Touchard [20]. John Riordan told me about this. However, the paper did play a certain role in publicizing the two lemmas it contains. But now there is an earlier anticipation to report. In 1902 Muir [18] proved a stronger result than Lemma 1: The number of ways of choosing k things from n in a line, with a separation of at least r between the selected objects, is

$$\binom{n - rk + r}{k}.$$

This was fifteen years before I was born. (I am grateful to Jacques Dutka for telling me about Muir's paper, in a letter dated September 22, 1986.)

Here is another anticipation: The symbolic multiplication idea in my first published paper [13] appears in [17] (Note: This is L. Lindelöf, not to be confused with his son, the better-known E. Lindelöf). John Riordan, a master of the literature on combinatorics, pointed this one out to me as well.

Humbled by these experiences, I have resolved never again to publish anything without a thorough search of the literature.

I close with four remarks:

1. I appreciate the honor of having my paper appear in the Gessel-Rota collection of combinatorial papers [8].
2. The neatest proof of Lemma 1 is that of Marshall Hall [9, p. 3]; it works just as well to prove Muir's generalization.
3. If the n objects are arrayed in a circle, the number of ways of choosing k, with a separation of at least r, is

$$\frac{n}{n - rk}\binom{n - rk}{k}.$$

The passage from a line to a circle that works so easily for $r = 1$ in the paper fails miserably for $r > 1$. So another idea is needed. I worked out a proof based on an idea I got from Bogart and Doyle [3]. Clark and Bonan have a nice proof in [5, Lemma 3, p. 221], as does David Singmaster in a preprint [19] dated August 10, 1989.
4. Dutka [6] has a large bibliography. I have added some additional references: [1, 2, 4, 7, 10, 11, 12, 14, 15, 16].

References

1. H. D. Abramson, On selecting separated objects from a row, *Amer. Math. Monthly* 76(1969), 1130–31.
2. Gregory F. Bachelis, The "problème des ménages à deux tables," preprint, 9 pp.
3. Kenneth P. Bogart and Peter G. Doyle, Non-sexist solution of the ménage problem, *Amer. Math. Monthly* 93(1986), 514–18.

4. Wen Chang Chu, On the number of combinations without k-separations, *J. Math. Res. Exposition* 7(1987), 511–20 (Chinese; English summary).
5. Dean S. Clark and Stanford C. Bonan, Experimental gambling system, *Math. Mag.* 60(1987), 216–22.
6. Jacques Dutka, On the problème des ménages, *Math. Intelligencer* 8(1986), no. 3, 18–25, 33.
7. B. S. El-Desouky, A generalization of selecting adjacent balls without separations, *Proc. Pakistan Acad. Sci.* 27(1990), 111–18.
8. Ira Gessel and Gian-Carlo Rota, eds., *Classic Papers iin Combinatorics*, Birkhäuser, Boston 1987.
9. Marshall Hall, *Combinatorial Theory*, 2d ed., Wiley, New York, 1986.
10. F. K. Hwang, Selecting k objects from a cycle with p pairs of separation s, *J. Comb. Theory* A 37(1984), 197–99.
11. F. K. Hwang, J. Korner, and V. K.-W. Wei, Selecting non-consecutive balls arranged in many lines, *J. Comb. Theory* A 37(1984), 327–36.
12. F. K. Hwang and Y.-C. Yao, A direct argument for Kaplansky's theorem on a cyclic arrangement and its generalization, *Oper. Res. Lett.* 10(1991), 241–43.
13. Irving Kaplansky, On a generalization of the "problème des rencontres," *Amer. Math. Monthly* 46(1939), 159–61.
14. Peter Kirschenhofer and Helmut Prodinger, Two selection problems revisited, *J. Comb. Theory* A 42(1986), 310–16.
15. John Konvalina, On the number of combinations without unit separation, *J. Comb. Theory* A 31(1981), 101–7.
16. John Konvalina and Yi-Hsin Liu, Subsets without unit separation and binomial products of Fibonacci numbers, *J. Comb. Theory* A 57(1991), 306–10.
17. L. Lindelöf, Un problème du calcul des probabilités, *Helsingf. Vet. Soc. Acta* 42(1899), 9 pp. (reviewed in Jahrbuch, vol. 30, p. 211).
18. T. Muir, Note on selected combinations, *Proc. Royal Soc. Edinburgh* 24(1902), 102–4 (reviewed in Jahrbuch, vol. 33, p. 229).
19. David Singmaster, Direct derivations for the number of unordered selections with repetitions and the ménage lemmas, preprint, 8 pp.
20. J. Touchard, Sur un problème de permutations, *C. R. Acad. Sci. Paris*, 198(1934), 631–33.

ANNALS OF MATHEMATICS
Vol. 46, No. 3, July, 1945

A CONTRIBUTION TO VON NEUMANN'S THEORY OF GAMES

By Irving Kaplansky

(Received February 1, 1945)

1. Introduction

For the two-person zero-sum game von Neumann's principal result is the existence of a well-defined *value* which both players can achieve by suitable strategies.[1] The simplest non-trivial case is where each player has a choice of precisely two moves; for this case von Neumann gives a complete treatment.[2] In this paper we attempt a generalization of von Neumann's results to an arbitrary two-person zero-sum game.

The idea which we use is to study the so-called "completely mixed" case, where each player must make some use of all of his possible moves. For this case a complete treatment is given. When the game is not completely mixed, there is at least one move which is superfluous, and an inductive procedure becomes possible. In this way we obtain a method of finding the value of a game in a finite number of steps. Unfortunately it has not yet been possible to evolve a corresponding Entscheidungsverfahren for the players' strategies.

2. Main results

A two-person zero-sum game G can be formulated as follows. The first player A selects an integer i ($i = 1, \cdots, m$); simultaneously the second player B selects an integer j ($j = 1, \cdots, n$). Then B pays A an amount a_{ij} (which may be positive, zero, or negative). Thus G is completely determined by the matrix $\| a_{ij} \|$.

A strategy for A consists of a vector $(p_1, \cdots, p_m)$ with $p_i \geq 0$, $\sum p_i = 1$. The idea is of course that he will play i with probability p_i. From von Neumann's results we know that there exist strategies $(p_1, \cdots, p_m)$ and $(q_1, \cdots, q_n)$ and a real number v' such that

$$(1) \qquad \sum_i p_i a_{ij} \geq v', \qquad (j = 1, \cdots, n)$$

$$(2) \qquad \sum_j q_j a_{ij} \leq v'. \qquad (i = 1, \cdots, m)$$

The number $v' = v'(G)$ is the unique *value* of G. Strategies satisfying (1) and (2) are called *good* strategies; in general they are not unique.

A strategy is *pure* if it has the form $(0, \cdots 0, 1, 0, \cdots 0)$; otherwise it is *mixed*. If each $p_i > 0$ we call the strategy $(p_1, \cdots, p_n)$ *completely mixed*. If the only good strategies are completely mixed, we shall call the game completely mixed.

---

[1] *Theory of Games and Economic Behaviour* by John von Neumann and Oskar Morgenstern, Princeton University Press, 1944. The theorem in question is proved in pp. 153–155.

[2] Loc. cit., pp. 169–173.

THEOREM 1. *If A has a completely mixed good strategy $(p_1 \cdots p_m)$, then any good strategy $(q_1, \cdots q_n)$ for B satisfies*

$$\sum_j a_{ij} q_j = v'. \qquad (i = 1, \cdots, m)$$

PROOF. If for example $\sum_j a_{1j} q_j > v'$ then[3] we have $p_1 = 0$, contrary to the hypothesis that $(p_1, \cdots, p_m)$ is completely mixed. The common sense interpretation is that if A can be punished whenever he plays 1, then his good strategy must exclude the playing of 1.

THEOREM 2. *Let $v' = 0$ and suppose all of A's good strategies are completely mixed. Then if r is the rank of $\| a_{ij} \|$,*

(i) $r \leqq n - 1$

(ii) $r \geqq m - 1$

(iii) *If $r = m - 1$, A has only one good strategy $(p_1, \cdots, p_m)$ and it satisfies*

$$(3) \qquad \sum_i a_{ij} p_i = 0. \qquad (j = 1, \cdots, n)$$

PROOF. (i) Since A has at least one completely mixed strategy $(p_1, \cdots, p_m)$, by Theorem 1 any good strategy $(q_1, \cdots, q_n)$ for B satisfies

$$\sum_j a_{ij} q_j = 0 \qquad (i = 1, \cdots, m)$$

whence $r \leqq n - 1$.

(ii) Suppose $r \leqq m - 2$. Then the system of equations

$$(4) \qquad \sum_i a_{ij} x_i = 0 \qquad (j = 1, \cdots, n)$$

has at least two linearly independent solutions. One of these solutions, say $(\pi_1, \cdots, \pi_m)$ must be independent of $(p_1, \cdots, p_m)$.

CASE I. $\sum \pi_i \neq 0$. We may assume $\sum \pi_i = 1$. Then the vector $(P_1, \cdots, P_m)$ given by

$$P_i = P_i(\lambda) = (1 + \lambda) p_i - \lambda \pi_i \qquad (\lambda \geqq 0)$$

has $\sum P_i = 1$ and

$$\sum_i a_{ij} P_i \geqq 0 \qquad (j = 1, \cdots, n)$$

Since $p_i > 0$, we have $P_i(\lambda) > 0$ for sufficiently small λ. On the other hand for large λ, at least one P_i is negative, since p_i and π_i are distinct. Let λ_0 be the smallest positive λ for which not all the P_i are positive; then $P_i(\lambda_0) \geqq 0$ and for at least one value of i, $P_i(\lambda_0) = 0$. Hence $P_i(\lambda_0)$ is a good strategy for A which is not completely mixed, contradicting our hypothesis.

CASE II. $\sum \pi_i = 0$. We repeat the same argument, using $P_i = p_i - \lambda \pi_i$.

(iii). Finally suppose $r = m - 1$. Then the equations (4) have a unique solution (up to a factor of proportionality). This solution must be proportional

---

[3] Loc. cit., Theorem 17:D, p. 161.

to any good strategy $(p_i, \cdots, p_m)$, for otherwise the proof given for (ii) could be repeated verbatim. Hence the good strategy is unique and satisfies (3).

THEOREM 3. *If $m > n$, then A has a good strategy which is not completely mixed.*

PROOF. We may assume $v' = 0$ without loss of generality. Then if the conclusion of Theorem 3 fails, we have by Theorem 2

$$m - 1 \leqq r \leqq n - 1,$$

a contradiction.

THEOREM 4. *If $m = n$, and the game is not completely mixed, then both A and B have good strategies which are not completely mixed.*

PROOF. Suppose that on the contrary A has only completely mixed good strategies. We assume $v' = 0$. Then by Theorem 2, $r = n - 1$ and A has a unique good strategy $(p_1 \cdots p_n)$ satisfying (3). Now B has a good strategy $(q_1, \cdots, q_n)$ which is not completely mixed; say $q_1 = 0$. If A_{ij} denotes the cofactor of a_{ij} then

$$\frac{q_1}{A_{i1}} = \frac{q_2}{A_{i2}} = \cdots = \frac{q_n}{A_{in}}$$

whence $A_{i1} = 0$ $(i = 1, \cdots, n)$. Hence the matrix

$$\begin{pmatrix} a_{12} & \cdots & a_{1n} \\ \vdots & & \vdots \\ a_{n2} & \cdots & a_{nn} \end{pmatrix}$$

obtained by deleting the first column from $\| a_{ij} \|$, has rank less than $n - 1$. The equations

$$\sum_i a_{ij} x_i = 0 \qquad\qquad (j = 2, \cdots, n)$$

have at least two linearly independent solutions: one of them our previous solution $(p_1, \cdots, p_n)$ and another $(\pi_1, \cdots, \pi_n)$. We must have

(5) $$a_{11}\pi_1 + a_{21}\pi_2 + \cdots + a_{n1}\pi_n \neq 0$$

for otherwise (4) would have two linearly independent solutions, contradicting $r = n - 1$. According as $\sum \pi_i \neq 0$ or $\sum \pi_i = 0$ we form

$$P_i = (1 + \lambda)p_i - \lambda\pi_i$$

or

$$P_i = p_i - \lambda\pi_i$$

and for suitable choice of λ (positive or negative according as (5) is positive or negative) we will obtain a good strategy $(P_1, \cdots, P_n)$ with

$$a_{11}P_1 + a_{21}P_2 + \cdots + a_{n1}P_n > 0.$$

This contradicts the uniqueness of A's good strategy.

THEOREM 5. *A game of value zero is completely mixed if and only if*
(1) *Its matrix is square ($m = n$) and has rank $n - 1$,*
(2) *All cofactors are different from zero and have the same sign.*

PROOF. *Necessity.* The necessity of (1) is evident from Theorem 2. Let A_{ij} denote the cofactor of a_{ij}. For any fixed j,

$$\frac{p_1}{A_{1j}} = \frac{p_2}{A_{2j}} = \cdots = \frac{p_n}{A_{nj}}.$$

Since the p's all are positive, $A_{1j}, \cdots, A_{nj}$ all have the same sign (or are all zero). A similar remark applies to $A_{i1}, \cdots, A_{in}$ for any fixed i. We exclude the possibility that each $A_{ij} = 0$, for then $r < n - 1$. Hence all A_{ij} are different from zero and have the same sign.

Sufficiency. Conversely suppose that $\|a_{ij}\|$ is square and of rank $n - 1$, and that all the cofactors A_{ij} have the same sign. Then if we select $(p_1, \cdots, p_n)$ proportional to $(A_{1j}, \cdots, A_{nj})$ we satisfy (3), and we may choose $p_i > 0$, $\sum p_i = 1$. Similarly we choose $(q_1, \cdots, q_n)$ with $q_j > 0$, $\sum q_j = 1$ and

$$(4) \qquad\qquad \sum_j q_j a_{ij} = 0. \qquad\qquad (i = 1, \cdots, n)$$

It is now clear that the game has value zero, and that both A and B possess completely mixed good strategies. From Theorem 1 it follows that any good strategies are solutions of (3) and (4). But the fact that $\|a_{ij}\|$ has rank $n - 1$ assures us that (3) and (4) have no solutions other than the completely mixed ones given above.

THEOREM 6. *The value v' of the completely mixed game $\|a_{ij}\|$ is given by*

$$(6) \qquad\qquad v' = \frac{|a_{ij}|}{\sum\limits_{i,j} A_{ij}}$$

where A_{ij} is the cofactor of a_{ij}. The denominator is always different from zero.

PROOF. If $\|a_{ij}\|$ has value v', then the game given by $b_{ij} = a_{ij} - v'$ has value zero and is obviously also completely mixed. In particular by Theorem 4, $|b_{ij}| = 0$. But $|b_{ij}| = |a_{ij}| - v'\sum A_{ij}$. It remains to be shown that $\sum A_{ij} \neq 0$. Let B_{ij} be the cofactor of b_{ij}. Then $|a_{ij}| = |b_{ij} + v'| = v'\sum B_{ij}$.

By Theorem 5, $\sum B_{ij} \neq 0$. The case $v' = 0$ obviously need not be considered. Hence we have $|a_{ij}| \neq 0$ and $\sum A_{ij} \neq 0$, as desired.

3. Determination of the value of a game

We can now state an explicit Entscheidungsverfahren for determining in a finite number of steps the value of an arbitrary two-person zero-sum game G.

First if the matrix of G is rectangular, say $m > n$, we are entitled by Theorem 3 to delete from G a suitable row without changing the value of the game. Of course we will not know in advance which row can be deleted but we can try all of them and take the best result. This can be continued until we reach a square matrix. The result can be stated as

THEOREM 7. *If $m > n$, $v(G) = \text{Max}_i \, v(G_i)$ where $\{G_i\}$ runs over the n by n sub-matrices of G. If $m < n$, $v(G) = \text{Min}_i \, v(G_i)$ where $\{G_i\}$ runs over the m by m sub-matrices.*

Next let a square game G be presented for inspection. To determine whether G is completely mixed, we find by (6) its supposed value v'. (If the denominator vanishes, we forthwith abandon the hypothesis that G is completely mixed). Then we reduce every element of G by v' and apply the test of Theorem 5. Actually in inspecting the cofactors it suffices to examine one row and one column. Alternatively, we may attempt to solve the equations

$$(7) \qquad \sum a_{ij}p_i = v'$$
$$(8) \qquad \sum a_{ij}q_j = v'.$$

If we obtain a solution with the p's and q's all positive, we have sustained the hypothesis that G is completely mixed, otherwise we have defeated it.

If it is settled that G is not completely mixed, we apply Theorem 4 and learn that a suitable row can be deleted. After this is done, the matrix is rectangular and by Theorem 3 a suitable column can be deleted. This can be summarized as

THEOREM 8. *If the square game G is not completely mixed, we have*

$$v'(G) = \underset{i}{\text{Max}} \, \underset{j}{\text{Min}} \, v'(G_{ij})$$

where G_{ij} is the game obtained by deleting from G its ith row and j^{th} column.

Note. It is clear, *mutatis mutandis*, that also

$$v'(G) = \underset{j}{\text{Min}} \, \underset{i}{\text{Max}} \, v'(G_{ij}).$$

Indeed the inequality of these two expressions is another necessary and sufficient condition for G being completely mixed. It is a curious oddity that when G is not completely mixed, the matrix $v'(G_{ij})$ corresponds to a strictly determined game.

4. Examples

$$(a) \qquad \begin{pmatrix} 5 & 3 & 0 \\ -2 & -5 & 2 \\ -5 & 7 & 2 \end{pmatrix}$$

From (6) we find $v' = 1$. Then by solving (7) and (8) we determine the unique good strategies $(1/2, 1/3, 1/6)$ for A and $(4/23, 1/23, 18/23)$ for B.

$$(b) \qquad \begin{pmatrix} 2 & -2 & 0 \\ -2 & 2 & 0 \\ 0 & 1 & 2 \end{pmatrix}$$

The determinant is zero, but one of the cofactors is also zero, so that this game is not completely mixed. The nine 2 by 2 sub-games have values given by the matrix

$$\begin{pmatrix} 4/3 & 0 & 0 \\ 1 & 1 & 2/5 \\ 0 & 0 & 0 \end{pmatrix}$$

and we find $v' = 2/5$. The corresponding good strategies are $(1/5, 0, 4/5)$ for A and $(3/5, 2/5, 0)$ for B.

(c) *Diagonal games.* The game

$$\begin{pmatrix} a_1 & 0 & 0 & \cdots & 0 \\ 0 & a_2 & 0 & \cdots & 0 \\ \cdots & \cdots & \cdots & \cdots & \cdots \\ 0 & 0 & 0 & \cdots & a_n \end{pmatrix}$$

is readily seen to be non-completely mixed (and of value zero) unless the a's all have the same sign. In the latter case (6) gives

$$\frac{1}{v'} = \frac{1}{a_1} + \frac{1}{a_2} + \cdots + \frac{1}{a_n},$$

i.e., nv' is the harmonic mean of the a's. This can be generalized without difficulty to games which decompose as in the classical theory of matrices:

$$\begin{pmatrix} G_1 & & & 0 \\ & G_2 & & \\ & & \ddots & \\ 0 & & & G_k \end{pmatrix}$$

Here $v'(G) = 0$ unless the $v'(G_i)$ have the same sign, in which case

$$\frac{1}{v'(G)} = \frac{1}{v'(G_1)} + \frac{1}{v'(G_2)} + \cdots + \frac{1}{v'(G_k)}.$$

(d) *Symmetric games.* The game G is symmetric if and only if its matrix is skew-symmetric. Of course $v'(G) = 0$, so the only interest lies in locating good strategies. Since the rank of a skew-symmetric matrix is even, we have the curious fact that an n by n symmetric game can be completely mixed only if n is odd.

For $n = 3$, the game

$$\begin{pmatrix} 0 & c & -b \\ -c & 0 & a \\ b & -a & 0 \end{pmatrix}$$

is completely mixed if and only if a, b, c are different from zero and have the same sign. The unique good strategy is then $(a/(a + b + c), b/(a + b + c), c/(a + b + c))$. (This is a sort of generalization of "Stone, Paper and Scissors").[4]

The symmetric game $\| a_{ij} \|$ of order 5 is completely mixed if and only if the five expressions

$$a_{25}a_{34} - a_{35}a_{24} + a_{45}a_{23}$$
$$-a_{15}a_{34} + a_{35}a_{14} - a_{45}a_{13}$$
$$a_{15}a_{24} - a_{25}a_{14} + a_{45}a_{12}$$
$$-a_{15}a_{23} + a_{25}a_{13} - a_{35}a_{12}$$
$$a_{14}a_{23} - a_{24}a_{13} + a_{34}a_{12}$$

have the same sign and the unique good strategy is then proportional to them.

NEW YORK CITY

---

[4] Loc. cit., p. 111.

Afterthought

When [2] appeared in 1944, I read it with fascination. I had not been aware of the launching of the theory of games in [1], and of course only years later did I learn about the partial anticipation by Borel [3].

The book's formula $(ad - bc)/(a + d - b - c)$ for the value of certain 2×2 games caught my eye. I felt that there had to be a generalization to appropriate $n \times n$ games. The numerator needs no comment. What about the denominator? Anyone who instantly guesses that it generalizes to the sum of the cofactors can go to the head of the class.

In retrospect I suspect that von Neumann knew the formula but decided to be generous to a naive newcomer.

I have two afterthoughts:

(1) In example (b) on page 478 I should have added that for both players there is only one good strategy (this of course often happens even if the game is not completely mixed).

(2) In the final paragraph of the paper I should have identified the five expressions as the principal 4×4 Pfaffians of the matrix and made the obvious conjecture for larger matrices (or better still, I should have proved it).

References

1. John von Neumann, Zur Theorie der Gesellschaftsspiele, *Math. Annalen* 100(1928), 295–320.
2. _____ and Oskar Morgenstern, *Theory of Games and Economic Behavior*, Princeton Univ. Press, 1944. (2d ed. 1947, 3d ed. 1953, paperback reprint 1980.)
3. *Econometrica* 21(1953), 95–127. This contains translations of the relevant Borel papers, along with comments by Fréchet and von Neumann.

LATTICES OF CONTINUOUS FUNCTIONS

IRVING KAPLANSKY

1. Introduction. Let X be a compact (=bicompact) Hausdorff space and $C(X)$ the set of real continuous functions on X. By defining addition and multiplication pointwise, we convert $C(X)$ into a ring. With the norm $\|f\| = \sup |f(x)|$, $C(X)$ becomes a Banach space. Finally, we may introduce an ordering by defining $f \geqq g$ to mean $f(x) \geqq g(x)$ for all x; this makes $C(X)$ a lattice.

Gelfand and Kolmogoroff [6][1] showed that, as a ring alone, $C(X)$ characterizes X. More precisely, if $C(X)$ and $C(Y)$ are isomorphic rings, then X and Y are homeomorphic. Banach [3, p. 170] proved that $C(X)$ as a Banach space characterizes X, if X is compact metric. Stone [5, p. 469] generalized this to any compact Hausdorff space, and Eilenberg [5] and Arens and Kelley [2] have since given other proofs. Finally, Stone [9] has shown that as a lattice-ordered group, $C(X)$ characterizes X. A negative result is that $C(X)$ as a topological linear space fails to characterize X [3, p. 184].

In this paper we shall prove the following result: as a *lattice alone* $C(X)$ characterizes X. This theorem is shown in §5 to subsume all the earlier results cited above. Moreover in this context we can replace the reals by an arbitrary chain, granted a suitable separation axiom. In §4 it is shown that the connectedness of X is equivalent to the indecomposability of $C(X)$ as a lattice.

I am greatly indebted to Professor A. N. Milgram for suggestions which led to a substantial simplification of my proof of Theorem 1.

2. Main theorem. Let R be a chain (simply ordered set). Until §6 it will be assumed that R has neither a minimal nor maximal element. There is a natural way of topologizing R [4, p. 27] which can be described as follows: for any $\alpha \in R$ let $U(\alpha)$ be the set of all $\beta \in R$ with $\beta > \alpha$, $L(\alpha)$ the set of all β with $\beta < \alpha$; then the U's and L's form a subbase of the open sets.

LEMMA 1. *If $\alpha, \beta \in R$ and $\alpha > \beta$, then there exist neighborhoods M, N of α, β such that $\gamma > \delta$ for all $\gamma \in M$, $\delta \in N$.*

PROOF. If there exists ξ with $\alpha > \xi > \beta$ we take $M = U(\xi)$, $N = L(\xi)$. If not, we take $M = U(\beta)$, $N = L(\alpha)$.

Received by the editors January 2, 1947.
[1] Numbers in brackets refer to the bibliography at the end of the paper.

Let X be a topological space and $C = C(X)$ the set of all continuous functions from X to R. We order C by defining $f \geqq g$ to mean $f(x) \geqq g(x)$ for all $x \in X$. Then C becomes a lattice, and in fact a distributive lattice.

DEFINITION. X is *R-separated* if for any x, $y \in X$ ($x \neq y$) and α, $\beta \in R$ there exists a continuous function f with $f(x) = \alpha$, $f(y) = \beta$. X is *R-normal*, if for any disjoint closed sets F, $G \subseteq X$ and any α, $\beta \in R$ there exists a continuous function f equal to α on F and β on G.

Any R-separated space is necessarily a Hausdorff space. If R is the real number system, it is known that conversely any compact Hausdorff space is R-normal, and this is likewise true if R is the set of reals with $\pm \infty$ adjoined (the latter is the same as a bounded closed interval on the real line). If R is disconnected, it is clear that X must be totally disconnected if it is to be R-separated, and conversely a totally disconnected space is R-separated for every R. In the extreme case where R consists of two elements, $C(X)$ is a Boolean algebra in its natural ordering, and the results in this paper are subsumed in Stone's theory of Boolean spaces and rings [7].

For any R we have the following result, which can be proved in exactly the same way as the lemma on p. 487 of [1].

LEMMA 2. *Suppose R has neither a maximal nor minimal element. Then if X is compact and R-separated, it is R-normal.*

By a *prime ideal* P in C [4, p. 78] we mean the set of antecedents of 0 in some lattice homomorphism of C onto the two-element lattice $(0, I)$. A prime ideal P is a sublattice containing with any element all smaller ones, and its complement $C - P$ has the dual property.

We can construct prime ideals in C as follows. Let Z be the lower half of a Dedekind cut in R, and for a fixed point $x \in X$, let P consist of all f with $f(x) \in Z$. Such prime ideals can be characterized as follows: $f \in P$ and $g(x) \leqq f(x)$ imply $g \in P$. We shall see below (§3) that these do not in general exhaust all prime ideals. However, we can prove a certain weakened version of this property.

DEFINITION. A prime ideal P in C is *associated* with a point $x \in X$ if $f \in P$ and $g(x) < f(x)$ imply $g \in P$.

LEMMA 3. *If X is compact, then any prime ideal P in C is associated with some point. If X is R-separated, the point is unique.*

PROOF. Suppose P is associated with no point of X. Then for every $x \in X$ we have functions f, g with $g(x) < f(x)$ and $f \in P$, $g \in Q = C - P$. By Lemma 1 the inequality $g < f$ extends to a neighborhood of x. A finite number of these neighborhoods cover X. Let $f_1, \cdots, f_n$ and $g_1, \cdots, g_n$ be the corresponding functions, and define $h = f_1 \cup \cdots \cup f_n$, $k = g_1 \cap \cdots \cap g_n$. Then $h > k$, but $h \in P$, $k \in Q$, a contradiction.

Suppose now that X is R-separated and that P is associated with both x and x'. Let m, n be any functions in P, Q respectively. There exists a function r with $r(x) < m(x)$, $r(x') > n(x')$. But these inequalities require r to be in both P and Q.

LEMMA 4. *If two prime ideals in C are associated with the same point of X, then their intersection contains a prime ideal.*

PROOF. Suppose P and P' are both associated with x. Let $f \in P$, $g \in P'$ and take any $\alpha \in R$ smaller than both $f(x)$ and $g(x)$. Let P'' be the set of all h with $h(x) \leq \alpha$; then $P'' \subset P \cap P'$.

LEMMA 5. *Suppose X is compact and R-separated. Let P, P', P'' be prime ideals in C with $P'' \subset P \cap P'$. Then P and P' are associated with the same point.*

PROOF. Suppose on the contrary that P and P' are associated with distinct points x and y. Let P'' be associated with z, where z is for definiteness different from x. Choose any f in P'', g in $C - P$, and then h with $h(z) < f(z)$, $h(x) > g(x)$. Then h is in P'' but not in P, a contradiction.

For use in §4 we insert at this point the following result.

LEMMA 6. *Suppose X is compact. If P is a prime ideal associated with x, $f \in P$, and $g \leq f$ on an open and closed set U containing x, then $g \in P$.*

PROOF. Suppose $g \in Q = C - P$. We note that P cannot be associated with a point y in the complement V of U, for we can construct a function m with $m(x) < f(x)$, $m(y) > g(y)$ and obtain a contradiction. Hence at every $y \in V$ we have functions h, k with $h > k$ in a neighborhood of y and $h \in P$, $k \in Q$. A finite number of neighborhoods cover the compact space V; if $h_1, \cdots, h_n, k_1, \cdots, k_n$ are the corresponding functions we have $f \cup h_1 \cup \cdots \cup h_n$ in P but exceeding $g \cap k_1 \cap \cdots \cap k_n$ in Q.

LEMMA 7. *Suppose X is compact and R-separated. Let f_0 be a fixed function in C, and for any subset S of X define $A(S)$ to be the intersection of all prime ideals containing f_0 which are associated with points of S. Then a point x is in the closure of S if and only if $A(S)$ is contained in a prime ideal associated with x.*

PROOF. Suppose x is in the closure of S. Choose $\alpha > f_0(x)$ and let P

consist of all f with $f(x) \leq \alpha$. Now $g \in A(S)$ implies $g \leq f_0$ at all points of S and hence $g(x) \leq f_0(x)$. It follows that $A(S) \subset P$. (For later use we remark that this half of the proof did not use R-separation.)

Suppose x is not in the closure of S. Then $A(S)$ cannot be contained in any prime ideal P associated with x. For let f be any function in $C - P$ and let $\alpha \in R$ be chosen with $f_0 > \alpha$ on S. By Lemma 2, a continuous function g exists which equals α on S and exceeds f at x. Then g is in $A(S)$ but not in P, a contradiction.

We can now prove our principal theorem.

THEOREM 1. *Let R be a chain with neither a minimal nor maximal element, and let it be endowed with its order topology. Let X be a compact R-separated space, and C the set of continuous functions from X to R. Then as a lattice, C characterizes X.*

PROOF. We say that two prime ideals in C are equivalent if their intersection contains a third prime ideal. Lemmas 3, 4, and 5 show that there is a one-one correspondence between points of X and classes of equivalent prime ideals. Lemma 7 shows that the topology of X can be expressed in terms of inclusion relations among the prime ideals, and this completes the proof.

3. **An example.** We shall give an example to show that not all prime ideals in C are simply based on a point in X and a Dedekind cut in R.

Let R be the reals, X the unit interval $[0, 1]$. Let A be the set of functions f for which $f(x) \leq -x$ in a neighborhood of 0 (the neighborhood depending on f). A is an ideal in C, that is, a sublattice containing with any element all smaller ones. Similarly let B be the set of all g with $g(x) \geq x$ in a neighborhood of 0; B is a dual ideal disjoint from A. By [8, Theorem 6] we can expand A, B to a prime ideal P and its complementary dual prime ideal Q. All functions f with $f(0) < 0$ go into P, all with $f(0) > 0$ go into Q, but those vanishing at 0 are split between P and Q.

Examples of this type can be constructed provided neither X nor R is discrete. They apparently require the axiom of choice.

4. **Connectedness and direct products.** If the space X splits into two open and closed sets X_1 and X_2, it is evident that the lattice $C = C(X)$ is the direct product [4, p. 13] of the lattices $C(X_1)$ and $C(X_2)$. We shall now prove the less trivial converse.

THEOREM 2. *Let R be a chain with neither a maximal nor minimal element, and let it be endowed with its order topology. Let X be a compact space and $C = C(X)$ the lattice of all continuous functions from X to R.*

Suppose C is a direct product of lattices, $C \cong C_1 \times C_2$. Then X splits into two open and closed sets X_1, X_2 in such a fashion that $C_i \cong C(X_i)$.

Proof. We first remark that any prime ideal P in C is of the form $P_1 \times C_2$ or $C_1 \times P_2$, where P_i is a prime ideal in C_i. This follows readily from Theorem 10 of [8], or it can be proved by repeating the argument of Lemma 3, if we replace X by a two-point space and R by C_1 at one point and C_2 at the other. Let us write Z_1 for the class of prime ideals of the form $P_1 \times C_2$, Z_2 for those of the form $C_1 \times P_2$.

Now take any point x in X and any prime ideal P associated with it. We define X_i to be the set of those x for which $P \in Z_i$. It follows from Lemma 4 that this definition makes sense, since any two prime ideals associated with the same point must fall into the same class.

Next we show that X_1 is closed. If not, there is a point $y \in X_2$ which is in the closure of X_1. By the first half of Lemma 7, $A(X_1)$ is contained in a prime ideal P associated with y. But P is of the form $C_1 \times P_2$ while $A(X_1)$ is of the form $B \times C_2$, so this is impossible. Similarly, X_2 is closed, and each is thus open and closed.

We shall now set up a lattice isomorphism between C_1 and $C(X_1)$. Given $c_1 \in C_1$, we pair it with any c_2 in C_2, obtaining a function $f \in C$, whose specialization to X_1 we shall call f_1. We prove that the correspondence is order-preserving, which incidentally shows it to be well defined. Suppose we have $d_1 \in C_1$ with $d_1 \leq c_1$ and that d_1 has given rise to the function g on X specializing to g_1 on X_1. If $g_1(x) > f_1(x)$ at some point $x \in X_1$, we can build a prime ideal P associated with x such that $f \in P$, $g \in C - P$. But P is of the form $P_1 \times C_2$ and hence $f \in P$ implies $g \in P$, since the C_1-component of f exceeds that of g. This contradiction shows that $f_1 \geq g_1$, as desired.

To set up the correspondence the other way we take a function f_1 on X_1, pair it with any f_2 on X_2, to obtain a function f whose C_1-component is, say, c_1. Suppose $g_1 \leq f_1$ and that g_1 gives rise to $d_1 \in C_1$. We assert that $d_1 \leq c_1$. For if not, we note that C_1 is a distributive lattice, whence [4, Theorem 5.8] there exists a prime ideal $P_1 \subset C_1$ with $c_1 \in P_1$, $d_1 \in C_1 - P_1$. For the corresponding prime ideal $P = P_1 \times C_2$ we have $f \in P$, $g \in C - P$. But P is associated with a point in X_1; by Lemma 6 we have a contradiction.

There remains the trivial observation that the two correspondences we have defined are inverse to each other, and we have completed the proof of Theorem 2.

Remark. It is to be observed that no separation axiom is required for Theorem 2. In case R is the reals, we can even drop the assumption of compactness by making use of the Stone-Čech compactification

$\beta(X)$. (Cf. [5, p. 578].) But for the general case no such device is available, and the validity of Theorem 2 for non-compact spaces remains undecided.

5. Rings and Banach spaces. Suppose we are given $C(X)$ as the *ring* of real continuous functions. Now $f \geq g$ is equivalent to the statement that $f - g$ is a square. Hence we are also given the lattice; the lattice characterization subsumes the ring characterization.

Suppose we are given $C(X)$ as a *Banach space*, with X a compact Hausdorff space. Let e be an extreme point on the unit sphere, that is, a point which is not an interior point of a segment lying in the unit sphere. It is easy to see that e can assume only the values 1 and -1; suppose it is 1 on Y_1 and -1 on Y_2, where Y_1, Y_2 are open and closed disjoint sets whose sum is X. Now a function $f \neq 0$ which satisfies $\|f/\|f\| - e\| \leq 1$ must be non-negative on Y_1 and non-positive on Y_2. Hence if we write $f \geq 0$ whenever it fulfills this condition, we preserve order on Y_1 and invert order on Y_2, as compared with the natural order in $C(X)$. This gives us a lattice which is at any rate isomorphic to $C(X)$. Hence $C(X)$ as a Banach space determines $C(X)$ as a lattice, and the lattice characterization subsumes the Banach space characterization.

Finally we remark that in our Theorem 2 we have likewise subsumed the analogous ring and Banach space theorems, in the following strong sense: if we have $C(X)$ expressed as a direct sum of rings or Banach spaces, then that very decomposition expresses the lattice as a direct product (it being understood in the Banach space case that the lattice being decomposed is the one obtained after inversion of order on Y_2). We omit the straightforward proof of this fact.

6. Chains with 0 and I. If R is a chain which has a minimal or maximal element or both, Theorems 1 and 2 remain valid, with the proviso that R-separation in the hypothesis of Theorem 1 is replaced by R-normality. In several of the proofs, however, there occur technical complications. We have preferred to avoid these complications, which add nothing to the fundamental idea. We shall merely mention the two main modifications that are necessary: (1) The definition of "equivalence" of prime ideals is revised to read: either their intersection contains a prime ideal, or the intersection of their complements contains a dual prime ideal. (2) In the characterization of the topology, x is in the closure of S if and only if $A(S)$ is contained in a prime ideal, or the object dual to $A(S)$ is contained in a dual prime ideal.

BIBLIOGRAPHY

1. R. F. Arens, *A topology for spaces of transformations*, Ann. of Math. vol. 47 (1946) pp. 480–495.

2. R. F. Arens ond J. L. Kelley, *Characterizations of the space of continuous functions over a compact Hausdorff space*, To be published in Trans. Amer. Math. Soc.

3. S. Banach, *Théorie des opérations linéaires*, Warsaw, 1932.

4. G. Birkhoff, *Lattice theory*, Amer. Math. Soc. Colloquium Publications, vol. 25, New York, 1940.

5. S. Eilenberg, *Banach space methods in topology*, Ann. of Math. vol. 43 (1942) pp. 568–579.

6. I. Gelfand and A. N. Kolmogoroff, *On rings of continuous functions on a topological space*, C. R. (Doklady) URSS. vol. 22 (1939) pp. 11–15.

7. M. H. Stone, *Applications of the theory of Boolean rings to general topology*, Trans. Amer. Math. Soc. vol. 41 (1937) pp. 375–481.

8. ———, *Topological representations of distributive lattices and Brouwerian logics*, Časopis Matematiky a Fysiky vol. 67 (1937) pp. 1–25.

9. ———, *A general theory of spectra II*, Proc. Nat. Acad. Sci. U.S.A. vol. 27 (1941) pp. 83–87.

UNIVERSITY OF CHICAGO

Afterthought

In the 1930s and 1940s many mathematicians became fascinated with $C(X)$, the continuous real or complex functions on a compact Hausdorff space X. Just about every aspect of its structure was put under a microscope.

A popular pastime was to prove that, under the structure in question, one could recover X from $C(X)$. My objective in this little paper was to offer a treatment that (hopefully) would be definitive for the real case, arguing that recovering X from $C(X)$, regarded as a partially ordered set, subsumed all previous results.

Probably every attempt to be definitive is ultimately doomed to defeat. In this case the defeat came quickly in Milgram's ingenious paper [3]. He recovered X from the structure of $C(X)$ as a multiplicative semigroup. There is no apparent connection (in either direction) with $C(X)$ as a partially ordered set.

REMARKS.

1. Milgram's proof works equally well for the complex or quaternion $C(X)$. One can even use certain other topological division rings.

2. One's first attempt at a proof in the semigroup setting would probably be to try the functions vanishing at a point as a way to recover the points of X. This fails (at least when I tried it). Instead Milgram successfully used the functions vanishing in a (variable) neighborhood of a point.

3. One can describe what Milgram did as a recovery of a ring from its multiplicative semigroup. There is a large literature on recovering a ring from its multiplicative group of invertible elements, usually in a noncommutative setting. But in the present context this fails. For instance, if X is connected, the multiplicative group of the real $C(X)$ is the direct product of a group of order two and a divisible torsionfree Abelian group, and thus is determined by one cardinal number.

4. See Shirota [4] for an extension to noncompact X.

5. Henriksen [1] made a gallant effort to minimize duplication.

6. My own followup paper [2] is not included in this collection. I would say that the only thing in it of possible enduring interest is the observation that a lattice automorphism of a real $C(X)$ is continuous if X satisfies the first axiom of countability, but may fail to be continuous otherwise.

References

1. Melvin Henriksen, On the equivalence of the ring, lattice, and semigroup of continuous functions, *Proc. Amer. Math. Soc.* 7(1956), 959–60.
2. Irving Kaplansky, Lattices of continuous functions, II, *Amer. J. of Math.* 70(1948), 626–34.
3. Arthur N. Milgram, Multiplicative semigroups of continuous functions, *Duke Math. J.* 16(1949), 377–83.
4. Taira Shirota, A generalization of a theorem of I. Kaplansky, *Osaka Math. J.* 4(1952), 121–32.

LOCALLY COMPACT RINGS.*

By IRVING KAPLANSKY.

---

1. Introduction. In this paper an account is given of some results in the structure theory of locally compact (= bicompact) rings. We begin by recapitulating, in somewhat generalized form, the known results on locally compact connected rings (Theorem 1). The theory of locally compact rings without nilpotent ideals is shown to be reducible to the totally disconnected case (Theorem 2). In **3** the hypothesis of boundedness is added, and a complete result given for the semi-simple case (Theorem 4). This section concludes with remarks on a class of rings including both compact rings and discrete rings with descending chain condition; it is shown that the known structure theorems for these two cases can thus be unified. The next two sections are devoted to maximal ideals, the existence and continuity of inverses, and the effect of chain conditions. A principal tool in this investigation is the fact (Lemma 4) that a locally compact totally disconnected ring has compact open subrings; this makes it possible to apply the structure theory for compact rings given in [11]. The final section assembles some results on locally compact primitive rings.

2. The component of 0. We shall use the definition of boundedness in [11]: a subset S of a topological ring A is *right bounded* if for any neighborhood U of 0 there exists a neighborhood V such that $V \cdot S \subset U$ (by $V \cdot S$ we mean the set of all products of elements of V by elements of S). Boundedness means both right and left boundedness, the latter being analogously defined.

The following theorem contains various known results as special cases; cf. [2], [5], [8], [11, Th. 8], [14], [15]. The proof is virtually the same as that in [11] but we repeat it for completeness.

THEOREM 1. *If A is a locally compact ring with C the component of 0, and B is a right bounded subgroup of A, then $CB = 0$.*

Proof. For any fixed character f of A, let $I(f)$ denote the set of all $a \in A$ with $f(aB) = 0$. Then $I(f)$ is clearly a subgroup of A; if we show it to be open it will contain C and we will have $f(CB) = 0$ for all f, so that

---

* Received August 12, 1947.

$CB = 0$. Choose neighborhoods U, V of 0 in A with $f(U) < 1/2$, $V \cdot B \subset U$. Then for $x \, \varepsilon \, V$ we have $nxB \subset U$ for any integer n, and hence $f(nxb) = nf(xb) < 1/2$ for all $b \, \varepsilon \, B$. Hence $f(xb) = 0$, $V \subset I(f)$, $I(f)$ is open.

We now quote the structure theorem [18, p. 110] which asserts that A is the direct sum (in the topological group sense) of a vector group N and a group in which the component P of 0 is compact. Since P is compact it is bounded [11, Lemma 10]. Hence we have the following corollary.

COROLLARY. *With the above notation* $P^2 = PN = NP = 0$.

Before proceeding to the next theorem, we make a remark on direct sums. If A is a topological ring and B, C are closed ideals with $B \cap C = 0$, $B + C = A$, then A is the direct sum of B and C in the ring-theoretic but not necessarily in the topological sense. However in a more special case we are able to assert that we have a direct sum in both senses: if B has a unit element e. For if the directed set a_i approaches a, then ea_i approaches ea, i. e., the B-components of a_i approach the B-component of a. By subtraction we get the same result for the C-components, and this verifies that we have the Cartesian product topology in $B + C$.

THEOREM 2. *A locally compact ring with no algebraically nilpotent* [1] *ideals is the direct sum of a connected ring and a totally disconnected ring. The former is a semi-simple algebra of finite order over the real numbers.*

Proof. In the notation of the Corollary to Theorem 1 we must have $P = 0$, since P is a nilpotent ideal. Then N becomes the component of 0 and is an ideal. Any nilpotent ideal in N would generate a nilpotent ideal in A; hence N has no nilpotent ideals. Also N is not only a vector space but an algebra over the real numbers; this requires only the verification of the appropriate associativity condition [cf. 8]. Thus N is a semi-simple algebra over the reals and has a unit element. We form the Peirce decomposition with respect to the latter, and we have $A = N + S$ where $S = A - N$ is totally disconnected.

3. Bounded rings. We begin with certain remarks valid for any bounded rings.

---

[1] A set S is algebraically nilpotent if for some n, $S^n = 0$. We have prefixed the term " algebraic " in order to distinguish between this and the topological nilpotence defined on p. 162 of [11].

LEMMA 1. *In a bounded ring the quasi-inverse* [2] *is uniformly continuous.*

Proof. We use a' to denote the quasi-inverse of a. The identity

$$a' - b' = (1 + a')(b - a)(1 + b')$$

shows that in a bounded ring, $a' - b'$ can be made arbitrarily small by choosing $a - b$ sufficiently small.

THEOREM 3. *In a complete bounded ring, the quasi-regular elements form a closed set.*

Proof. Suppose the directed set a_i converges to a and that a'_i exists for every i. Lemma 1 shows that a'_i is a Cauchy directed set. Its limit a' is the quasi-inverse of a.

From [11, Th. 4] we derive the following corollary.

COROLLARY. *The radical of a complete bounded ring is closed.*

The following lemma gives a new proof of Jacobson's result [10, Th. 26] that a two-sided ideal in a semi-simple ring is semi-simple.

LEMMA 2. *If B is a two-sided ideal in a ring A, we have $R(B) = R(A) \cap B$, where R denotes the radical in each case.*

Proof. First on the mere assumption that B is a right ideal, we prove $R(B) \supset R(A) \cap B$. Let $x \,\varepsilon\, R(A) \cap B$; then x has a quasi-inverse y which is in B since $y = -x - xy$. Hence $R(A) \cap B$ consists of elements which are quasi-regular in B, and since it is an ideal in B, it is contained in $R(B)$. Conversely suppose $x \,\varepsilon\, R(B)$, where B is now a two-sided ideal in A. Then for any $a \,\varepsilon\, A$, $-(xa)^2$ is in $R(B)$ and is quasi-regular, say with quasi-inverse z. Hence $-xa$ has the quasi-inverse $xaoz$, and $x \,\varepsilon\, R(A)$.

The following structure theorem reduces the study of locally compact bounded semi-simple rings to the discrete case, since compact semi-simple rings are completely known [11, Th. 16].

THEOREM 4. *A locally compact bounded semi-simple ring is the direct sum of a compact semi-simple ring and a discrete semi-simple ring.*

Proof. It follows from Theorem 1 that A is totally disconnected. Since A is bounded and locally compact, it has a system of ideal neighborhoods of 0 [11, Lemma 9]. In particular we have a compact open ideal B, which

---

[2] The terms quasi-inverse, radical, semi-simple, and primitive are used as defined by Jacobson in [10]. We also use the notation $xoy = x + y + xy$.

is semi-simple by Lemma 2. B has a unit element [11, Th. 16], and we may use the Peirce decomposition to write $A = B + C$, with $C = A - B$ discrete since B is open. That C is semi-simple follows from another application of Lemma 2.

The presence of a compact open ideal may be used to obtain results without the assumption of semi-simplicity. The following is an example.

THEOREM 5. *A commutative locally compact totally disconnected bounded ring is the direct sum of a compact ring with unit, and a ring which modulo its radical is discrete.*

Proof. Our assumptions imply the existence of a compact open ideal I, which in turn [11, Th. 17] is the direct sum of a compact ring B with unit e, and a radical ring D. We form the Peirce decomposition $A = B + C$, with $B = Ae$ and C the annihilator of e. Now D is an ideal in I, which is an ideal in A. By two applications of Lemma 2 we have $R(D) = R(A) \cap D$, i. e. $D \subset R(A)$. Since $R(A) = R(B) + R(C)$ and $D \subset C$, we have $D \subset R(C)$. Now $C - D = (B + C) - (B + D) = A - I$ is discrete. Hence $C - R(C)$ is discrete.

We shall conclude this section with some remarks on a still more special class of rings. These rings satisfy the following three conditions: (1) local compactness, (2) boundedness, (3) the descending chain condition for right ideals which contain a fixed open two-sided ideal. Special cases are compact rings and discrete rings with the descending chain condition on right ideals. We shall now briefly indicate how the classical results in the latter case, and the results for compact rings in [11], may be thus unified and generalized.

First we consider the semi-simple case. Theorem 4 is applicable and condition (3) shows that the discrete summand has the descending chain condition, and accordingly is a direct sum of a finite number of matrix rings over division rings. The result may be restated thus: a semi-simple ring satisfying (1), (2), and (3) is the Cartesian direct sum of matrix rings over division rings, all but a finite number of the components being finite.

In the general case let R be the radical of a ring A satisfying (1), (2) and (3). By the Corollary to Theorem 3, R is closed. The arguments of [11, p. 163] may now be repeated virtually verbatim, and we have that R is the union of all (topologically) nil left and right ideals, and in the totally disconnected case is itself topologically nilpotent. The idempotents of $A - R$ may be transferred to A as in [11, Lemma 12] and this yields the analogues of Theorems 17 and 18: if A is commutative it is the direct sum of primary rings and a radical ring, and a primary ring is the ring of matrices over a completely primary ring.

4. Quasi-inverses. We repeat from [11] the following definition: a topological ring is a Q_r-ring if the right quasi-regular elements form an open set, a Q-ring if the quasi-regular elements form an open set. It suffices [11, Lemma 2] that there be a neighborhood of 0 consisting of (right) quasi-regular elements. Moreover [11, Lemma 4] in a Q-ring, the quasi-inverse is continuous wherever defined if it is continuous at 0.

LEMMA 3. *Let B be a closed two-sided ideal in the topological ring A. Then A is a Q-ring with continuous quasi-inverse if and only if both B and A — B are.*

Proof. Suppose that A is a Q-ring with continuous quasi-inverse. If x is near 0 in B, then x has a quasi-inverse $y = -x - xy$ necessarily in B, which, being near 0 in A is also near 0 in B. Again, for any element z near 0 in $A - B$, we may pick an inverse image in A near 0 in A. The image of the latter's quasi-inverse provides us with a quasi-inverse of z near 0 in $A - B$.

Conversely suppose that both B and $A - B$ are Q-rings with continuous quasi-inverse. For x near 0 in A the coset $x + B$ is near 0 in $A - B$. The quasi-inverse of $x + B$ is a coset having a representative y near 0 in A. Then xoy is in B, and being small, has a small quasi-inverse z. Then yoz is the desired quasi-inverse of x.

We shall now apply Lemma 3 to locally compact rings. First we dispose of the connected case.

THEOREM 6. *A connected locally compact ring is a Q-ring with continuous quasi-inverse.*

Proof. In the notation of the Corollary to Theorem 1, P is a closed two-sided ideal which is certainly a Q-ring with continuous quasi-inverse in virtue of $P^2 = 0$. Next $A - P$ is an algebra of finite order over the reals. It can be normed [1, Th. 4] and hence [11, Lemmas 3, 5] is a Q-ring with continuous quasi-inverse. The result now follows from Lemma 3.

The treatment of the totally disconnected case is based upon the following lemma.

LEMMA 4. *A locally compact totally disconnected ring A has a system of neighborhoods of 0 which are compact open subrings.*

Proof. We know that A has subgroup neighborhoods of 0. Let U be such a compact open subgroup, and select an open subgroup V such that $V \subset U$, $VU \subset U$ (this is possible since U is bounded). Define $W = V + V^2 + V^3 + \cdots$. Then W is a compact open subring of A and $W \subset U$.

For later use we note the following refinement.

LEMMA 5. *Let A be a locally compact totally disconnected ring having a unit element 1 such that the closed subring C generated by 1 is compact. Then A has a compact open subring containing 1.*

Proof. If U is any compact open subring, the desired subring is $U + C$.

We remark that the hypothesis of Lemma 5 is of course satisfied if A has finite characteristic. Another case where the hypothesis holds is where A has a compact open subgroup U containing an element a such that the mapping $x \rightarrow xa$ is a homeomorphism (for example, A may be any non-discrete locally compact totally disconnected division ring); for then C is homeomorphic to Ca which is compact. An example where the hypothesis fails is the ring of integers under the discrete topology.

LEMMA 6. *A compact ring A which is not a Q-ring contains a set of non-zero idempotents with cluster point 0.*

Proof. If the radical R of A is open, A is clearly a Q-ring. Hence we suppose that R is not open, which means that $A - R$ is infinite. By [11, Th. 16] we may find in $A - R$ an infinite set of idempotents having only 0 as a cluster point. We may build [11, Lemma 12] idempotents e_i in A mapping on these. The set e_i has a cluster point e, necessarily an idempotent, and necessarily in R since it maps on 0 mod R. But the negative of an idempotent is right quasi-regular only if it is 0, and so $e = 0$.

We now prove the principal result of this section. It provides for the locally compact case an affirmative answer to the question raised in [11] as to whether any Q_r-ring is a Q-ring.

THEOREM 7. *A locally compact ring A is a Q-ring with continuous quasi-inverse if and only if A is a Q_r-ring.*

Proof. In the light of Lemma 3 and Theorem 6, it will suffice to treat the totally disconnected case. Then by Lemma 4, A has a compact open subring B. We assert that B is a Q-ring, for otherwise by Lemma 6 there are negatives of non-zero idempotents arbitrarily near 0, which is impossble in a Q_r-ring. It then follows that A is a Q-ring. That the quasi-inverse is continuous is a consequence of Lemma 1.

We may now prove Theorems 21 and 22 of [11] without countability assumptions.

THEOREM 8. *A locally compact ring without divisors of 0 is a Q-ring with continuous quasi-inverse.*

Proof. The proof is like that of Theorem 7, except that the presence of idempotents near 0 is now ruled out by the remark that the only possible idempotents are 0 and 1.

We prove one further result of this kind.

THEOREM 9. *A locally compact ring with the ascending or descending chain condition on closed right ideals is a Q-ring with continuous quasi-inverse.*

Proof. Assuming that the usual compact open subring is not a Q-ring we again find an infinite set of idempotents; but this time we arrange, as is clearly possible in the light of [11, Th. 16], that $e_i e_j = e_j e_i = e_j$ for $i \geqq j$ or $i \leqq j$ according as we have the ascending or descending chain condition. In the former case $\{e_i A\}$ is an ascending set of closed right ideals. But if $e_i A = e_j A$ for $i > j$ then $e_i = e_j a$, and left-multiplying by e_j gives $e_j = e_j a = e_i$, a contradiction. The argument is similar for the descending case.

In Theorems 7, 8 and 9 we have generalized Otobe's result [16] that the inverse is continuous in a locally compact division ring. These do not exhaust the possible hypotheses that are adequate. We mention one more: assume no nilpotent ideals (so that Theorem 2 is applicable), and assume outright that there is a neighborhood of 0 free of non-zero idempotents.

At the end of **5** we shall give an example of a locally compact ring in which the quasi-inverse is not continuous.

We conclude this section with the following supplement to Theorem 9.

THEOREM 10. *A semi-simple Q-ring with the descending chain condition on closed right ideals has the descending chain condition on all right ideals. Hence a locally compact semi-simple ring with the descending chain condition on closed right ideals is the direct sum of a finite number of matrix rings over locally compact division rings.*

Proof. We shall use Jacobson's structure theory of semi-simple rings

[10, pp. 310-312], sharpened by the use of Segal's notion of regular ideals.[3] We may summarize the facts that we need as follows: the regular maximal right ideals M_i in A have intersection 0. We form for each i the ideal $P_i = (M_i : A) =$ the set of all $x \, \varepsilon \, A$ with $Ax \subset M_i$. By the use of right multiplications the primitive ring $A - P_i = Q_i$ may be represented as a dense ring of endomorphisms in the vector space $A - M_i$.

Now in our topological context we note further that M_i is closed [17, Th. 1.6], and from this it follows readily that P_i is closed. Let $x_1, x_2, \cdots$ be linearly independent elements in the vector space $A - M_i$ in which Q_i acts, and let I_n denote the set of elements (i. e. linear transformations) annihilating $x_1, \cdots, x_n$. It is easy to see that the I's form a properly descending chain of closed right ideals. Since this chain must terminate $A - M_i$ is finite-dimensional. Next we note that a finite number of the M's already have intersection 0. From this we can conclude that A is the direct sum of a finite number of simple rings (cf. the argument at the foot of p. 314 of [10]). In the locally compact case we use Theorem 9 to get that A is a Q-ring, and thus verify the second sentence of the theorem. We remark finally that the direct sum and matrix representations involved hold in the topological as well as in the algebraic sense, as can be seen by persistent use of idempotents.

5. Maximal ideals. The results in this section will be obtained by further exploitation of the existence of compact open subrings. We begin with two preliminary results.

LEMMA 7. *Let e be an idempotent in a topological ring A, such that eAe is a Q_r-ring. Then any maximal right ideal M containing $\{a - ea\}$ is closed.*

Proof. Suppose on the contrary that the closure of M is A. Then M contains $e + x$ with x arbitrarily small, say so small that exe is right quasi-regular, $exe \circ y = 0$. Then M contains $z = (e + x)(e + ey)$, also $ez - z$, and hence ez. A computation shows that $ez = e$, so that M contains e and all of A, a contradiction.

LEMMA 8. *Let e be an idempotent in a topological ring A, such that eAe is a Q_r-ring. Then if x lies in every closed regular maximal right ideal of A, ex is in the radical of A.*

---

[3] A right ideal I is regular if there exists a left unit e modulo I, that is, an element e such that $ex - x \, \varepsilon \, I$ for all x.

Proof. Consider the right ideal I generated by $\{a - ea\}$ and $e + xe$. If I is a proper ideal it can be expanded to a maximal right ideal M excluding e, which is closed by Lemma 7, and regular since e is a left unit. Then $x \varepsilon M$, $xe \varepsilon M$, $e \varepsilon M$, $M = A$, a contradiction. Hence $I = A$, and we must have

$$b + (e + xe)(e + d) = e$$

where $b \varepsilon \{a - ea\}$, $d \varepsilon A$. Left-multiplying by e we find $exeoed = 0$. Thus $(ex)e$ is right quasi-regular and hence so is $e(ex) = ex$ (cf. the last identity on p. 154 of [11]). Since xc is also in every closed regular maximal right ideal for every c, we likewise have exc right quasi-regular, and hence ex is in the radical.

THEOREM 11. *A locally compact ring which is not a radical ring has a closed regular maximal right ideal.*[4]

Proof. First we consider the totally disconnected case. We find a compact open subring B. If B is a Q-ring, so is A, and (unless A is a radical ring) we know that A has regular maximal right ideals which are closed. Otherwise, let R be the radical of B. In $B - R$ we select a primitive idempotent, and we find in B an idempotent e mapping on f. Now we know that f annihilates a neighborhood of 0 in $B - R$. Hence for a suitable neighborhood U of 0 in B we have $eUe \subset R$. This shows that eAe is a Q-ring, and Lemma 7 then provides us with the desired closed regular maximal right ideal, namely any right ideal containing $\{x - ex\}$ and maximal with respect to exclusion of e.

In the general case let C be the component of 0 in A. If $A - C$ is a radical ring it is a Q-ring, so is A (Lemma 3 and Theorem 6), and we are finished. Otherwise by the first part of the proof $A - C$ has a closed regular maximal right ideal. We take its inverse image in A.

THEOREM 12. *Let A be a locally compact totally disconnected ring with a unit element 1 such that the closed subring generated by 1 is compact. Then the intersection S of the closed regular maximal right ideals in A is the closure R' of the radical R of A.*

Proof. That S contains R' is clear, since S is closed and contains R. Conversely suppose $x \varepsilon S$. We select (Lemma 5) a compact open subring B containing 1; let T be the radical of B. In $B - T$ we may find a directed set f_i of idempotents of finite rank approaching 1 (an idempotent of finite

---

[4] Concerning this question of the existence of closed maximal ideals, R. Arens has remarked that his ring L^ω [3] has no closed maximal ideals at all.

rank is the sum of a finite number of primitive idempotents); this is an immediate consequence of [11, Th. 16]. We build idempotents e_i in B mapping on f_i [11, Lemma 12]. Then $\{e_i\}$ has the cluster point 1, for such a cluster point is necessarily an idempotent mapping on the unit element of $B - T$, and 1 is the only element fulfilling these conditions. By Lemma 8, $e_i x \, \varepsilon \, R$ and it follows that $x \, \varepsilon \, R'$.

The significance of a result like Theorem 12 is as follows. Suppose that A is a locally compact semi-simple ring. Then $R = R' = 0$, and under the hypothesis of Theorem 12, $S = 0$. We then immediately deduce a representation of A as a subdirect sum of locally compact primitive rings, the point being that only with *closed* maximal ideals is a topology inherited by the components of the sum. It need hardly be added that if A is a Q-ring, the assumption about the unit element is superfluous. Whether it is in any event superfluous I have not been able to determine.

We give an example now of the kind of subdirect sum that can arise in locally compact rings.[5] Let F_p denote the p-adic numbers, I_p the p-adic integers. Let A be the set of all sequences $\{a_i\}$ of elements of F_p, with all but a finite number of components in I_p; let B be the subset with all components in I_p. The topology is as follows: neighborhoods of 0 in B are taken to be neighborhoods of 0 in A, and B is given the Cartesian product topology. Then B is compact so that A is locally compact. Continuity of addition and multiplication in A are readily verified. A is semi-simple: in fact the sequences with 0 at a designated place form a closed maximal ideal, and these ideals have intersection zero. A noteworthy feature of this ring is that the (multiplicative) inverse is not continuous. The elements b_i with p in the i-th place and 1 elsewhere approach the unit element of A (the sequence of all 1's), but b_i^{-1} does not approach the unit element since the difference does not even enter B.

6. Primitive rings. As was observed earlier, Theorem 12 provides a partial reduction of locally compact semi-simple rings to the primitive case. Thus an important step in the theory would be accomplished if locally compact primitive rings were classified. It follows from Theorem 2 and [10, Lemma 4] that only the totally disconnected case need be studied.

The most important result so far obtained is the structure theorem of Jacobson [9]: any non-discrete locally compact division ring admits a valuation and is an algebra of finite order over its center. There have been three

---

[5] This kind of construction is closely related to certain group-theoretic results of Braconnier and Dieudonné [7].

later contributions: Otobe [16] removed the countability assumption in Jacobson's proof, Braconnier [4] announced a proof based on Haar measure, and a new proof is given in [12]. The latter proof leans strongly on the following result.

LEMMA 9. *A locally compact topological linear space over a non-discrete locally compact division ring is finite-dimensional.*

The connected case is a classical result of F. Riesz. The general case was announced in [6] and a proof is given in [12]. In the characteristic zero case a proof was already given by Jacobson in [9], and we wish to record here that the characteristic zero case also follows from a theorem of Mackey [13].

We turn now from division rings to primitive rings. Jacobson [9, Th. 5.3] gave the following result: a non-discrete locally compact simple [6] ring S of characteristic zero is an algebra of finite order over the p-adic numbers. However there appears to be a gap in the proof that is not easy to supply: the tacit assumption (especially on p. 440) that multiplication by p^{-1} is continuous, so that $p^n G$ is open where G is a compact open subgroup of S. The more general statement is made on p. 442 that if S is a locally compact p-group [7] which is an algebra over the rational numbers, then S is an algebra of finite order over the p-adic numbers. That this statement is in error is shown by the following example: let S be the set of all sequences of p-adic numbers with the proviso that in each sequence there is a bound to the negative powers of p. Let T be the subset of S consisting of those sequences with all components p-adic integers. We give T the Cartesian product topology and make it open in S. Multiplication is ruled out of the picture by setting $S^2 = 0$. Then S is locally compact and a p-group, but it is infinite-dimensional over the p-adic numbers. It goes without saying that multiplication by p^{-1} is not continuous, and consequently S is not a *topological* linear space over the p-adic numbers.

If we add to the hypothesis of simplicity the existence of a unit, then p^{-1} becomes a member of S and multiplication by it will be continuous. However a somewhat shorter proof is made possible by the assumption of a unit.

THEOREM 13. *A non-discrete locally compact simple ring A of characteristic zero with unit element is an algebra of finite order over its center.*

We first note the following lemma.

---

[6] A simple ring here means one with no proper ideals, and not merely no proper closed ideals.

[7] A p-group is one in which every element satisfies $p^n a \to 0$.

LEMMA 10. *For any compact commutative totally disconnected group G, there exists a prime p and an element $a \neq 0$ in G such that $p^n a \to 0$.*

The assumption of the first axiom of countability in Jacobson's proof [9, Lemma 4.4] can be avoided without difficulty. A brief proof can also be based on character theory: the character group H of G is a discrete group with all its elements of finite order. The elements of H with order a power of p form a direct summand H_p which must be non-zero for some p. The annihilator G_p of $H - H_p$ is easily seen to be a p-group.

Proof of Theorem 13. The center of any simple ring with unit is a field. Hence the center Z of A is a locally compact field. If we show that Z is non-discrete, then by Lemma 9 $[A:Z]$ will be finite. Now for any p the set of elements a in A such that $p^n a \to 0$ forms an ideal A_p. By Lemma 10 this ideal is non-void for some p, and $A_p = A$ since A is simple. In particular $p^n \cdot 1 \to 0$, Z is non-discrete.

We present another result that can be proved in a similar way.

THEOREM 14. *Let A be a non-discrete locally compact primitive ring and suppose that A is of characteristic zero and has minimal ideals. Then A is an algebra of finite order over its center.*

Proof. Let eA be a minimal right ideal. Then eA is a locally compact topological linear space over the locally compact division ring eAe, and A is represented as a dense ring of endomorphisms on this linear space. Again we have only to prove that eAe is non-discrete, for this will make eA finite-dimensional. By Lemma 10 we have a non-zero ideal I consisting of all a with $p^n a \to 0$. By [10, Th. 29], e is in I. Hence eAe is not discrete.

We shall conclude with an example showing that in the characteristic p case, it is possible for a locally compact primitive ring to be infinite-dimensional over its center. Let K be a finite field and A the set of all infinite matrices $a = (a_{ij})$, $a_{ij} \varepsilon K$, which are "ultimately triangular": for any a there exists $N = N(a)$ such that $a_{ij} = 0$ for $j > i > N$. Let B be the subset of matrices which are actually triangular: $a_{ij} = 0$ for $j > i$. We topologize B with the weak topology, a general neighborhood of 0 in B consisting of all matrices with the first n rows zero. This makes B compact, and by declaring that B is open in A we make A locally compact. That A is primitive is clear and incidentally it has a unit element and minimal ideals. The one point needing serious verification is the continuity of multiplication, a verification which we omit.

This ring is not simple: the linear transformations with finite-dimensional range form a proper ideal in A. However A is simple in the sense that there are no proper *closed* ideals. It seems unlikely that a locally compact ring with minimal ideals could be simple in the strong sense, since completeness appears to require the presence of linear transformations with infinite-dimensional range.

UNIVERSITY OF CHICAGO.

BIBLIOGRAPHY.

1. A. A. Albert, "Absolute valued real algebras," *Annals of Mathematics*, vol. 48 (1947), pp. 495-501.

2. H. Anzai, "On compact topological rings," *Proceedings of the Imperial Academy of Tokyo*, vol. 19 (1943), pp. 613-615.

3. R. Arens, "The space L^ω and convex topological rings," *Bulletin of the American Mathematical Society*, vol. 52 (1946), pp. 931-935.

4. J. Braconnier, "Groupes d'automorphismes d'un groupe localement compact," *Comptes Rendus de l'Académie des Sciences, Paris*, vol. 220 (1945), pp. 382-384.

5. J. Braconnier, "Sur les modules localements compacts," *Comptes Rendus de l'Académie des Sciences, Paris*, vol. 222 (1946), pp. 527-529.

6. J. Braconnier, "Sur les espaces vectoriels localements compacts," *Comptes Rendus de l'Académie des Sciences, Paris*, vol. 222 (1946), pp. 777-778.

7. J. Braconnier and J. Dieudonné, "Sur les groupes abeliens localements compacts," *Comptes Rendus de l'Académie des Sciences, Paris*, vol. 218 (1944), pp. 577-579.

8. N. Jacobson and O. Taussky, "Locally compact rings," *Proceedings of the National Academy of Sciences*, vol. 21 (1935), pp. 106-108.

9. N. Jacobson, "Totally disconnected locally compact rings," *American Journal of Mathematics*, vol. 58 (1936), pp. 433-449.

10. N. Jacobson, "The radical and semi-simplicity for arbitrary rings," *American Journal of Mathematics*, vol. 67 (1945), pp. 300-320.

11. I. Kaplansky, "Topological rings," *American Journal of Mathematics*, vol. 69 (1947), pp. 153-183.

12. I. Kaplansky, "Topological methods in valuation theory," *Duke Mathematical Journal*, vol. 14 (1947), pp. 527-541.

13. G. Mackey, "A remark on locally compact abelian groups," *Bulletin of the American Mathematical Society*, vol. 52 (1946), pp. 940-944.

14. Y. Otobe, "On quasi-evaluations of compact rings," *Proceedings of the Imperial Academy of Tokyo*, vol. 20 (1944), pp. 278-282.

15. Y. Otobe, "Note on locally compact simple rings," *Proceedings of the Imperial Academy of Tokyo*, vol. 20 (1944), p. 283.

16. Y. Otobe, "On locally compact fields," *Japanese Journal of Mathematics*, vol. 19 (1945), pp. 189-202.

17. I. E. Segal, "The group algebra of a locally compact group," *Transactions of the American Mathematical Society*, vol. 61 (1947), pp. 69-105.

18. A. Weil, "L'Intégration dans les Groupes Topologiques et ses Applications," *Actualités Scientifiques et Industrielles*, 869, Paris, 1938.

Afterthought

I can pinpoint the moment I became interested in locally compact rings. It was in the summer of 1938 at the University of Chicago. I was a student and one of the courses for which I registered was really a seminar run by Albert, with mostly guest speakers. Nathan Jacobson ("Jake") gave a talk, presenting his recent results on locally compact division rings.

Seven years later I was a newly appointed instructor at Chicago. World War II had just ended. I was eager to launch new projects. One of them was to follow Jake and see if I could add something to the study of locally compact rings.

Let me note that Jake had published three papers on the subject: [4], [5], and [6]; the first was joint with Olga Taussky. The historical background is nicely sketched on page 6 of volume I of [7].

I decided first to study compact rings closely [8], arguing that this should be useful later, since the interesting locally compact rings have lots of compact open subrings. The present paper was followed by two sequels: [10] and [11]. I feel that the one selected for this volume is an adequate sample.

Looking back, I find it particularly pleasing that Jake's general structure theory of rings meshed nicely (it was quite recent at the time). For instance, the Jacobson radical of a locally compact ring is closed; better yet, it is an intersection of closed regular maximal left ideals. This result appears in [10].

I was defeated by the case of simple locally compact rings in characteristic p. I salute Skornyakov [13] for settling this. The way it worked out suggests using the terminology "tame" and "wild." The tame ones are matrix rings over division rings (i.e., simple Artinian). The others are wild, and wild ones exist. Here is a pretty result: Wildness is characterized by the presence of nontrivial open one-sided ideals. Warner [15, pp. 276–281] presented an account of Skornyakov's construction.

I conclude with five remarks:

1. With Skornyakov's work the program was finished, at least as I saw it. Nevertheless I find it interesting that in the Small collection of reviews in ring theory there is a whole subheading for locally compact rings (29.01). Pages 847–854 in the volume covering the years from 1940 to 1979 fall under this heading. Then there are more papers with "compact" generalized to "linearly compact." In the 1980 to 1984 collection all papers on topological rings are lumped together.

2. Right at the start there was a remarkable dichotomy. Take a locally compact division ring D, for instance. If the topology is discrete, we have a purely algebraic object on our hands and there is nothing to be said. But the moment we add nondiscreteness to the hypotheses we get a tight structure theory.

Goldman and Sah [1], [2] pointed out the way to match this dichotomy in the general case: Assume that the given locally compact ring R has no open two-sided ideals other than R itself.

3. There is a flaw in [9, p. 814] that I point out belatedly (it was called to my attention long ago by Malcolm Smiley). For nondiscrete locally compact alternative division rings, the targets do not exist! I blush to admit that I was apparently

56

unaware that five-dimensional quadratic forms over the fields in question are null forms, and so there are no Cayley division algebras.

4. Applications of the theory of locally compact rings have not proliferated. But I did find two. In [12, p. 676], Siegel makes a fleeting reference. In [3] Iwasawa uses the theory in a more substantial way, to prove the existence of closed maximal ideals. To be sure, he does point out that this is much easier in his commutative case.

5. Much of Seth Warner's definitive treatise on topological rings [15] is devoted to locally compact rings; I recommend it enthusiastically to the reader. The preceding book [14] on topological fields is an important prelude.

References

1. Oscar Goldman and Chih-han Sah, On a special class of locally compact rings, *J. of Algebra* 4(1966), 71–95.
2. _____, Locally compact rings of special type, *J. of Algebra* 11(1969), 363–454.
3. Kenkichi Iwasawa, On the rings of valuation vectors, *Ann. of Math.* 57(1953), 331–56.
4. Nathan Jacobson and Olga Taussky, Locally compact rings, *Proc. Natl. Acad. Sci. USA* 21(1935), 106–8.
5. Nathan Jacobson, Totally disconnected locally compact rings, *Amer. J. of Math.* 58(1936), 433–49.
6. _____, A note on topological fields, *Amer. J. of Math.* 59(1937), 889–94.
7. _____, *Collected Mathematical Papers*, 3 volumes, Birkhäuser, Boston, 1989.
8. Irving Kaplansky, Topological rings, *Amer. J. of Math.* 69(1947), 153–83.
9. _____, Topological rings, *Bull. Amer. Math. Soc.* 54(1948), 809–26.
10. _____, Locally compact rings II, *Amer. J. of Math.* 73(1951), 20–24.
11. _____, Locally compact rings III, *Amer. J. of Math.* 74(1952), 929–35.
12. Carl Ludwig Siegel, Discontinuous groups, *Ann. of Math.* 44(1943), 674–89 (*Ges. Abh.* vol. II, 390–405).
13. L. A. Skornyakov, Einfache lokal bikompakte Ringe, *Math. Zeit.* 87(1965), 241–51.
14. Seth Warner, *Topological Fields*, North-Holland, New York, 1989.
15. _____, *Topological Rings*, North-Holland, New York, 1993.

RINGS WITH A POLYNOMIAL IDENTITY

IRVING KAPLANSKY

1. Introduction. In connection with his investigation of projective planes, M. Hall [2, Theorem 6.2][1] proved the following theorem: a division ring D in which the identity

$$(1) \qquad (xy - yx)^2 z = z(xy - yx)^2$$

holds is either a field or a (generalized) quaternion algebra over its center F. In particular, D is finite-dimensional over F, something not assumed a priori. The main result (§2) in the present paper is the following: if D satisfies *any* polynomial identity it is finite-dimensional over F. There are connections with other problems which we note in §§3, 4.

2. Proof of finite-dimensionality. Let A be an algebra (no assumption of finite order) over a field F. We denote by $F[x_1, \cdots, x_r]$ the free algebra generated by r indeterminates over F. We say that A satisfies a polynomial identity if there exists a nonzero element f in $F[x_1, \cdots, x_r]$ such that $f(a_1, \cdots, a_r) = 0$ for all a_i in A.

LEMMA 1.[2] *If A satisfies any polynomial identity, then it satisfies a polynomial identity in two variables.*

PROOF. Suppose A satisfies the equation $f(x_1, \cdots, x_r) = 0$. Replacing x_i by $u^i v$ we obtain the equation $g(u, v) = 0$, with g a polynomial which is not identically zero.

LEMMA 2. *If A satisfies any polynomial identity, it satisfies a polynomial identity which is linear in each variable.*

PROOF. Suppose A satisfies $f(x_1, \cdots, x_r) = 0$ and that f is not linear in x_1. Then

$$f(y + z, x_2, \cdots, x_r) - f(y, x_2, \cdots, x_r) - f(z, x_2, \cdots, x_r) = 0$$

is satisfied by A. This is a polynomial (in $r+1$ variables), not identically zero, and with degree in y and z lower than the degree of f in x_1. By successive steps of this kind we reach a polynomial linear in all variables.

Presented to the Society, September 3, 1947; received by the editors August 20, 1947.

[1] Numbers in brackets refer to the bibliography at the end of the paper.

[2] Cf. [7, Satz 2].

575

LEMMA 3. *Suppose that A satisfies an identity g = 0 where g is linear in all its variables, and let K be a field over F. Then the Kronecker product $A \times K$ also satisfies g = 0.*

PROOF. We have to show that $g(x_1, \cdots, x_r)$ vanishes when the x's are replaced by elements $\sum a_i k_i$ of $A \times K (a_i \in A, k_i \in K)$. This follows from the linearity of g and the fact that the k's commute with everything.

LEMMA 4.[3] *If A satisfies an identity f = 0 which is of degree not greater than n in each of its variables, and if F has at least $n+1$ elements, then A satisfies an identity which is homogeneous in each of its variables.*

PROOF. We write $f = \sum f_i$ where f_i is of degree i in x_1. Replacing x_1 by λx_1 where $\lambda \in F$, we find $\sum \lambda^i f_i = 0$. We do this for n different nonzero scalars. Using the resulting Vandermonde determinant, we obtain $f_i = 0$. We repeat this procedure with each of the variables.

LEMMA 5. *If $f(x, y)$ is any nonzero element of $F[x, y]$, we can find two matrices a, b with elements in F such that $f(a, b) \neq 0$.*

PROOF. A field of degree k over F may be represented by $k \times k$ matrices over F. Thus by taking matrices split into suitable smaller blocks, we may arrange for a scalar field of any desired (finite) size.

Suppose that on the contrary f vanishes for all matrices with elements in F. Then by Lemma 4 we may pass to a polynomial $g(x, y)$ which also vanishes for any matrices over F, and which is homogeneous in y, say of degree t. Take a to be a diagonal matrix: $a = \mathrm{diag} (u_1, \cdots, u_n)$, with n greater than the degree of f, and the u's as yet undefined. For b we choose an n by n matrix which permutes the elements of a diagonal matrix cyclically. Thus

(2) $b^i a b^{-i} = \mathrm{diag} (u_{i+1}, u_{i+2}, \cdots, u_n, u_1, \cdots, u_i).$

We consider $g(a, b)b^{-t}$. Using the relation (2), we systematically push b to the right of a, and every such operation induces a cyclic permutation. Corresponding to the typical monomial $x^i y^j x^k y^l x^m \cdots$ in g, we obtain the term

$$(u_1)^i (u_{1+j})^k (u_{1+j+l})^m \cdots$$

in the upper left corner of $g(a, b)b^{-t}$. Moreover given any such term we can unambiguously reconstruct the monomial from which it arose (we use here the fact that n exceeds the degree of g, so that there is no overlapping of the u's). Thus the upper left entry of $g(a, b)b^{-t}$ is a

---

[3] Cf. [7, Satz 1].

polynomial $h(u_1, \cdots, u_n)$ (that is to say, an ordinary commutative polynomial), which is not identically zero. By enlarging the field of scalars, if again necessary, we can find values of the u's which make h different from 0. For this choice we have $g(a, b) \neq 0$, a contradiction.

We prove the main theorem of the paper not only for division rings, but more generally for primitive algebras in the sense of Jacobson.

THEOREM 1. *A primitive algebra satisfying a polynomial identity is finite-dimensional over its center.*

PROOF. First we assert that A is a matrix algebra of finite order over a division algebra. For if not we would have[4] that for every k, A has a sub-algebra homomorphic to D_k where D is a division algebra over F, and D_k is the algebra of matrices of order k over D. The polynomial identity satisfied by A is inherited by D_k for all k, and therefore holds for all matrices with coefficients in F. By Lemma 1 we may assume that the polynomial in question has two variables, and this contradicts Lemma 5.

The problem is thus reduced to the case where A is a division algebra, say with center Z. By Zorn's lemma, select a maximal subfield K of A, and form the Kronecker product $A \times K$ over Z. By a theorem of Nakayama and Azumaya [7], $A \times K$ is a dense algebra of linear transformations in a vector space over K. By Lemma 2, A has a linear polynomial identity, and by Lemma 3 this identity survives in $A \times K$. By Lemma 1 we pass to an identity in two variables in $A \times K$. Repetition of the argument of the preceding paragraph shows that $A \times K$ is finite-dimensional over K, hence A is finite-dimensional over Z.

The following further result is virtually a restatement of Lemma 5.

THEOREM 2. *The free algebra in any number of indeterminates has a complete set of finite-dimensional representations.*

PROOF. Let $f(x_1, \cdots, x_r)$ be an element in the free algebra. We make the replacement $x_i = u^i v$ of Lemma 1. For the resulting polynomial $g(u, v)$, we can find, by Lemma 5, matrices a and b such that $g(a, b) \neq 0$. The mapping induced by sending x_i into $a^i b$ $(i = 1, \cdots, r)$ and all other x's into 0 is a homomorphism of the free algebra into the algebra generated by a and b, and in this representation $f(x_1, \cdots, x_r)$ is not sent into 0.

3. **Algebraic algebras.** It is true conversely that any algebra of finite order satisfies a polynomial identity. For algebras over the real numbers, this was shown by Wagner [8, pp. 531–532]; his identity

---

[4] [3, Th. 3]. Jacobson has since shown that a unit element is not necessary for this theorem (oral communication).

for n by n matrices has degree $(n^2-n+2)(n^2-n+4)/4$. E. R. Kolchin has remarked that any algebra of order n satisfies an identity of degree $n+1$:

$$\sum \pm x_1 x_2 \cdots x_{n+1} = 0$$

where the sign is positive or negative according as the permutation is even or odd. Incidentally the existence of these identities furnishes a brief proof of Malcev's result [6, Theorem 9] that the free algebra in two or more indeterminates cannot be faithfully represented by matrices of finite order.

A wider class of algebras than those of finite order is covered by the following result.[5]

THEOREM 3. *An algebraic algebra of bounded degree satisfies a polynomial identity.*

PROOF. To avoid complications of notation, we give the proof for the case where the degrees of the elements are bounded by 3. Thus every element x satisfies an equation

(3) $$x^3 + \alpha x^2 + \beta x = 0$$

where α, β are scalars which of course depend on x. (We do not assume a unit element, and so no constant appears in (3).) We take the commutator with y, obtaining

$$[x^3, y] + \alpha [x^2, y] + \beta [x, y] = 0$$

where $[x, y]$ denotes $xy - yx$. Then we take the commutator with $[x, y]$ and finally with $[[x, y], [x^2, y]]$. The result is:

$$[[[x, y], [x^2, y]], [[x, y], [x^3, y]]] = 0,$$

a polynomial of degree 11. The polynomial moreover is not identically zero, since for example there is only one term

$$xyx^2 yxyx^3 y.$$

In the general case where the degree is bounded by n, we get an identity of degree $2^{n+1} - 2^{n-1} - 1$. The case $n=2$ yields an identity equivalent to (1), with z replaced by x.

Combining Theorems 1 and 3, we obtain a new proof of a theorem due to Jacobson [3, Theorems 5 and 7].

THEOREM 4. *A primitive algebraic algebra of bounded degree is finite-dimensional over its center.*

---

[5] I am indebted to Prof. Jacobson for Theorem 3 and its proof.

An outstanding question in the theory of algebraic algebras is Kurosch's analogue of Burnside's problem: is an algebraic algebra necessarily locally finite?[6] Affirmative answers have been contributed in two cases: (1) algebraic algebras of bounded degree [3, 4, and 5], and (2) representable[7] algebraic algebras [6]. A third (trivial) instance where the answer is affirmative is the commutative case. A hypothesis that is weaker than any of these is the assumption of a polynomial identity. It thus seems natural to try Kurosch's problem next for algebraic algebras with a polynomial identity. We shall now contribute an affirmative answer for the case of a nil algebra; in the light of [3, Theorem 15] the question is thus reduced to the semi-simple case.

THEOREM 5. *A nil algebra satisfying a polynomial identity is locally finite.*

PROOF. We use the terminology and results of Levitzki (cf. [5] and the references given there). Exactly as in [5] we reduce to the case where A is semi-regular. We choose an element $a \neq 0$ such that $a^2 = 0$. The algebra Aa is also semi-regular. We convert our identity into one, say $f(x_1, \cdots, x_r) = 0$, which is linear in each variable (Lemma 2). Suppose the variable x_1 actually appears first in at least one of the monomials comprising f. We gather all such terms and write

$$(4) \qquad f(x_1, \cdots, x_r) = x_1 g(x_2, \cdots, x_r) + h(x_1, \cdots, x_r).$$

Each monomial of h has a factor x_1 which appears later than the first term. It follows that if we substitute a for x_1, and any elements of Aa for $x_2, \cdots, x_r$, we introduce a factor a^2 in each term of h and thus make h vanish. Thus $ag(x_2, \cdots, x_r)$ must also vanish for $x_2, \cdots, x_r$ in Aa. This makes g a right annihilator of Aa. Since Aa is semi-regular it has no right annihilator except 0. Hence Aa satisfies the identity $g = 0$. By induction on the degree of the identity we have that Aa is locally finite, which contradicts its semi-regularity.

4. Further remarks. (a) Wagner's main theorem in [8] asserts that an ordered algebra over the reals satisfying a polynomial identity is necessarily commutative. Our results furnish a short proof of a special case: *any ordered primitive algebra satisfying a polynomial identity is commutative.* This is an immediate consequence of Theorem

---

[6] An algebra is locally finite if every finitely generated sub-algebra is of finite order.

[7] An algebra is representable if it is isomorphic to a ring of matrices of finite order over some extension field.

1 and Albert's theorem [1] to the effect that any ordered algebra of finite order is a field.

(b) The argument of Theorem 1 can be made to yield an explicit upper bound for the order of the algebra over its center, in terms of the degree and number of variables of the identity; the bound however is far too generous. In a special case like Hall's theorem there is no difficulty in getting the precise bound. Specifically it is only necessary to produce 3 by 3 matrices violating the linearized form of (1). One may in this way also verify the following result: a semi-simple algebra satisfying a polynomial identity of degree not greater than 3 is commutative.

(c) We have for simplicity given all the results in this paper for algebras, but they may be extended to rings as follows. Assume that the polynomials in question have as coefficients operators α such that $\alpha x = 0$ implies $x = 0$. It is to be observed that this holds in particular if the coefficients are ± 1. Thus we may assert that any primitive ring satisfying (1) is an algebra of finite order; and Theorem 5 thus extended subsumes Levitzki's theorem [5] that a finitely generated nil-ring of bounded index is nilpotent.

Added in proof (May 1, 1948). I am indebted to Dr. Harish-Chandra for the following brief proof of Lemma 5. Let n be the degree of $f(x, y)$ and I the ideal $F[x, y]$ generated by monomials of degree not less than $n+1$. Then $F[x, y]/I$ is an algebra, which can be faithfully represented by matrices since it is of finite order. This gives us matrices for which $f \neq 0$.

BIBLIOGRAPHY

1. A. A. Albert, *On ordered algebras*, Bull. Amer. Math. Soc. vol. 46 (1940) pp. 521–522.

2. M. Hall, *Projective planes*, Trans. Amer. Math. Soc. vol. 54 (1943) pp. 229–277.

3. N. Jacobson, *Structure theory for algebraic algebras of bounded degree*, Ann. of Math. (2) vol. 46 (1945) pp. 695–707.

4. I. Kaplansky, *On a problem of Kurosch and Jacobson*, Bull. Amer. Math. Soc. vol. 52 (1946) pp. 496–500.

5. J. Levitzki, *On a problem of A. Kurosch*, Bull. Amer. Math. Soc. vol. 52 (1946) pp. 1033–1035.

6. A. Malcev, *On the representations of infinite algebras*, Rec. Math. (Mat. Sbornik) N. S. vol. 13 (1943) pp. 263–286 (Russian with English summary).

7. T. Nakayama and G. Azumaya, *Über einfache distributive Systeme unendlicher Ränge* II, Proc. Imp. Acad. Tokyo vol. 20 (1944) pp. 348–352.

8. W. Wagner, *Über die Grundlagen der projektiven Geometrie und allgemeine Zahlensysteme*, Math. Ann. vol. 113 (1937) pp. 528–567.

UNIVERSITY OF CHICAGO

Afterthought

In his collected papers [2, vol. II, p. 1, and vol. III, starting on p. 475] Nathan Jacobson has recounted the early history of polynomial identities and the subsequent theory as we now know it. I shall not duplicate any of this except to describe again one scene.

On that summer day in 1947 I listened with great interest as he presented a proof of Marshall Hall's characterization of quaternion division algebras by the identity that says that the square of any commutator is central. Right after the lecture we talked about it. One of us speculated that maybe *any* identity would imply finite-dimensionality. Which of us? I don't remember for sure. I have checked with Jake and neither does he. The human race will apparently never know.

The scene switches to "the Point." This is a bit of land jutting out into Lake Michigan, near the University of Chicago. The swimming is great in the summer (although not exactly legal). I have a clear recollection of getting the idea for Lemma 5 on page 576, the key point in the paper (of course it subsequently was superseded by Harish-Chandra's remark). It happened while sunbathing between two swims.

I wasted no time getting it written up and submitted; the *AMS Bulletin* received it on August 20, 1947.

I turn next to central polynomials. In [2, vol. III, p. 487] Jake says that already in that summer of 1947 we discussed the possibility that central polynomials might exist for matrices larger than 2×2. In a draft of this note I wrote that this was at variance with my recollection and that I believed that for some reason or other the idea took a while to surface. In a letter commenting on the draft Jake wrote me that he stood by his statement. So here again the historical record will have to stand as uncertain. As for what appeared in writing, there is a relevant passage on page 237 of [3]. The context there concerned an effort to construct central elements in certain C^*-algebras (the effort succeeded by a different method); it was noted that 2×2 matrices were an "honorable exception," since the central polynomial $(xy - yx)^2$ was then available. It might be said that this passage implicitly asked whether central polynomials exist for larger matrices, but I would rather say that I failed to raise the question explicitly since I did not believe that they existed (see below). Incidentally, the *AMS Transactions* received this paper on December 10, 1949.

In [4] the problem of the existence of central polynomials was at last explicitly stated (although I messed it up and had to redo it in [5]).

I have heard it said that I conjectured the existence of central polynomials. Well, we are all cut from the same cloth. When a proposed theorem turns out to be true, we bask when credited with conjecturing it. When the opposite happens, we protest that we merely raised the question (when we can get away with it). But this time I am going to be honest. I did not conjecture it, certainly not in writing. In conversation, if anything, I was pessimistic. All the greater was my pleasure when Formanek [1] and Razmyslov [6] independently constructed central polynomials for matrices of any size. (This was another one of those amazingly frequent coincidences across the Iron Curtain.)

There is quite a literature on polynomial identities. In the Small collection of reviews in ring theory it gets a whole section to itself: 11. In the 1940–1979 volume this stretches from page 227 to page 260, and I counted 208 papers.

References

1. Edward Formanek, Central polynomials for matrix rings, *J. of Algebra* 23(1972), 129–32.
2. Nathan Jacobson, *Collected Mathematical Papers*, 3 volumes, Birkhäuser, Boston, 1989.
3. Irving Kaplansky, The structure of certain operator algebras, *Trans. Amer. Math. Soc.* 70(1951), 219–55.
4. _____, Problems in the theory of rings, pp. 1–3 of a report on a conference on linear algebras, June 1956, NAS-NRC, Washington, Publ. 502, 1957.
5. _____, "Problems in the theory of rings" revisited, *Amer. Math. Monthly* 77(1970), 445–54.
6. J. P. Razmyslov, A certain problem of Kaplansky, *Izv. Akad. Nauk SSSR Ser. Mat.* 37(1973), 483–501 (Russian).

THE WEIERSTRASS THEOREM IN FIELDS
WITH VALUATIONS

IRVING KAPLANSKY[1]

In [2, Theorem 32][2] the author showed that an analogue of the Weierstrass-Stone theorem holds in topological rings having ideal neighborhoods of 0. Earlier, Dieudonné [1] had proved the Weierstrass-Stone theorem for the field of p-adic numbers. Now the field of p-adic numbers has an open subring (the p-adic integers) with ideal neighborhoods of 0. It seems plausible, therefore, to expect that the method of [2] will apply, provided one has a supplementary device for "getting into" the p-adic integers. This is in fact the case, and the result is applicable to any division ring with a valuation of rank one.[3] The requisite lemma reads as follows:

LEMMA 1. *Let F be a division ring with a valuation of rank one, and B its valuation ring. Let a be any nonzero element in F, and K a compact subset of F. Then there exists a (non-commutative) polynomial f with coefficients in F, without a constant term, and satisfying* $f(a) = 1$, $f(K) \subset B$.

PROOF. Let P denote the maximal ideal in B, that is, the set of all elements x with $V(x) > 0$, where V denotes the valuation. Let K' denote the subset of K with values less than $V(a)$; K' will again be compact. For any c in K' there is a compact open subset of K' containing c and contained in $c(1+P)$. Take a finite covering of K' consisting of such neighborhoods: say $U_1, \cdots, U_r$ with U_i contained in $c_i(1+P)$. Suppose the c's numbered so that $V(c_i) \geq V(c_{i+1})$. Now $1 - c_i^{-1}U_i$ is a compact subset of P, and consequently the values of its elements have a positive lower bound α_i. Choose integers $n(1), \cdots, n(r)$ in succession large enough so that

$$V(a^{-1}c_i) + \sum_{j=1}^{i-1} n(j)V(c_j^{-1}c_i) + n(i)\alpha_i \geq 0$$

for $i = 1, \cdots, r$. Then the polynomial

$$f(x) = 1 - (1 - a^{-1}x)(1 - c_1^{-1}x)^{n(1)} \cdots (1 - c_r^{-1}x)^{n(r)}$$

satisfies the requirements of the lemma.

Received by the editors April 18, 1949.
[1] John Simon Guggenheim Memorial Fellow.
[2] Numbers in brackets refer to the bibliography at the end of the paper.
[3] That is to say, a valuation whose value group is a subgroup of the real numbers.

356

We shall restate [2, Theorem 32] in a slightly sharpened form.

LEMMA 2. *Let A be a topological ring with unit element and ideal neighborhoods of 0, let X be a totally disconnected locally compact Hausdorff space, and C the ring of all continuous functions from X to A vanishing at ∞. Topologize C by uniform convergence. Let D be a closed subring of C, containing for any distinct points x, $y \in X$ and any a, $b \in A$, a function f with $f(x) = a$, $f(y) = b$. Then $D = C$.*

PROOF. Let U be a fixed ideal neighborhood of 0, $1 \notin U + U$. Let K be a compact subset of X and x a point not in K. Then D contains a function f with $f(x) = 1$ and $f(y) = 0$ for a given y in K. The function f will take values in U in a suitable neighborhood of y. A finite number of these neighborhoods cover K; if g is the product of the corresponding f's, we have $g(x) = 1$, $g(K) \subset U$.

We start again with an arbitrary $z \in X$ and an h in D with $h(z) = 1$. There is a compact neighborhood L of z with $h(L) \subset 1 + U$, and a larger compact set M such that h is in U in the complement of M. For any given w in $M - L$ we can, by the preceding paragraph, find a function p in D with $p(L) \subset U$, $p(w) = h(w)$. Then the function $h(h - p)$ has the following properties: in L its values lie in $1 + U$, and it vanishes at w and accordingly lies in U in a neighborhood of w. A finite number of these neighborhoods cover the compact set $M - L$. The product of the corresponding elements $h(h - p)$ gives us an element q in D with $q(L) \subset 1 + U$, $q(L') \subset U$, L' the complement of L. By combining such elements we can get "within U" of the characteristic function of any compact open set. Since D is closed, it actually contains all characteristic functions of compact open sets, from which it follows readily that $D = C$.

As an immediate consequence of Lemmas 1 and 2, we have the following generalization of Dieudonné's theorem:

THEOREM. *Let F be a division ring with a valuation of rank one, X a totally disconnected locally compact Hausdorff space, C the ring of all continuous functions from X to F vanishing at ∞. Topologize C by uniform convergence. Let D be a closed subring of C, admitting left-multiplication by the constant functions, and containing for any two distinct points x, $y \in X$ a function vanishing at x but not at y. Then $D = C$.*

BIBLIOGRAPHY

1. J. Dieudonné, *Sur les fonctions continues p-adiques*, Bull. Sci. Math. (2) vol. 68 (1944) pp. 79–95.
2. I. Kaplansky, *Topological rings*, Amer. J. Math. vol. 69 (1947) pp. 153–183.

INSTITUTE FOR ADVANCED STUDY

Afterthought

I have been fascinated by the Weierstrass-Stone theorem ever since I learned about it as a student at Harvard. So when Dieudonné proved his p-adic Weierstrass theorem, it was natural for me instantly to think that this too should be generalized à la Stone. Then a pleasant thing happened. I recalled that I had already proved a Weierstrass-Stone–type theorem for compact rings, and it turned out that with one extra idea this worked.

Moral: You never can tell where an oddball piece of mathematics might some day be useful.

The reference below is a survey article. It contains several references to developments subsequent to my paper.

Reference

J. B. Prolla, On the Stone-Weierstrass theorem for modules over non-Archimedean valued fields, pp. 413–32 in *Functional Analysis, Holomorphy, and Approximation Theory* (Rio de Janeiro, 1980), North-Holland, 1982; *Math. Rev.* 84c, 46078.

ANNALS OF MATHEMATICS
Vol. 53, No. 2, March, 1951

PROJECTIONS IN BANACH ALGEBRAS[1]

By Irving Kaplansky

(Received March 2, 1950)

1. Introduction

There have been two main contributions to the theory of operator algebras on Hilbert space. In a series of five memoirs, Murray and von Neumann have made important strides toward the structure theory of the weakly closed case. In work begun by the Russian school, a study has been made of the more general case where merely uniform closure is assumed. In terminology suggested by Segal, we call these W^*-algebras and C^*-algebras respectively. A notable advantage of the C^*-case is the existence of an elegant system of intrinsic postulates due to Gelfand and Neumark [2]; so one can, and does, study C^*-algebras in an abstract fashion that pays no attention to any particular representation.

A corresponding characterization of W^*-algebras is not known, but nevertheless several substitutes have been suggested. In [6] von Neumann postulated from the start a second topology behaving like the weak topology. Steen [8] assumed completeness relative to a topology induced by positive functionals. In the present paper the entire burden will be thrown upon a more algebraic, and in some sense more elementary assumption; briefly put, our postulate is the assumption of least upper bounds in the partially ordered set of projections—the precise axioms are given in §2. We call the algebras in question AW^*-algebras (the "A" suggesting "abstract").

This work is in essence a continuation of the study that was begun by Rickart in [7]. The exact connection between AW^*-algebras and his B_p^*-algebras is explained in the appendix; the main idea is that it is necessary to strengthen his postulates suitably if they are to work well in the non-separable case.

The fundamental role in the theory is played by the Murray-von Neumann concept of the equivalence of projections. The first problem one encounters is that of proving the additivity of equivalence. The case where all projections involved are orthogonal is rapidly disposed of in Lemma 3.1. With the aid of this result we then develop a preliminary structure theory which at length enables us to prove unrestricted additivity (Theorem 5.5).

Even in the case of W^*-algebras the main theorems are formally new, since there is no assumption of separability, and since everything is done not just for factors but for algebras "in the large". In particular, the classification into types I, II and III can be extended to algebras which are not factors, and every AW^*-algebra is a unique direct sum of algebras of the three types. There is another classification cutting across this: the division into finite and infinite cases. In the present paper we carry the structure theory far enough to verify that the projections in a finite AW*-algebra form a continuous geom-

---

[1] Part of the work on this paper was done while the author was a fellow of the John Simon Guggenheim Memorial Foundation.

235

etry. By making use of the known results on continuous geometries, we thus establish the existence of a suitable dimension function in finite AW^*-algebras.

2. The lattice of projections

By a C^*-algebra we mean a uniformly closed self-adjoint algebra of operators on a Hilbert space. One may give an equivalent intrinsic set of axioms due to Gelfand and Neumark: a C^*-algebra is a Banach *-algebra in which, for every x, $\| x^*x \| = \| x \|^2$ and x^*x has a quasi-inverse.

A W^*-algebra is a weakly closed self-adjoint algebra of operators on a Hilbert space. No intrinsic axioms for a W^*-algebra are known. (One must be a little careful in formulating this problem, for a C^*-algebra may admit two faithful representations, in only one of which it is weakly closed. Presumably the desired axioms would characterize C^*-algebras having *some* faithful weakly closed representation.) The purpose of this paper is to study the consequences of two axioms which, while they fall short of characterizing W^*-algebras, do preserve many of their features.

In any C^*-algebra we may partially order the projections (= self-adjoint idempotents) by specifying that $e \geq f$ means $ef = f$ (the latter being equivalent to $fe = f$). We observe that if e and f commute, they have $e + f - ef$ as a least upper bound (LUB); this is apparently all that can be said in general concerning the lattice properties of this partially ordered set. However it is known that in a W^*-algebra the projections form a complete lattice. Postulate A asserts a portion of this, and postulate B assures us of a sufficient supply of projections (otherwise postulate A might be vacuous).

Definition. An AW^*-algebra is a C^*-algebra satisfying:

(A) In the partially ordered set of projections, any set of orthogonal projections has a LUB,

(B) Any maximal commutative self-adjoint subalgebra is generated by its projections (that is, it is equal to the smallest closed subalgebra containing its projections).

REMARK. That W^*-algebras satisfy postulate B is known; a maximal commutative subalgebra is weakly closed, and postulate B then follows from any version of the spectral theorem. In assessing the strength of these postulates, it is helpful to examine the commutative case. Then (A) and (B) are precisely equivalent to the assertion that the space of maximal ideals is the Boolean space of a complete Boolean algebra. On the other hand, for a commutative W^*-algebra the space of maximal ideals is the Boolean space of a measure algebra. Thus in the commutative case, AW^*-algebras generalize W^*-algebras to the same extent that complete Boolean algebras generalize measure algebras.

Our use of postulate (B) will for the present be confined to the derivation of the following lemma:

LEMMA 2.1. *Let A be a commutative C^*-algebra which is generated by its projections; let $x \in A$ and a positive ε be given. Then A contains a projection e which is a multiple of x and satisfies $\| x - ex \| < \varepsilon$.*

PROOF. By the known structure theorem we may represent A as the set of all continuous complex functions vanishing at infinity on a locally compact Hausdorff space T; write $x(t)$ for the function representing x. Our hypothesis implies that T is total disconnected. There exists a compact open set U, containing the (compact) set where $|x(t)| \geq \varepsilon$ and contained in the set where $|x(t)| \geq \varepsilon/2$. By taking e to be the characteristic function of U we satisfy the requirements of the lemma.

Our chief tool throughout the paper will be a sort of continuity property of the LUB of orthogonal projections.

LEMMA 2.2. *Let A be a C^*-algebra in which postulate* (B) *holds, and let $e_i \, \epsilon \, A$ be orthogonal projections with a LUB e in A. Then:* (a) $xe_i = 0$ *for all i implies $xe = 0$,* (b) $e_i x = x e_i$ *for all i implies $ex = xe$.*

PROOF. (a) Write $y = ex^*xe$, and embed y and e in a maximal commutative self-adjoint subalgebra C. By Lemma 2.1 we can choose a projection f in C, such that f is a multiple of y and $\|fy - y\| < \varepsilon$. Since $ye_i = 0$, we have $fe_i = 0$, whence the projection $e - fe$ satisfies $e - fe \geq e_i$ for every i. Hence $e - fe \geq e$, $fe = 0$. Since $\|e\| = 1$ and $y = (yf - y)e$, we find $\|y\| < \varepsilon$. This being true for every ε, we have $y = 0$, $xe = 0$.

(b) It suffices to consider the case where x is self-adjoint—for if e_i commutes with x it also commutes with x^*, and we may write $x = (x + x^*)/2 + (x - x^*)/2$. So we assume $x = x^*$, and observe that $e \geq e_i$ implies $(x - ex)e_i = 0$. By part (a), $(x - ex)e = 0$. Taking adjoints, we find $ex = xe$.

THEOREM 2.3. *Let S be any subset of an AW^*-algebra A. Then the right annihilator of S in A is a principal right ideal generated by a projection.*

PROOF. In the right annihilator I of S, select (Zorn's lemma) a maximal set $\{e_i\}$ of orthogonal projections, and write $e = \text{LUB } e_i$. By Lemma 2.2, $e \, \epsilon \, I$. To prove that $I = eA$, we take $y \, \epsilon \, I$ and have to prove that $x = y - ey = 0$. If x is non-zero, so is xx^*, and by Lemma 2.1 there exists a non-zero projection f which is a (two-sided) multiple of xx^*. Then f (along with x) is in I, $ex = 0$ implies $ef = 0$, $e_i f = 0$ for all i, contradicting the maximality of $\{e_i\}$.

COROLLARY 1. *In an AW^*-algebra, the right annihilator of a right ideal is a principal two-sided ideal generated by a central projection.*

PROOF. The right annihilator in question is of the form eA, and is a two-sided ideal. In any ring without nilpotent ideals, this implies that e is in the center.

COROLLARY 2. *An AW^*-algebra has a unit element.*

PROOF. Apply Corollary 1 to the annihilator of 0.

COROLLARY 3. *The projections in an AW^*-algebra form a complete lattice.*

PROOF. Given an arbitrary set of projections $\{e_i\}$, let fA be their right annihilator. Then $e_i f = 0$, whence $e_i \leq 1 - f$. Also if $e_i \leq g$, then $e_i(1 - g) = 0$, $1 - g \, \epsilon \, fA$, $1 - f \leq g$. In other words, $1 - f$ is the LUB of $\{e_i\}$.

THEOREM 2.4. *If A is an AW^*-algebra, so is the center of A, any maximal commutative self-adjoint subalgebra of A, and any subalgebra of the form eAe, where e is a projection in A.*

PROOF. From part (b) of Lemma 2.2 it follows that the LUB of central projections is again central. This tells us that postulate (A) is valid in the center Z of A. Next let x be any element in Z. By Corollary 1 of Theorem 2.3, the annihilator of x in A is of the form eA with e a projection in Z. Hence the annihilator of x in Z is ϵZ. This is enough to assure us that postulate (b) holds in Z. The proof for the other two cases is similar and is left to the reader.

Following the terminology of [3, p. 411], we define the C^*-sum of C^*-algebras $\{A_i\}$ to be the algebra of all bounded sequences $\{a_i\}$, $a_i \in A_i$, with the natural norm and *-operation. It is straightforward to verify that a C^*-sum of AW^*-algebras is again an AW^*-algebra. Conversely, C^*-sums arise when the center of an AW^*-algebra is decomposed.

LEMMA 2.5. *Let A be an AW^*-algebra, and $\{e_i\}$ a set of orthogonal central projections with LUB 1. Then A is isomorphic to the C^*-sum of $\{e_iA\}$.*

PROOF. There is a natural map of A into the C^*-sum: $x \rightarrow \{e_ix\}$. It is an isomorphism, and [3, p. 411] norm-preserving. We have to see that it is onto. To see that A contains a projection with prescribed components $f_i \in e_iA$, we simply take the LUB of $\{f_i\}$. We now have all the projections of the C^*-sum, and the latter is generated by its projections, since it is an AW^*-algebra.

3. Equivalence of projections

We say that the projections e and f are equivalent, written $e \sim f$, if there exists an element x with $xx^* = e$, $x^*x = f$; note that x is necessarily in eAf, and that the mapping $a \rightarrow x^*ax$ is a *-isomorphism of eAe onto fAf. We write $e \precsim f$ if $e \sim g$ with $g \leq f$. We shall take for granted the elementary properties of these relations. Our first main problem is to prove the additivity of equivalence: if $\{e_i\}$ are orthogonal projections with LUB e, $\{f_i\}$ orthogonal projections with LUB f, and $e_i \sim f_i$, we must prove $e \sim f$. The proof given by Murray and von Neumann [4, p. 142] depends on taking a weak limit of partially isometric elements, and so is not available to us; we must throw the entire burden on the process of taking the LUB of projections $\{e_i\}$. In the special case where e_i and f_i are orthogonal, we can link them by a sort of "bisecting" projection, as in the following proof. Unrestricted additivity of equivalence will be proved later (Theorem 5.5), but only after some preliminary theory has been developed.

LEMMA 3.1. *In an AW^*-algebra, suppose $e_i \sim f_i$, where $\{e_i\}$ are orthogonal, $\{f_i\}$ are orthogonal, and each e_i is orthogonal to each f_j. Let e, f be the LUB's of e_i, f_i. Then $e \sim f$.*

PROOF. Suppose $x_ix_i^* = e_i$, $x_i^*x_i = f_i$. Write $2g_i = e_i + x_i + x_i^* + f_i$. Then g_i is a projection satisfying

(1) $2e_ig_ie_i = e_i$, $2g_ie_ig_i = g_i$,   and same with f_i replacing e_i.

Moreover $\{g_i\}$ are orthogonal. Let g denote their LUB, and write $x = e(2g - 1)$. Then $xx^* = e$, and it remains to prove

(2) $x^*x = (2g - 1)e(2g - 1) = f$.

Let us write e_i' for the LUB of $\{e_j, j \neq i\}$. By Lemma 2.2, e_i' is orthogonal to g_i. Moreover we have $e = e_i + e_i'$, and so $eg_i = e_ig_i$. Similarly $ge_i = g_ie_i$. Putting these together, we get $2ege_i = 2e_ig_ie_i$, and the latter is e_i by (1). So $(2eg - 1)e_i = 0$ and hence, by Lemma 2.2, $(2eg - 1)e = 0$. Similar arguments establish

$$(3) \qquad 2ege = e, \qquad 2geg = g, \qquad \text{and same with } f \text{ instead of } e.$$

Let us write $h = (2g - 1)e(2g - 1)$, $k = (2g - 1)f(2g - 1)$. Using (3) and the orthogonality of e and f (two applications of Lemma 2.2), we find $fhf = 4fgegf = 2fgf = f$. Similarly $eke = e$. Again $hfh = (2g - 1)eke(2g - 1) = h$. From $fhf = f$ we get $(fh - f)(fh - f)^* = 0$, $fh = f$. From $hfh = h$ we similarly get $hf = h$. Hence $h = f$, and this proves (2).

From Lemma 3.1 we can deduce the analogue of the Schröder-Bernstein theorem; the proof by Murray and von Neumann [4, p. 152] can be quoted nearly verbatim and we accordingly omit it.

THEOREM 3.2. *If $e \precsim f$ and $f \precsim e$ in an AW^*-algebra, then $e \sim f$.*

We next take up the question of comparability of projections, again with an added assumption of orthogonality; the assumption in Lemma 3.4 that e and f are orthogonal will be dropped later (Theorem 5.6). We first prove a preliminary lemma.

LEMMA 3.3 *Let A be a C^*-algebra satisfying posiulute (B) of §2, and let e and f be projections in A such that $eAf \neq 0$. Then A contains non-zero projections e' and f' such that $e' \leq e$, $f' \leq f$ and $e' \sim f'$.*

PROOF. Let x be a non-zero element of eAf, and embed xx^* in a maximal commutative self-adjoint subalgebra C. In the functional representation for C, xx^* is represented by a non-negative function; from this, and the total disconnectedness of the space of maximal ideals, we see that there exists a self-adjoint element y in C such that xx^*y^2 is a non-zero projection e'. We have $e' \leq e$, since $ex = x$. Write $z = yx$, so that $zz^* = e'$, and z^*z is an equivalent projection f'. Since $xf = x$, we have $f' \leq f$.

LEMMA 3.4. *Let e and f be orthogonal projections in an AW^*-algebra. Then there exist central projections g, h such that $g + h = 1$, $eg \succsim fg$, $eh \precsim fh$.*

PROOF. Consider sets $\{e_i\}$, $\{f_i\}$ of orthogonal projections with $e_i \leq e$, $f_i \leq f$, $e_i \sim f_i$. Zorn permits us to choose among these a maximal set, and for this choice we let k, m be the respective LUB's of $\{e_i\}$, $\{f_i\}$. By Lemma 3.1, $k \sim m$. Write $k' = e - k$, $m' = f - m$. If $k'Am' \neq 0$, then by Lemma 3.3 the sets $\{e_i\}$, $\{f_i\}$ can be enlarged, a contradiction. So $k'Am' = 0$. Consider the annihilators of $Ak'A$ and $Am'A$. By Corollary 1 of Theorem 2.3, these are generated by central projections, say g_1 and h_1. Then $Am'A \subset h_1A$, and taking annihilators we find $g_1A \supset (1 - h_1)A$. This implies $g + h = 1$ where $g = g_1$, $h = h_1(1 - g_1)$. That g and h fulfill the requirements of the lemma follows from the further facts: $gh = 0$, $gm' = 0$, $hk' = 0$.

For later use we derive a further lemma.

LEMMA 3.5. *In the AW^*-algebra A suppose $\{e_i\}$ is a maximal set of equivalent* *orthogonal projections, say with cardinal number* $\aleph$ *($\aleph$ infinite). Then there exists a non-zero central projection h such that h is the LUB of $\aleph$ orthogonal projections equivalent to he_1.*

PROOF. Write e for the LUB of $\{e_i\}$. We apply Lemma 3.4 to e_1 and $1 - e$. It is impossible that $e_1 \precsim 1 - e$, for this would contradict the assumed maximality of $\{e_i\}$. Hence there is a non-zero direct summand hA in which $h(1 - e) \precsim he_1$. Now he is the LUB of projections he_i each of which is equivalent to he_1. It will therefore suffice for us to prove $h \sim he$.

For this purpose we split the projections $\{he_i\}$ into two subsets of equal cardinal number: say $\{f_i\}$ and $\{g_i\}$, and write f and g for their respective LUB's. We have $he = f + g$ and $f \sim g$. Next we argue that the projection k, given by $k = g + h(1 - e)$, satisfies $k \sim f$. For $h(1 - e)$, being $\precsim he_1$, may be swallowed up by one of the projections comprising $\{f_i\}$, and the remaining f_i may still be paired off with the g_i, both sets being infinite. By Lemma 3.1 we have $k \precsim f$. Since on the other hand $f \sim g \leq k$, it follows from Theorem 3.2 that $k \sim f$, and then $k \sim g$. Further $f + k \sim f + g$, that is, $h \sim he$.

4. Some structure theory

The fundamental concept in the classification of AW^*-algebras is that of *finiteness* of projections [4, p. 155]; a projection e is finite if $f \sim e$ and $f \leq e$ imply $f = e$, otherwise infinite. It is to be observed that along with a finite projection, all smaller ones are likewise finite. We say that an AW^*-algebra is finite if its unit element is finite (and hence all projections finite); and we shall call A *purely infinite* if every non-zero central projection is infinite.

LEMMA 4.1. *The LUB of finite central projections is finite.*

PROOF. Let $\{e_i\}$ be a set of orthogonal finite central projections, with LUB e; and suppose $e \sim f \leq e$. Then for every i, $e_i = ee_i \sim fe_i$. The finiteness of e_i implies $e_i = fe_i$, $f \geq e_i$, hence $f \geq e$, $f = e$.

It follows from this lemma that an AW^*-algebra has a unique maximal finite central projection; on decomposing the algebra with respect to it, we obtain:

THEOREM 4.2. *Any AW^*-algebra is a direct sum of a finite algebra and a purely infinite algebra.*

We proceed to develop a fragment of structure theory for infinite algebras.

LEMMA 4.3. *An infinite AW^*-algebra contains an infinite set of equivalent orthogonal non-zero projections.*

PROOF. Suppose $xx^* = 1$, $x^*x \neq 1$; write $e_i = (x^*)^i x^i$, $y_i = (x^*)^i x^{i+1}$. Then $\{e_i - e_{i+1}\}$ are orthogonal non-zero projections and $y_i - y_{i+1}$ implements an equivalence between $e_i - e_{i+1}$ and $e_{i+1} - e_{i+2}$.

We can take the set of projections just constructed and expand it (Zorn's lemma) to a maximal set. By Lemma 3.5 we can then drop down to a direct summand hA such that h is the LUB of an infinite set of orthogonal equivalent projections. By further splitting these projections into $\aleph_0$ subsets of the same cardinal number, we exhibit h as the LUB of $\aleph_0$ equivalent orthogonal projections. Now assuming that A is purely infinite, we may repeat this construction on $(1 - h)A$; by a transfinite induction we obtain:

LEMMA 4.4. *In a purely infinite AW*-algebra, 1 can be exhibited as the LUB of $\aleph_0$ equivalent orthogonal projections.*

One could proceed to study the decomposition of AW^*-algebras according to the cardinal number of sets of equivalent orthogonal projections, obtaining in this way a refinement of Theorem 4.2. We shall not go into such matters, and shall for the present confine ourselves to deriving a corollary of Lemma 4.4.

LEMMA 4.5. *In a purely infinite AW^*-algebra there exists a projection e such that $1 \sim e \sim 1 - e$.*

PROOF. Split the projections given by Lemma 4.4 into four subsets of $\aleph_0$ each, say with LUB's $f_1, \cdots, f_4$. Then the projections in $f_2 + f_3$ can be paired off with those in f_1, and so (Lemma 3.1) $f_2 + f_3 \sim f_1$; similarly $f_1 + f_4 \sim f_2$. Adding we get $1 \sim e$, where $e = f_1 + f_2$, and further $e \sim 1 - e$.

We turn next to a decomposition theory which cuts across the above; this is the splitting into types analogous to Murray and von Neumann's types I, II and III. The appropriate formulation for algebras which are not necessarily factors is as follows.

DEFINITION. A non-zero projection e is abelian if eAe is commutative. An AW^*-algebra is of type I if every direct summand has an abelian projection; type II if it has no abelian projections and every direct summand contains a finite projection; type III if all projections are infinite. An AW^*-algebra which is both finite and of type II will be said to be of type II_1.

Types I, II and III are readily seen to be mutually exclusive, and moreover they are preserved under the taking of LUB's of central projections. Hence:

THEOREM 4.6. *Any AW^*-algebra is uniquely expressible as a direct sum of three algebras, respectively of types I, II and III.*

The following lemmas can be regarded as a preliminary crude version of dimension theory.

LEMMA 4.7. *Let e be an abelian projection in an AW^*-algebra, and f a projection with $f \leq e$. Then f is of the form $f = he$, where h is a central projection.*

PROOF. Consider the right annihilator of fA; by Corollary 1 of Theorem 2.3, this can be written $(1 - h)A$ where h is a central projection. We claim that $f = he$. For $f \leq h$ and $f \leq e$, whence $f \leq he$. We write $k = he - f$ and observe $fAk = f(eAe)k = 0$, since eAe is commutative and $f, k \in eAe$. It follows that $k \in (1 - h)A, k = 0$.

LEMMA 4.8. *Let e be an abelian projection in an AW^*-algebra A. Then A contains a non-zero central h such that h is the LUB of orthogonal projections equivalent to he.*

PROOF. Let $e_1 = e$, and let $\{e_i\}$ be a maximal set of orthogonal equivalent projections; write $f = $ LUB of $\{e_i\}$. It is impossible that $e \precsim 1 - f$; hence by Lemma 3.4 there exists a non-zero central projection g such that $(1 - f)g \sim k \leq eg$. If $k = 0$, g satisfies the requirements of the lemma; otherwise, by Lemma 4.7, $k = he$ with h a non-zero central projection, and h fulfills the requirements.

LEMMA 4.9. *In any C^*-algebra A, $x^*xAy = 0$ implies $xx^*Ay = 0$.*

PROOF. The left annihilator of Ay is a closed two-sided ideal I. By [3, Th. 7.2], $x^*x \in I$ implies $xx^* \in I$.

LEMMA 4.10. *Let A be a finite AW^*-algebra for which $1 = e_1 + \cdots + e_n$, with $\{e_i\}$ equivalent orthogonal abelian projections. Then any set of equivalent orthogonal non-zero projections in A has at most n members.*

REMARKS. 1. The assumption that A is finite is actually redundant (Theorem 6.2). 2. This, and more general results, can be proved by classical considerations on matrices, but it seems preferable to prove it by methods of the present theory.

PROOF. Let $f_1, \cdots, f_{n+1}$ be equivalent orthogonal projections. If $e_1 A f_1 = 0$, then (Lemma 4.9) $e_i A f_1 = 0$, $f_1 = 0$. So we may assume $e_1 A f_1 \neq 0$, and it follows from Lemma 3.3 that e_1 contains a projection g which is equivalent to a projection contained in f_1. Then A contains $n + 1$ orthogonal projections equivalent to g. But by Lemma 4.7, $g = he_1$ with h a central projection, and h is the sum of n orthogonal projections equivalent to g. This contradicts the finiteness of h.

LEMMA 4.11. *Let A be a finite AW^*-algebra of type I, and $\{e_i\}$ a set of orthogonal projections in A. Then there exists a direct summand of A in which all but a finite number of the e's vanish.*

PROOF. We cite Lemma 4.8; and after a change of notation we may assume that 1 is the sum of n equivalent orthogonal abelian projections (n is necessarily finite). Choose equivalent abelian projections f_i such that, the e's being appropriately numbered, we have $f_1 \leq e_1, \cdots, f_k \leq e_k$ with k as large as possible (by Lemma 4.10, $k \leq n$). Then if e_j is other than $e_1, \cdots, e_k$ we must have $f_1 A e_j = 0$; otherwise by Lemma 3.3 we could get the above set-up with $k + 1$ instead of k. If we write $(1 - h)A$ for the right annihilator of $f_1 A$, with h a central projection by Corollary 1 of Theorem 2.3, we have $he_j = 0$, and the direct summand hA fulfills the requirements.

LEMMA 4.12. *Let A be an AW^*-algebra with no abelian projections, and e a projection in A. Then e can be written $e = f + g$, with f and g equivalent orthogonal projections.*

PROOF. Consider sets of projections $\{f_i\}$, $\{g_i\}$, all mutually orthogonal, all $\leq e$, and with $f_i \sim g_i$. Choose a maximal set by Zorn's lemma, and let f, g be the respective LUB's. Then f and g are orthogonal (Theorem 2.3) and equivalent (Lemma 3.1). If $h = e - f - g$ is not zero, we observe that by hypothesis hAh is not commutative, and so there exists a projection $k \leq h$ such that k is not in the center of hAh. This says that $kA(h - k) \neq 0$. By Lemma 3.3 we can find non-zero equivalent projections contained in k and $h - k$, and these may be used to enlarge $\{f_i\}$ and $\{g_i\}$, a contradiction.

LEMMA 4.13. *Let e be a finite projection in an AW^*-algebra; suppose $e \lesssim 1 - e$, and let f, g be equivalent projections contained in e. Then $e - f \sim e - g$.*

PROOF. The equivalence between e and a portion of $1 - e$ induces an equivalence between $e - f$ and a projection h orthogonal to e. By Lemma 3.4, the algebra may be split into two summands in each of which $e - g$ and h are comparable. Thus after dropping down to a direct summand, we may assume, say, that $e - f \lesssim e - g$. If equivalence does not hold here, then on adding the equivalent projections f and g we contradict the finiteness of e.

LEMMA 4.14. *Let A be a finite AW^*-algebra, and $\{e_i\}$ a countably infinite set of orthogonal projections in A, with LUB e. Suppose $e \lesssim 1 - e$, and write $f_n = e_1 + \cdots + e_n$. Then it cannot be the case that $f_n \lesssim e - f_n$ for all n.*

PROOF. We shall suppose the contrary and derive a contradiction. Since $e \lesssim 1 - e$, there exists a projection g orthogonal to e and equivalent to $e - e_1$. We proceed to build a sequence of orthogonal projections $\{h_i\}$ with the properties: $h_i \leq g$ and $h_i \sim e_i$. Suppose this has been done as far as $n - 1$. We have

$$(4) \qquad f_n \lesssim e - f_n \leq e - e_1 \sim g.$$

In the equivalence induced by (4) of f_n into a portion of g, suppose that f_{n-1} corresponds to g'. Then

$$(5) \qquad e_n \lesssim g - g' \sim g - (h_1 + \cdots + h_{n-1}),$$

the last equivalence following from Lemma 4.13. Equation (5) shows that we may continue the process to h_n. Now let k be the LUB of the h's. By Lemma 3.1, $e \sim k$, and this yields $e \lesssim e - e_1$, in contradiction to finiteness.

5. Equivalence of projections, continued

We take up again the problem of the additivity of equivalence, and in the following lemma we nearly complete the solution.

LEMMA 5.1. *Let A be an AW^*-algebra, $\{e_i\}$ orthogonal projections with LUB e, $\{f_i\}$ orthogonal projections with LUB f. Suppose $e_i \sim f_i$ for every i. Then we can conclude $e \sim f$ in any of the following cases:* (a) $e \lesssim 1 - f$, (b) A *is purely infinite,* (c) A *is finite of type I,* (d) A *is arbitrary and* $\{e_i\}$, $\{f_i\}$ *are countable.*

PROOF. (a) We have an equivalence between e and a projection g orthogonal to f; suppose this induces an equivalence of e_i and g_i. Then we have $g_i \sim f_i$, and by Lemma 3.1, $g \sim f$. Hence $e \sim f$.

(b) By Lemma 4.5, there exists a projection g with $1 \sim g \sim 1 - g$. The equivalence between 1 and g induces equivalences between e_i, e, f_i, f and say e_i', e', f_i', f'. Then $e' \leq g \sim 1 - g \leq 1 - f'$. Hence part (a) is applicable and shows that $e' \sim f'$. This implies $e \sim f$.

(c) We cite Lemma 4.11. Since we know the additivity of equivalence for finite sums, we obtain an equivalence between the components of e and f in the appropriate direct summand. A transfinite induction completes the proof.

(d) Theorem 4.2 and part (b) reduce the problem to the case where A is finite. Then by Theorem 4.6, A is the direct sum of an algebra of type I and one of type II_1; and by part (c), the problem is entirely reduced to the II_1 case. For the present, we treat this final case only under the added countability assumption.

We begin by applying Lemma 4.12; we split each e_i, and correspondingly each f_i "in half". It suffices for us to study one of the two halves thus created. After a change of notation, this means that we can assume $e \lesssim 1 - e$. We next cite Lemma 4.14 and Lemma 3.4. They tell us that, after dropping down to a direct summand and changing notation, we have $e - g \lesssim g$ for a suitable n and $g = e_1 + \cdots + e_n$. If we write $h = f_1 + \cdots + f_n$, we have $g \sim h$ by finite additivity of equivalence, and so it will suffice to prove $e - g \sim f - h$. Now

$$e - g \lesssim g \sim h \leq 1 - (f - h),$$

and $e - g \sim f - h$ follows from part (a).

With countable additivity thus established, we proceed to develop another fragment of equivalence theory. Let x be an arbitrary element of an AW^*-algebra. By Theorem 2.3, the right annihilator of x may be written as $(1 - e)A$, e a projection. We shall call e the *right projection* for x, and there is a dual definition of *left projection*; this terminology agrees with that of Rickart [7, p. 534].

THEOREM 5.2. *The right and left projections of any element in an AW^*-algebra are equivalent.*

PROOF. Let e, f be the right and left projections for x. Let $y = x^*x$, and embed y in a maximal commutative self-adjoint subalgebra C; suppose y is represented by $y(t)$ in the functional representation for C. Let U_n be a compact open set containing the set where $| y(t) | \geq 1/n$, and contained in the set where $| y(t) | \geq 1/(n + 1)$. Write e_n for the characteristic function of U_n and $g = $ LUB of $\{e_n\}$; note that $e_n \leq e_{n+1}$. Since e_n is a multiple of y, and $y(1 - e) = 0$, we have $e_n(1 - e) = 0$, $e_n \leq e$, $g \leq e$. On the other hand

$$\| y(1 - g) \| = \| y(1 - e_n)(1 - g) \| \leq \| y(1 - e_n) \| \leq 1/n,$$

whence $y(1 - g) = 0$, $(1 - g)x^*x(1 - g) = 0$, $x(1 - g) = 0$, $1 - g \epsilon (1 - e)A$, $g \geq e$. We have proved $g = e$.

Next we note that there exists a self-adjoint element z_n in C such that $yz_n^2 = e_n$; write $w_n = xz_n$, and $w_n^*w_n$ is then equal to e_n, whence $w_nw_n^*$ is an equivalent projection f_n. From $(1 - f)x = 0$ we deduce $(1 - f)f_n = 0$, $f_n \leq f$. It is readily verified that $w_n^*w_{n+1} = e_n$, $w_nw_{n+1}^* = f_n$, from which it follows further that $f_n \leq f_{n+1}$, and that $w_{n+1} - w_n$ implements an equivalence between $e_{n+1} - e_n$ and $f_{n+1} - f_n$. Moreover $\{e_{n+1} - e_n\}$ are orthogonal projections with LUB e, and $\{f_{n+1} - f_n\}$ are orthogonal projections each $\leq f$. By part (d) of Lemma 5.1, we have $e \lesssim f$. Similarly $f \lesssim e$, and by Theorem 3.2, $e \sim f$.

LEMMA 5.3. *Let e and f be projections in an AW^*-algebra. Then the right projection for $e(1 - f)$ is $(e \cup f) - f$, and the left projection is $e - (e \cap f)$.*

PROOF. Let $x = e(1 - f)$, g be its right projection, and $h = f + g$. Then $xf = 0$ implies $f \epsilon (1 - g)A$, $fg = 0$. So h is a projection, and our task is to prove $h = e \cup f$. Of course $h \geq f$. To prove $h \geq e$, we observe $x(1 - g) = 0$, whence $e(1 - h) = 0$. Suppose that $k \geq e \cup f$. Then $1 - k$ annihilates both e and f, hence right-annihilates x, and so $1 - k \epsilon (1 - g)A$, $k \geq g$. Together with $k \geq f$, this gives $k \geq h$. We have proved $h = e \cup f$, and $g = (e \cup f) - f$. A dual argument proves that the left projection of x is $e - (e \cap f)$.

By combining the two previous results we obtain:

THEOREM 5.4. *For any projections e, f in an AW^*-algebra, $(e \cup f) - f \sim e - (e \cap f)$.*

We can now conclude the problem of the additivity of equivalence.

THEOREM 5.5. *Let A be an AW*-algebra, $\{e_i\}$ orthogonal projections with LUB e, $\{f_i\}$ orthogonal projections with LUB f, and suppose $e_i \sim f_i$ for every i. Then $e \sim f$.*

PROOF. Reference to Lemma 5.1 reveals that we need only treat the case where A is of type II_1. By two successive applications of Lemma 4.12, we chop each e_i, and correspondingly each f_i, into four equivalent orthogonal parts. It suffices for us to discuss one of the four parts thus created. After a change of notation, this means that we can assume that A contains four orthogonal projections equivalent to e, and also four orthogonal projections equivalent to f.

Write $g = (e \cup f) - f$. Since g and f are orthogonal, we see by Lemma 3.4 that A splits into two summands in which g and f are comparable. In the usual changed notation, we may treat two cases. Case I. $f \lesssim g$. By Theorem 5.4, $g \lesssim e$, and $f \lesssim e \lesssim 1 - e$. We apply part (a) of Lemma 5.1. Case II. $g \lesssim f$. Write $h = 1 - (e \cup f)$. We claim that $h \gtrsim f$. Otherwise by Lemma 3.4, we pass to a direct summand where $h \lesssim f$. Then $1 = h + g + f$ is the union of three orthogonal projections $\lesssim f$, while on the other hand it contains four orthogonal projections equivalent to f; this contradicts finiteness. Hence $f \lesssim h \leq 1 - e$, and we may apply part (a) of Lemma 5.1. We have proved $e \sim f$ in all cases.

On the basis of Theorem 5.5, we can now establish unrestricted versions of Lemma 3.4 and Lemma 4.13. The proofs are left to the reader.

THEOREM 5.6. *Let e and f be arbitrary projections in an AW*-algebra. Then there exist central projections g, h with $gh = 0$, $g + h = 1$, and $eg \gtrsim fg$, $eh \lesssim fh$.*

REMARK. This decomposition can be refined further into three parts, where e is equivalent to, "strictly less than", or "strictly greater than" f; the decomposition is then unique.

THEOREM 5.7. *Let e, f be equivalent projections in a finite AW*-algebra. Then $1 - e \sim 1 - f$, and moreover there exists a unitary element carrying e into f.*

6. The lattice of projections, continued

We begin by establishing the additivity of finiteness.

LEMMA 6.1. *Let e, f, g be projections in an AW*-algebra with $ef = 0$, $g \leq e + f$. Write $h = e + f - g$. Then A may be split into two direct summands in which $g \lesssim e$ and $h \lesssim f$ respectively.*

PROOF. We apply Theorem 5.6 (or Lemma 3.4) and split A according to the comparability of $g \cap f$ and $h \cap e$. Case I. $g \cap f \lesssim h \cap e$. Then by Theorem 5.4,

$$g = (g \cap f) + [g - (g \cap f)] \lesssim (h \cap e) + [(g \cup f) - f].$$

The last two terms are orthogonal projections, both $\leq e$. Hence $g \lesssim e$. Case II. $h \cap e \lesssim g \cap f$. We have

$$h = (h \cap e) + [h - (h \cap e)] \lesssim (g \cap f) + [(h \cup e) - e]$$

and argue similarly that $h \lesssim f$.

THEOREM 6.2. *If e, f are finite projections in an AW*-algebra, so is $e \cup f$.*

PROOF. We may assume $e \cup f = 1$. If 1 is infinite, we apply Theorem 4.2, and after dropping down to a direct summand, we may assume that 1 is purely infinite. By Lemma 4.5 there then exists a projection g such that $1 \sim g \sim 1 - g$. By Theorem 5.4,

$$1 - e = (e \cup f) - e \sim f - (e \cap f) \leqq f,$$

and so $1 - e$ is finite. By Lemma 6.1 we may split the algebra into two direct summands in which $h \lesssim e$ and $1 - h \lesssim 1 - e$ respectively; but both of these statements are contradictory.

We proceed to discuss the modular law. Let e, f, g be projections with $e \leqq g$. We form the two expressions relevant for the modular law: $h = (e \cup f) \cap g$ and $k = e \cup (f \cap g)$. Now both h and k have the property that their union with f is $e \cup f$, and their intersection with f is $f \cap g$. By two applications of Theorem 5.4:

$$h - (f \cap g) \sim (e \cup f) - f \sim k - (f \cap g),$$

whence $h \sim k$. In any lattice we have $h \geqq k$; if we now throw in the assumption that h is finite, we deduce $h = k$. On putting this together with Theorem 6.2, we have proved the following theorem.

THEOREM 6.3. *The finite projections in an AW*-algebra form a modular lattice.*

We proceed to examine the connection between AW^*-algebras and continuous geometry. A preliminary lemma is needed.

LEMMA 6.4. *Let A be a finite AW^*-algebra, and $\{e_\rho\}$ a well ordered ascending set of projections in A, with LUB e. Suppose $e_\rho \lesssim f$ for every ρ. Then $e \lesssim f$.*

PROOF. We may assume without loss of generality that when λ is a limit ordinal, e_λ is the LUB of the preceding e's. We proceed to construct a well ordered ascending set of projections $\{f_\rho\}$, with the properties: (1) $e_\rho \sim f_\rho$, (2) $f_\rho \leqq f$, (3) when λ is a limit ordinal, f_λ is the LUB of the preceding f's. Suppose f_ρ has been constructed for $\rho < \alpha$. *Case* I. α a limit ordinal. We take $f_\alpha =$ LUB of the preceding f's. Then properties (2) and (3) are clear. To verify (1), we observe that f_α is the LUB of f_1 and $\{f_{\rho+1} - f_\rho\}$, and e_α likewise is the LUB of e_1 and $\{e_{\rho+1} - e_\rho\}$. By Theorem 5.5, $e_\alpha \sim f_\alpha$. *Case* II. α not a limit ordinal. By hypothesis there is an equivalence between e_α and a portion of f; suppose this induces an equivalence of $e_{\alpha-1}$ and g. Then $e_\alpha - e_{\alpha-1} \lesssim f - g$. On the other hand, we have $f - f_{\alpha-1} \sim f - g$ by Theorem 5.7. Hence $e_\alpha - e_{\alpha-1} \lesssim f - f_{\alpha-1}$, that is, $e_\alpha - e_{\alpha-1} \sim h$ with $h \leqq f - f_{\alpha-1}$. We take f_α to be $f_{\alpha-1} + h$. When the construction is completed, we have, by Theorem 5.5, $e \sim$ LUB of $\{f_\rho\}$, and hence $e \lesssim f$.

THEOREM 6.5. *The projections in a finite AW^*-algebra form a continuous geometry (not necessarily irreducible).*

PROOF. Of the postulates given in [5], the only one that remains for us to check is the continuity of the lattice operations. By duality, we need only consider the LUB. Let then $\{e_\rho\}$ be a well ordered ascending set of projections with LUB e; let the LUB of $\{e_\rho \cap f\}$ be g. Our task is to prove $h = (e \cap f) - g = 0$. Write $k_\rho = (e \cap f) \cup e_\rho$. Then by Theorem 5.4,

(6) $$h \leq (e \cap f) - (e_\rho \cap f) \sim k_\rho - e_\rho \leq e - e_\rho .$$

We claim that $e_\rho \lesssim e - h$; otherwise, by Theorem 5.6, there would be a direct summand where $e - h$ is equivalent to a proper part of e_ρ. On adding this to (6), we contradict the finiteness of e. Hence $e_\rho \lesssim e - h$ for every ρ. By Lemma 6.4, $e \lesssim e - h$, and by finiteness, $h = 0$.

In a continuous geometry, the center is defined as the set of elements with unique complements, and equivalence is defined as the possession of a common complement. In the following theorem we identify these concepts with the corresponding ones for AW^*-algebras.

THEOREM 6.6. *In any AW^*-algebra:* (a) *a projection is central if and only if it has a unique complement,* (b) *two projections with a common complement are equivalent,* (c) *two finite equivalent projections have a common complement.*

PROOF. (a) If e is a central projection, then for any projection f we have $e \cap f = ef$, $e \cup f = e + f - ef$. Hence if f is a complement of e, we must have $ef = 0, f = 1 - e$. Conversely, suppose the projection e has $1 - e$ as its unique complement. By Theorem 5.6, we may drop to a direct summand where e and $1 - e$ are comparable. We may assume $e \sim f \leq 1 - e$; say $x^*x = e$, $xx^* = f$. Write $2g = e + x + x^* + f$, $h = 1 - e - f$, $k = g + h$; we note that g is a projection orthogonal to h and so k is a projection. We shall prove that k is a complement of e. (1) Let $m = e \cup k$. Then $m \geq e, g$ and so $mg = g$, $me = e$, $mx^* = x^*$, and

$$x^*f + f = 2gf = m(2gf) = m(x^*f + f) = x^*f + mf.$$

Hence $mf = f$ and $m \geq f$; then $m \geq e + f + h = 1$. (2) Let $n = e \cap k$. From $en = n$ and $he = 0$ we get $hn = 0$; then $n = kn = (g + h)n = gn$. From $2ege = e$ we then get

$$2n = 2egen = en = n,$$

and $n = 0$. Thus both k and $1 - e$ are complements of e, and by hypothesis, $k = 1 - e$. This entails $g = f$ and then $f = e = 0$. We have proved that e is in the center, as desired.

(b) If e and f are complements, then by Theorem 5.4, $e \sim 1 - f$. Any projection having f as complement is likewise equivalent to $1 - f$ and hence to e.

(c) Let e and f be equivalent finite projections. Since we can work within $e \cup f$, we may assume that the whole algebra is finite (Theorem 6.2). Theorem 6.5 then permits us to apply the theory of continuous geometries. By [5, Vol. III, p. 24, Th. 2.7], after dropping to a direct summand, we can assume that e and f are comparable in the sense of continuous geometry; that is, there exists g with $g \leq f$ such that e and g have a common complement. By part (b), $e \sim g$, and so $f \sim g$. By finiteness, $f = g$.

We can now apply the known theory of continuous geometry as a short cut to establishing dimension theory in finite AW^*-algebras. The results may be set forth as follows. Let A be a finite AW^*-algebra, with center Z. There exists a function D, defined on the projections of A and taking values in Z, with the following properties: (1) $0 \leqq D(e) \leqq 1$ for every e, (2) $D(e) = e$ if $e \in Z$, (3) $D(e) = D(f)$ if and only if $e \sim f$, (4) $D(e + f) = D(e) + D(f)$ if $ef = 0$; (5) D is uniquely determined by the foregoing. The passage from this dimension function to a trace is the next problem, but we shall not go into it in this paper.

7. Appendix

We shall devote this final section to clarifying the connection between AW^*-algebras and Rickart's B_p^*-algebras.

In the first place, Rickart's starting point was not a C^*-algebra but a B^*-algebra; a B^*-algebra is a Banach *-algebra in which $\| x^*x \| = \| x \|^2$, the axiom that x^*x has a quasi-inverse being dropped. We shall now show that in Rickart's case (and somewhat more generally) this last axiom is an automatic consequence.

We need the fact, proved by Arens [1, p. 788], that in a B^*-algebra x^*x cannot be -1. The following brief proof is perhaps of independent interest. We know from commutative theory that $x^*x = xx^* = -1$ is impossible in a B^*-algebra; hence the result is a consequence of the following purely algebraic lemma.

LEMMA 7.1. *Let A be a ring with a unit element and an involution* * *such that* $y^*y = 0$ *implies* $y = 0$. *Then* $x^*x = -1$ *implies* $xx^* = -1$.

PROOF. Let $y = (1 + x)(1 + xx^*)$. One verifies that $y^*y = 0$. Hence $y = 0$, and from $x^*y = 0$ we find $xx^* = -1$.

We now generalize the result of Arens.

LEMMA 7.2. *If in a B^*-algebra $x^*x = -e$ with e an idempotent, then $x = 0$.*

PROOF. We note that $(x - xe)^*(x - xe) = 0$ and so $x = xe$. Write $y = x - ex$; then $y^*y = -e - x^*ex$. We have $\| x \|^2 = \| x^*x \| = \| e \| = 1$, and hence $\| x^*ex \| \leqq 1$. Also x^*ex is in the algebra eAe which has e as unit element. Hence the spectrum of y^*y is $\leqq 0$, and we may write $y^*y = -z^2$, where $z^* = z$, $z \in eAe$. From $z = ze$ and $ey = 0$ we get $zy = 0$. Then $(y^* + z)(y + z) = 0$, $y + z = 0$. On left-multiplying by z we get $z^2 = 0$, $z = 0$, $y = 0$. Hence $ex = x$ and $x \in eAe$. This reduces the problem to the case treated by Arens, and proved again above.

The fact that B_p^*-algebras are C^* is implied by the following theorem.

THEOREM 7.3. *Let A be a B^*-algebra satisfying postulate (B) of §2, that is, every maximal commutative self-adjoint subalgebra of A is generated by its projections. Then A is a C^*-algebra.*

PROOF. It is known that the problem is equivalent to proving that every x^*x has spectrum $\geqq 0$. Suppose on the contrary that $y = x^*x$ has negative numbers in its spectrum. Embed y in a maximal commutative self-adjoint subalgebra C. In the functional representation for C, the function $y(t)$ assumes some negative values. Our hypothesis implies that the space of maximal ideals of C is totally disconnected. Hence there exists a compact open set U such that $y(t)$ is negative in U. Let e be the characteristic function of U; then there exists a self-adjoint element z in C such that $yz^2 = -e$. If $w = xz$, we find $w^*w = -e$, in contradiction to Lemma 7.2.

To facilitate the ensuing discussion, we introduce a definition. We shall call

A an $\aleph$-AW^*-algebra if it is a C^*-algebra satisfying postulate (B) of §2, and satisfying postulate (A) for sets of projections with cardinal number less than $\aleph$. We are interested only in $\aleph \geqq \aleph_1$; note that in an $\aleph_1$-AW^*-algebra, a LUB is assumed for any countable set of orthogonal projections. We now assert that $\aleph_1$-AW^*-algebras coincide with B_p^*-algebras (no unit element being assumed for the latter). In the direction $B_p^* \rightarrow \aleph_1$-$AW^*$, the requisite arguments are given by Rickart—the lack of a unit element makes no essential difference. In the other direction, we must show that in an $\aleph_1$-AW^*-algebra right and left projections exist for any element; but this was in effect shown in the first paragraph of the proof of Theorem 5.2.

It is perhaps of interest to see how far one can carry the theory of $\aleph$-AW^*-algebras, and the following remarks can be made. (1) Lemma 5.3 can be interpreted as proving the existence of the LUB of two projections. (2) It then follows readily that the projections form an $\aleph$-lattice. (3) Lemma 3.1 holds, if confined to sets with cardinal less than $\aleph$. (4) Theorem 3.2 is then valid, only countable sets being involved in its proof. (5) It seems a plausible conjecture that Theorem 5.2 and its various consequences are valid, but I am unable to prove this. (6) On the other hand, there are definite limitations in the theory, above all because Theorem 5.6 may fail, as shown by the following example. Let H be a non-separable Hilbert space, and B the algebra of all bounded operators on H with separable range. Let A be the algebra obtained from $B \oplus B$ by adjunction of a unit. Then it is not difficult to verify that A is an $\aleph_1$-AW^*-algebra, that its center is the complex numbers and it is hence indecomposable, and that nevertheless there exist projections that are not comparable; indeed there exist non-zero projections e, f with $eAf = 0$. The pathology shown by this example indicates that it is probably fruitless to pursue the theory of $\aleph$-AW^*-algebras.

UNIVERSITY OF CHICAGO

BIBLIOGRAPHY

1. R. ARENS, *Approximation in, and representation of, certain Banach algebras*, Amer. J. Math. 71 (1949), 763–790.
2. I. GELFAND and M. NEUMARK, *On the imbedding of normed rings into the ring of operators in Hilbert space*, Rec. Math. (Mat. Sbornik) N. S. 12 (1943), 197–213.
3. I. KAPLANSKY, *Normed algebras*, Duke Math. J. 16 (1949), 399–418.
4. F. J. MURRAY and J. VON NEUMANN, *On rings of operators*, Ann. of Math. 37 (1936), 116–229.
5. J. VON NEUMANN, Continuous Geometry, Princeton, 1937.
6. J. VON NEUMANN, *On an algebraic generalization of the quantum mechanical formalism*, Rec. Math. (Math. Sbornik) N. S. 1 (1936), 415–484.
7. C. E. RICKART, *Banach algebras with an adjoint operation*, Ann. of Math. 47 (1946), 528–550.
8. S. W. P. STEEN, *An introduction to the theory of operators* IV, Proc. Cambridge Phil. Soc. 35 (1939), 562–578.

Afterthought

Roger Godement reviewed this paper for *Mathematical Reviews* (vol. 13, p. 48). He frowned. I paraphrase and translate a portion of his review: What is the point of this generalization from W^* to AW^*, except perhaps to offer simplified proofs?

I am pleased that he noticed the simplified proofs. As for the main charge, I plead guilty and throw myself on the mercy of the court.

Still, the idea of studying W^*-algebras or algebras resembling them intrinsically lives on fitfully. As Gert Pedersen said in a 1985 talk at UC Berkeley [5, p. 51], the subject refuses to die.

The paper [3] of Kadison is worthy of note. By assuming suitable least upper bounds not just for projections but for all self-adjoint elements, one intrinsically characterizes W^*-algebras.

In [6] I singled out some advances in the theory of AW^*-algebras that I found noteworthy.

Another thought in this vein: Without this preliminary skirmish with AW^*-algebras, I would not have conjectured and then proved that any orthocomplemented complete modular lattice is a continuous geometry.

As noted in the preface to [4] I later changed my mind about the proper way to axiomatize AW^*-algebras. Question: Why did I do it clumsily in the first place? Lame reply: The process of taking the least upper bound of a set of orthogonal projections was so fundamental and so heavily used that I slid into making it an axiom. However, more was needed since there might not be any projections other than 0 or 1. Making an assumption about maximal commutative subalgebras was unfortunate; Zorn's lemma had to be invoked every time the axiom was used.

Two expository comments are made in [4]. (a) Lemma 4.14 on page 243 could have been proved immediately without the countability assumption. The details are in [4, Theorem 51, p. 77]. (It is erroneously labeled Theorem 5.) (b) The "parallelogram law" (Theorem 5.4, p. 244) can be reached by a more direct and more algebraic path. This is spread through several pages of [4], winding up with Theorem 62 on page 96.

Berberian's book [1] is a comprehensive account of Baer *-rings, an algebraic generalization of AW^*-algebras. I advise a reader to have on hand the update [2]; it is available from the author.

References

1. Sterling K. Berberian, *Baer *-Rings*, Springer-Verlag, (Berlin), 1972.
2. _____, *Baer Rings and Baer *-Rings*, 130 pp. Latest corrected version, 1992.
3. Richard V. Kadison, Operator algebras with a faithful weakly-closed representation, *Annals of Math.* 64(1956), 175–81.
4. Irving Kaplansky, *Rings of Operators*, Benjamin, (New York), 1968.
5. _____, MSRI after three years, *Math. Intelligencer* 7, no. 4, 48–55.
6. _____, To Harold Widom on his 60th birthday, to appear in the proceedings of a September 1992 conference on Wiener-Hopf and Toeplitz operators at UC Santa Cruz.

A THEOREM ON DIVISION RINGS

IRVING KAPLANSKY

The object of this note is to prove the following theorem.

Theorem. *Let A be a division ring with centre Z, and suppose that for every x in A, some power (depending on x) is in Z: $x^{n(x)} \in Z$. Then A is commutative.*

This theorem contains as special cases three previously known results.

1. It includes Wedderburn's theorem that any finite division ring is commutative, and the generalization by Jacobson [3, Theorem 8] asserting that any algebraic division algebra over a finite field is commutative; for in such an algebra every non-zero element has some power equal to 1.

2. It includes a theorem of Emmy Noether, as generalized by Jacobson [3, Lemma 2], stating that any non-commutative algebraic division algebra contains an element separable over the centre; for otherwise a suitable p^mth power of every element would lie in the centre.

3. Hua [1, Theorem 7] has proved the special case of the theorem where the power n is independent of x, and the characteristic is at least n.

Although our theorem generalizes the two cited theorems of Jacobson, we are not giving a new proof of these theorems. In fact, we shall prove a preliminary lemma on fields which reduces the problem precisely to these two theorems.

Lemma. *Let K be a field and L an extension of K, $L \neq K$, with the property that for every x in L, some power (the power depending on x) lies in K. Then L has prime characteristic, and it is either purely inseparable over K, or algebraic over its prime subfield.*

Proof. If L is indeed purely inseparable over K, there is of course nothing to prove. So suppose L contains an element y, y non $\in K$, which is separable over K. By a suitable isomorphism leaving K elementwise fixed, y can be sent into an element $z \neq y$ (of course z need not be in L). We have, say, $y^r \in K$ and and so $z^r = y^r$, whence $z = \epsilon y$ with $\epsilon^r = 1$. Suppose $(1 + y)^s \in K$; then similarly $1 + z = \eta(1 + y)$ with $\eta^s = 1$. We cannot have $\epsilon = \eta$, for then $\epsilon = 1$, $z = y$. So we may solve for y:

$$(1) \qquad y = (1 - \eta)(\eta - \epsilon)^{-1}.$$

We see that y is algebraic over the prime subfield P of K. If k is any element of K, we can repeat this argument with $k + y$ instead of y, and thus deduce

Received September 26, 1950.

290

87

that $k + y$, and hence k, is algebraic over P. In short, K is algebraic over P. If P has prime characteristic, we have reached the other possibility stated in the conclusion of the lemma, so it remains only to exclude the possibility that P has characteristic 0 (which means that it is the field of rational numbers). This we do as follows. For any integer i we have an expression like (1) for $y + i$:

$$(2) \qquad\qquad y + i = (1 - \eta_i)(\eta_i - \epsilon_i)^{-1}.$$

Moreover, the definition of η_i and ϵ_i shows that they lie in the normal field, say Q, generated by y over P. But Q, being a finite-dimensional extension of P, contains only a finite number of roots of unity. This leaves us powerless to account for the infinite number of elements in (2).

Proof of the theorem. If $A \neq Z$, choose any element x not in Z, and let L be the field generated by Z and x. Then the hypothesis of the lemma is ful- filled (with Z playing the role of K). The possibility that Z has prime charac- teristic and is algebraic over its prime subfield is ruled out by the first theorem of Jacobson cited above. So it must be true that L is purely inseparable over Z. This is the case for every x, and we contradict the second theorem of Jacobson.

Theorem 7 of [1] actually states that a non-commutative division ring is generated by its nth powers. Our theorem can be given a corresponding extension as follows. For every x of a non-commutative division ring A, let there be given a positive integer $n(x)$ such that $n(x) = n(a^{-1}xa)$ for all $a \neq 0$; let B be the division subring generated by the elements $x^{n(x)}$; then $B = A$. For B is invariant under all inner automorphisms, and if $B \neq A$ then by the theorem of Cartan-Brauer-Hua [1, Theorem 2] B is contained in the centre of A, contradicting the above theorem.

In conclusion we discuss two possibilities of generalization. In the first place we might consider relaxing the requirement that A be a division ring. In fact, our theorem remains correct if we merely assume that A is semi- simple in the sense of Jacobson [2]. The manœuvre for proving this has become fairly standard since the appearance of Jacobson's paper. If P is a primitive ideal in A, our hypothesis is inherited by A/P; if we prove that each A/P is commutative we will know that A is commutative, and so we need only consider the case where A is primitive. We represent A as a dense ring of linear transformations in a vector space V over a division ring. We now in effect check our theorem for two-by-two matrices. In detail: if V is more than one-dimensional, let α and β be linearly independent vectors, and let x be an element of A sending α into itself and annihilating β. It is impossible for any power of x to be in the centre. So V is one-dimensional, and we are back to the division ring case of the theorem.

Another path along which to proceed is to have a polynomial more general than x^n. We shall not attempt more than the case where n is independent of

x, although it would be interesting to invent plausible "one-parameter families" generalizing $\{x^n\}$. We assume then that there exists a polynomial f with coefficients in Z (we can suppose it has no constant term) such that $f(x) \in Z$ for every x. Since A then satisfies the identity $f(x)y - yf(x) = 0$, it follows forthwith from [4, Theorem 1] that A is finite-dimensional over Z. But as a matter of fact it is again true that A is commutative. For suppose f has smallest possible degree among polynomials with $f(x) \in Z$. We can suppose there is an element u in Z no power of which is 1 (otherwise Z would be of prime characteristic and algebraic over its prime field, etc.). Consider the polynomial $g(x) = f(x) - u^n f(xu^{-1})$, n being the degree of f; the degree of g is less than n, and it again has the property $g(x) \in Z$ for every x. The only way out is for g to be identically zero, which means $f(x) = x^n$, and we are back to the old case.

One must step cautiously in attempting to generalize this last result beyond division rings: observe that the ring of two-by-two matrices over $GF(2)$ satisfies the identity $x^8 = x^2$.

REFERENCES

[1] L. K. Hua, *Some properties of a sfield*, Proc. Nat. Acad. Sci. USA, vol. 35 (1949), 533-537.
[2] N. Jacobson, *The radical and semi-simplicity for arbitrary rings*, Amer. J. of Math., vol. 67 (1945), 300-320.
[3] ———— *Structure theory for algebraic algebras of bounded degree*, Ann. of Math., vol. 46 (1945), 695-707.
[4] I. Kaplansky, *Rings with a polynomial identity*, Bull. Amer. Math. Soc., vol. 54 (1948), 575-580.

University of Chicago

Afterthought

Herstein [1] significantly generalized the theorem of this paper, and this generalization appeared in Jacobson's book [2, pp. 218–219]. I find that Herstein's theorem becomes a little neater when restated as follows.

THEOREM. *Let R be a ring in which every element has a central power. Then the nilpotent elements in R form an ideal I and R/I is commutative.*

Although the accounts of Herstein and Jacobson are models of clarity, I have decided to present yet another. There are two reasons: The two preliminary lemmas on semigroups may be of some iinerest, and later in the proof I cite an underused theorem that I think deserves publicity.

LEMMA 1. *If xy is central in a semigroup then $(xy)^n = x^n y^n$ for all n.*

Proof. Pull xy repeatedly out of the middle and place it (say) on the left.

LEMMA 2. *In a semigroup with 0 suppose that x is nilpotent and that some power of xy is central. Then xy is nilpotent.*

Proof. Say $(xy)^m$ is central. The case for $m = 1$ follows from Lemma 1. By applying this to x and $y(xy)^{m-1}$, we get the general case.

Proof of the theorem. Right multiples of nilpotent elements are covered by Lemma 2 and left multiples follow by symmetry. To complete the proof that I is an ideal, we have to show that I is closed under addition. So suppose that x and y are nilpotent and that $(x + y)^r$ is central. Write $a = x(x + y)^{r-1}$, $b = y(x + y)^{r-1}$. Then a and b are nilpotent (since they are multiples of nilpotent elements), $a + b$ is central, a and b commute, so a and b are nilpotent, and $x + y$ is nilpotent.

Now (changing notation) we assume that R has no nonzero nilpotent elements and we have to prove that R is commutative. The first step is to reduce to the case where R has no zero-divisors. The underused theorem referred to above asserts that if P is a minimal prime ideal in R then R/P has no zero-divisors. Since the minimal prime ideals of R intersect in 0, this first reduction is done. Then we move to the case of a division ring by forming the localization S of R relative to all nonzero central elements. That S is a division ring follows readily from the fact that elements of R have central powers, and one easily sees that S inherits the property of central powers. Once we reach the case of a division ring, I (naturally enough) have nothing to add.

References

1. I. N. Herstein, A theorem on rings, *Can. J. of Math.* 5(1953), 238–41.
2. N. Jacobson, Structure of Rings, *Amer. Math. Soc. Coll. Publ.* vol. 37, revised ed., 1964.

A THEOREM ON RINGS OF OPERATORS

IRVING KAPLANSKY

1. Introduction. The main result (Theorem 1) proved in this paper arose in connection with investigations on the structure of rings of operators. Because of its possible independent interest, it is being published separately.

The proof of Theorem 1 is closely modeled on the discussion in Chapter I of [3]. The connection can be briefly explained as follows. Let N be a factor of type II_1; then in addition to the usual topologies on N, we have the metric defined by $[[A]]^2 = T(A^*A)$, T being the trace on N. Now it is a fact that in any *bounded* subset of N, the $[[\ ]]$-metric coincides with the strong topology—this is the substance of Lemma 1.3.2 of [3]. In the light of this observation, it can be seen that Theorem 1 is essentially a generalization (to arbitrary rings of operators) of the ideas in Chapter I of [3].

Before stating Theorem 1, we collect some definitions for the reader's convenience. Let R be the algebra of all bounded operators on a Hilbert space H (of any dimension). In R we have a natural norm and *-operation. A typical neighborhood of 0 for the strong topology in R is given by specifying $\epsilon > 0, \xi_1, \cdots, \xi_n \in H$, and taking the set of all A in R with $\|A\xi_i\| < \epsilon$; for the weak topology we specify further vectors $\eta_1, \cdots, \eta_n \in H$ and take the set of all A with $|(A\xi_i, \eta_i)| < \epsilon$. By a *-algebra of operators we mean a self-adjoint subalgebra of R, that is, one containing A^* whenever it contains A; unless explicitly stated, it is not assumed to be closed in any particular topology. For convex subsets of R, and in particular for subalgebras, strong and weak closure coincide [2, Th. 5]. An operator A is self-adjoint if $A^* = A$, normal if $AA^* = A^*A$, unitary if $AA^* = A^*A =$ the identity operator I.

2. The main result. We shall establish the following result.

THEOREM 1. *Let M, N be *-algebras of operators on Hilbert space, $M \subset N$, and suppose M is strongly dense in N. Then the unit sphere of M is strongly dense in the unit sphere of N.*

Received October 6, 1950. This paper was written in connection with a research project on spectral theory, sponsored by the Office of Naval Research.

Pacific J. Math. 1 (1951), 227–232.

227

We shall break up the early part of the proof into a sequence of lemmas. Lemma 1 is well known and is included only for completeness.

LEMMA 1. *In the unit sphere of* R, *multiplication is strongly continuous, jointly in its variables; and any polynomial in* n *variables is strongly continuous, jointly in its arguments.*

Proof. It is easy to see that multiplication is strongly continuous separately in its variables, even in all of R. Consequently [1, p. 49] we need only check the continuity of AB at $A = B = 0$. Since $||A|| \leq 1$, this is a consequence of

$$||AB\xi|| \leq ||A|| \; ||B\xi|| \leq ||B\xi|| .$$

Since addition and scalar multiplication are continuous (in all of R), the continuity of polynomials follows.

The precaution taken in the next lemma, in defining the mapping on the pair (A, A^*), is necessary since $A \longrightarrow A^*$ is not strongly continuous.

LEMMA 2. *Let* $f(z)$ *be a continuous complex-valued function, defined for* $|z| \leq 1$. *Then the mapping* $(A, A^*) \longrightarrow f(A)$ *is strongly continuous on the normal operators of the unit sphere of* R.

Proof. We are given a normal operator A_0 with $||A_0|| \leq 1$, a positive ϵ, and vectors ξ_i in H $(i = 1, \cdots, n)$. We have to show that by taking A, A^* to be normal with norm ≤ 1, and in suitable strong neighborhoods of A_0, A_0^*, we can achieve

$$(1) \qquad\qquad || [f(A) - f(A_0)] \xi_i || < \epsilon .$$

By the Weierstrass approximation theorem, there exists a polynomial g in two variables such that

$$(2) \qquad\qquad |g(z, z^*) - f(z)| < \epsilon/3 ,$$

for $|z| \leq 1$, z^* denoting the conjugate complex of z. By elementary properties of the functional calculus for normal operators, we deduce from (2):

$$(3) \qquad\qquad ||g(A, A^*) - f(A)|| < \epsilon/3 ,$$

$$(4) \qquad\qquad ||g(A_0, A_0^*) - f(A_0)|| < \epsilon/3 .$$

By Lemma 1, if we take A, A^* in appropriate neighborhoods of A_0, A_0^*, we have

(5) $$\left\| \left[g(A, A^*) - g(A_0, A_0^*) \right] \xi_i \right\| < \epsilon/3 .$$

By combining (3), (4), and (5) we obtain (1).

The next lemma follows from Lemma 2 as soon as it is admitted that $*$ is strongly continuous on unitary operators. This can, for example, be deduced from two known facts: (a) the strong and weak topologies coincide on the set of unitary operators, and (b) $*$ is weakly continuous.

LEMMA 3. *Let f be a continuous complex-valued function defined on the circumference of the unit circle. Then the mapping $U \longrightarrow f(U)$ is strongly continuous on the set of unitary operators.*

The *Cayley transform* is the mapping $A \longrightarrow (A - i)(A + i)^{-1}$; it is defined for any self-adjoint operator and sends it into a unitary operator.

LEMMA 4. *The Cayley transform is strongly continuous on the set of all self-adjoint operators.*

Proof. We have the identity

(6) $$(A - i)(A + i)^{-1} - (A_0 - i)(A_0 + i)^{-1} = 2i(A + i)^{-1}(A - A_0)(A_0 + i)^{-1} .$$

When A is self-adjoint, we have $\left\| (A + i)^{-1} \right\| \leq 1$. In order to make the left side of (6) small on a vector ξ, it therefore suffices to make $A - A_0$ small on the vector $(A_0 + i)^{-1} \xi$.

We shall prove a stronger form of Lemma 5 below (Corollary to Theorem 2).

LEMMA 5. *Let h be a real-valued function defined on the real line, and suppose that h is continuous and vanishes at infinity. Then the mapping $A \longrightarrow h(A)$ is strongly continuous on the set of all self-adjoint operators.*

Proof. Define

$$f(z) = h\left[-i(z + 1)(z - 1)^{-1} \right] \qquad \text{for } |z| = 1, \ z \neq 1,$$
$$= 0 \qquad \text{for } z = 1.$$

Then f is continuous on the circumference of the unit circle. Moreover,

$$h(A) = f\left[(A - i)(A + i)^{-1} \right] .$$

The mapping $A \longrightarrow h(A)$ is thus the composite of two maps: the Cayley transform, and the mapping on unitary operators given by f. By Lemmas 4 and 3, these latter two maps are strongly continuous. Hence so is $A \longrightarrow h(A)$.

Proof of Theorem 1. There is clearly no loss of generality in assuming M and N to be uniformly closed, for the unit sphere of M is even uniformly dense in the unit sphere of its uniform closure.

Let us write Z for the set of self-adjoint elements in M, and Z_1 for the unit sphere of Z. Let B be a given self-adjoint element in N, $\|B\| \leq 1$. By hypothesis, B is in the strong closure of M. We shall argue in two successive steps that B is actually in the strong closure of Z_1. We begin by remarking that B is in the weak closure of M, since the latter coincides with the strong closure of M. Now * is weakly continuous, and hence so is the mapping $A \longrightarrow (A + A^*)/2$. This mapping leaves B fixed, and sends M onto Z; hence B is in the weak closure of Z. Since Z is convex, this coincides with the strong closure of Z.

Let $h(t)$ be any real-valued function of the real variable t which is continuous and vanishes at infinity, satisfies $|h(t)| \leq 1$ for all t, and satisfies $h(t) = t$ for $|t| \leq 1$. We have that $h(B) = B$. Also h can be meaningfully applied within Z, since we have assumed M to be uniformly closed, and in fact $h(Z) = Z_1$. By Lemma 5, the mapping $A \longrightarrow h(A)$ is strongly continuous on self-adjoint operators. Hence B is in the strong closure of Z_1.

This accomplishes our objective as far as self-adjoint operators are concerned. To make the transition to an arbitrary operator, we adopt the device of passing to a matrix algebra.[1] Let N_2 be the algebra of two-by-two matrices over N. In a natural way, N_2 is again a uniformly closed *-algebra of operators on a suitable Hilbert space (compare §2.4 of [3]). It contains in a natural way M_2, the two-by-two matrix algebra over M. The strong topology on N_2 is simply the Cartesian product of the strong topology for the four replicas of N; thus M_2 is again strongly dense in N_2. Now let C be any operator in N, $\|C\| \leq 1$. We form

$$D = \begin{pmatrix} 0 & C \\ C^* & 0 \end{pmatrix}$$

and we note that $D \in N_2$, $D^* = D$, $\|D\| \leq 1$. Let U be any proposed strong neighborhood of D. By what we have proved above, there exists in U a self-adjoint element F,

---

[1] I am indebted to P. R. Halmos for this device, which considerably shortened my original proof of Theorem 1.

$$F = \begin{pmatrix} G & H \\ H^* & K \end{pmatrix}$$

with $F \in M_2$, $\|F\| \leq 1$. By suitable choice of U we can make H lie in a given strong neighborhood of C. Also $\|F\| \leq 1$ implies $\|H\| \leq 1$. This proves that C lies in the strong closure of the unit sphere of M, and concludes the proof of Theorem 1.

3. **Remarks.** (a) Since strong and weak closure coincide for convex sets, we can, in the statement of Theorem 1, replace "strongly" by "weakly" at will.

(b) From Theorem 1 we can deduce that portion of [2, Th. 8] that asserts that a *-algebra of operators is strongly closed if its unit sphere is strongly closed; but it does not appear to be possible to reverse the reasoning.

(c) As Dixmier has remarked [2, p. 399], Theorem 1 fails if M is merely assumed to be a subspace (instead of a *-subalgebra).

4. **Another result.** In concluding the paper we shall return to Lemma 5 and show that the hypothesis can be weakened to the assumption that h is bounded and continuous. It should be noted that we cannot drop the word "bounded," since for example it is known that the mapping $A \longrightarrow A^2$ is not strongly continuous.

Actually we shall prove a still more general result, which may be regarded as a generalization of Lemma 4.2.1 of [3].

THEOREM 2. *Let $h(t)$ be a bounded real-valued Baire function of the real variable t, and A_0 a self-adjoint operator. Let S be the spectrum of A_0, and T the closure of the set of points at which h is discontinuous; suppose S and T are disjoint. Then the mapping on self-adjoint operators, defined by $A \longrightarrow h(A)$, is continuous at $A = A_0$.*

Proof. We may suppose that

(7)
$$|h(t)| \leq 1$$

for all t. Given $\epsilon > 0$, and vectors ξ_i, we have to show that for A in a suitable strong neighborhood of A_0, we have

(8)
$$\|[h(A) - h(A_0)]\xi_i\| < \epsilon.$$

Choose a function $k(t)$ which satisfies: (a) k is continuous and vanishes at infinity, (b) $k(t) = 1$ for t in S, (c) $k(t) = 0$ for t in an open set containing T. Define

$p = hk$, $q = 1 - k + hk$. Then $p = q = h$ on S, and so

(9) $$p(A_0) = q(A_0) = h(A_0).$$

Also p and $q - 1$ are continuous and vanish at infinity. Hence Lemma 5 is applicable, and for a certain strong neighborhood of A_0 we have

(10) $$\| [p(A) - p(A_0)] \xi_i \| < \epsilon/4, \qquad \| [q(A) - q(A_0)] \xi_i \| < \epsilon/2.$$

The following is an identity:

(11) $$h = (1 - h)p + hq.$$

From (9) and (11) we get

(12) $$h(A) - h(A_0) = [1 - h(A)][p(A) - p(A_0)] + h(A)[q(A) - q(A_0)].$$

From (7), (10), and (12), we deduce (8), as desired.

If in particular h is continuous, then T is void and we get a simplified corollary.

COROLLARY. *Let $h(t)$ be a continuous bounded real-valued function of the real variable t. Then the mapping $A \longrightarrow h(A)$ is strongly continuous on the set of all self-adjoint operators.*

REFERENCES

1. N. Bourbaki, *Éléments de Mathématique*, Livre III, *Topologie Générale*, Actualités Sci. Ind., No. 916, Paris, 1942.

2. J. Dixmier, *Les fonctionelles linéaires sur l'ensemble des opérateurs bornés d'un espace de Hilbert*, Ann. of Math. **51** (1950), 387-408.

3. F. J. Murray and J. von Neumann, *On rings of operators IV*, Ann. of Math. **44** (1943), 716-808.

UNIVERSITY OF CHICAGO

Afterthought

The theorem turned out to be much more useful than I anticipated. This goes to show once more that the applicability of a piece of mathematics is not easy to predict.

I venture to quote from page 25 of the book referenced below:

The density theorem is Kaplansky's great gift to mankind. It can be used every day, and twice on Sunday.

REMARK. Gert Pedersen is equally witty in real life.

Reference

Gert K. Pedersen, *C*-Algebras and Their Automorphism Groups*, Academic Press, London, 1979.

THE STRUCTURE OF CERTAIN OPERATOR ALGEBRAS

BY

IRVING KAPLANSKY[1]

1. Introduction. While the structure of commutative Banach algebras is today fairly well understood, the study of the noncommutative case is as yet in its infancy. It therefore appears to be in order to devote some attention to special classes of noncommutative Banach algebras. In this paper we have singled out a class which turns out to be comparatively accessible. These algebras (we call them *CCR*-algebras) are in the first place *C**-algebras (uniformly closed self-adjoint algebras of operators on Hilbert space); secondly we impose a hypothesis most briefly stated as follows: all the irreducible *-representations consist of completely continuous operators (the actual definition of *CCR* in §5 is given in a slightly different form, which is later shown to be equivalent).

Any commutative *C**-algebra is *CCR*, for the irreducible *-representations are then one-dimensional. The structure problem in the commutative case has been completely solved: one gets all the continuous complex functions vanishing at ∞ on a locally compact Hausdorff space. The main point of the present paper is that this kind of representation by continuous functions survives in a *CCR*-algebra, in a somewhat attenuated form. As a fundamental tool in setting up the representation, we employ the structure space introduced by Jacobson [8][2]: the space X of all primitive ideals, topologized by making the closure of $\{P_i\}$ the set of all primitive ideals containing $\cap P_i$. At every point P of X we thus have a primitive *C**-algebra A/P, and we may associate with any $a \in A$ the "function" whose value at P is the homomorphic image $a(P)$ of a mod P. Naturally it is hopeless to speak of the continuity of $a(P)$, for we cannot compare elements in the unrelated algebras A/P. In special cases where the various algebras are not too badly unrelated, it was possible in [2] and [12] to get around this difficulty by utilizing a space other than the structure space. It does not seem to be possible to extend this device to the algebras studied in the present paper. But in any event there are some advantages in another attack which we now explain. Instead of $a(P)$ we consider $\|a(P)\|$; this is a real-valued function on X whose continuity is perfectly meaningful. For many purposes this weakened continuity appears to suffice. It turns out (Theorem 4.1) that $\|a(P)\|$ is a continuous function for every $a \in A$ if and only if X is a Hausdorff space.

Received by the editors December 10, 1949.

[1] Most of the work on this paper was done while the author was a fellow of the John Simon Guggenheim Memorial Foundation.

[2] Numbers in brackets refer to the bibliography at the end of the paper.

219

Now the structure space of a CCR-algebra A is not necessarily Hausdorff, so this result is not directly applicable. We next introduce the idea of a composition series, confining ourselves exclusively to series which are well-ordered ascending, and we are able to prove (Theorem 6.2) that A has a composition series of closed ideals I_ρ such that each $I_{\rho+1}/I_\rho$ has a Hausdorff structure space. This is our main structure theorem, and though it is by no means conclusive, it is strong enough to enable us to carry out (in §7) a fairly complete study of ideals and subalgebras of a CCR-algebra. Actually the results can be pushed slightly beyond CCR-algebras to a class that we call GCR: a GCR-algebra is one having a composition series whose factor algebras are CCR.

The proof of Theorem 6.2 proceeds in several stages. First (Theorem 4.2) we treat the case where all A/P are finite-dimensional and of the same order (or equivalently where all irreducible *-representations of A are finite-dimensional and of the same degree). In Theorem 5.1 we show that the structure space of a CCR-algebra, while not necessarily Hausdorff, is at least of the second category. This information is used to make two further advances; first to the case where all A/P are finite-dimensional, and then to the case of an arbitrary CCR-algebra.

Because of the intimate relation known to exist between locally compact groups and C^*-algebras, these results have applications to groups, which will be presented in a subsequent paper. The relevant groups (one might call them CCR-groups) have the property that for any irreducible unitary representation, the extension to the L_1-algebra consists of completely continuous operators. While it is not clear at this writing how far-reaching a class of groups this is, it is appropriate to note that the results of Gelfand and Neumark [5] show that the Lorentz group is a CCR-group.

In §2 some preliminary results are derived concerning algebras of completely continuous operators. The results in §3 are perhaps of independent interest; they are obtained by a systematic use of the technique of partitions. In Theorem 3.1 we prove, quite briefly, a general result concerning ideals in algebras of continuous functions; this theorem contains all previous results of this kind known to the author. Theorems 3.2–3.4 are devoted to non-commutative generalizations of the Weierstrass-Stone theorem. These considerations incidentally yield a new proof of the commutative Weierstrass-Stone theorem, the proof being of course based on partitions rather than the usual lattice-theoretic methods. The two commutative proofs are of about the same length; but partitions have the advantage of working in the non-commutative case, while lattices do not.

In §8 we prove some results indicating that CCR-algebras behave in certain respects as if they had a unit element, and in §9 the weakly closed case is considered. In the final section (§10) we turn from C^*-algebras to a non-topological analogue: algebraic algebras. Parallel, though not identical, arguments are applicable and yield a similar structure theory. An application is made to the problem raised by Kurosch as to whether every algebraic algebra is locally finite: it is shown (Theorem 10.4) that the answer is affirmative if it is affirmative for primitive algebras and nil algebras.

2. **Dual C^*-algebras.** We shall be speaking exclusively of Banach algebras with complex scalars, although nearly everything can be extended to the real case.

A C^*-algebra may be defined in either of the following ways: (1) a uniformly closed self-adjoint subalgebra of the algebra of all bounded operators on a Hilbert space, (2) a Banach algebra with a conjugate-linear involution *, satisfying the hypotheses that $\|x^*x\| = \|x\|^2$ and x^*x has a quasi-inverse. The equivalence of these two definitions is due to Gelfand and Neumark [4], at least in the presence of a unit element; in [12, Theorem 7.1] the case where there is no unit was treated.

By a *-representation of a C^*-algebra we mean a *-preserving homomorphism into the algebra of bounded operators on a Hilbert space; a *-representation is necessarily continuous [19, p. 77]. A *-representation is trivial if every element is sent into the zero operator; it is irreducible if it is nontrivial and there are no proper closed invariant subspaces.

For the reader's convenience we restate two results [12, Theorems 6.4 and 7.2] for which we shall have frequent use; indeed the second of these will be constantly used without further reference. (We shall not be concerned in this paper with B^*-algebras in the sense of [12], and so we have restricted the following statements to C^*-algebras.)

LEMMA 2.1. *Let ϕ be an algebraic *-preserving isomorphism of a C^*-algebra A into a dense subalgebra of a C^*-algebra B. Then ϕ maps A onto B and is an isometry.*

LEMMA 2.2. *Let I be a closed two-sided ideal in a C^*-algebra A. Then A/I (in its natural norm and *-operation) is again a C^*-algebra.*

We next recall a definition given in [12, p. 411]. If A_i is a set of C^*-algebras, we define the C^*-sum of $\{A_i\}$ to be the set of all $\{a_i\}$, $a_i \in A_i$, with $\|a_i\|$ bounded; this is in a natural way again a C^*-algebra. More important is the $C^*(\infty)$-sum, which is the subset of all sequences vanishing at ∞, that is, all sequences $\{a_i\}$ such that for any $\epsilon > 0$ all but a finite number of $\|a_i\|$ are less than ϵ.

In [10] a dual ring was defined to be a topological ring in which for every closed right (left) ideal I we have $R[L(I)] = I$ and $L[R(I)] = I$ respectively, where L and R denote the left and right annihilators. In [12, Theorem 8.3] the structure of a dual C^*-algebra was determined as follows:

LEMMA 2.3. *A dual C^*-algebra is the $C^*(\infty)$-sum of simple dual C^*-algebras (simple in the sense of having no proper closed two-sided ideals); a simple dual*

C*-algebra is the algebra of all completely continuous operators on a suitable Hilbert space.

The converse of Lemma 2.3 is also valid. The first of the statements in Lemma 2.4 follows from [12, Theorem 8.4] and the second from [10, Theorems 6 and 7][3].

LEMMA 2.4. *The algebra of all completely continuous operators on a Hilbert space is dual. The $C^*(\infty)$-sum of dual C^*-algebras is dual.*

The definition of a dual ring by the properties of annihilators will not be at all relevant in what is to follow; however the designation "dual" has the merit of brevity, and we shall stick to it. The following two alternative characterizations of dual C^*-algebras are more useful.

THEOREM 2.1. *The following statements are equivalent for a C^*-algebra A: (1) A is dual, (2) A has a faithful *-representation by completely continuous operators, (3) the socle of A is dense (the socle is the union of the minimal right ideals).*

Proof. That (1) implies (2) follows from Lemma 2.3: by taking the (Hilbert space) direct sum of the Hilbert spaces on which the simple components of A are represented, we get in a natural way a faithful representation of A by completely continuous operators.

(2)→(3). Suppose that A is a C^*-algebra of completely continuous operators and let S be the closure of its socle. The spectral theorem for completely continuous operators asserts that a self-adjoint element x in A is a limit, in the uniform topology, of linear combinations of finite-dimensional projections. The latter are in A, being uniform limits of polynomials in x, and moreover they are in S (the right ideal generated by a finite-dimensional projection is the union of a finite number of minimal right ideals). Hence $x \in S$ and $S = A$.

(3)→(1). Suppose that the socle of A is dense. Let $\{I_n\}$ denote the distinct minimal two-sided ideals of A, J_n the closure of I_n, K_n the closure of the union of the remaining I's, B the $C^*(\infty)$-sum of the J's. By [12, Theorem 7.3], J_n is the algebra of all completely continuous operators on a Hilbert space. By Lemma 2.4, B is dual. We shall prove that A is dual by establishing its isomorphism with B. We know that $I_m I_n = 0$ $(m \neq n)$ and consequently $J_n K_n = 0$. Then $J_n \cap K_n = 0$ (a C^*-algebra has no nil ideals). Next $J_n + K_n$ is dense in A since it contains at least the socle. By [12, Lemma 8.1] $J_n + K_n$ is all of A, and A is the direct sum of J_n and K_n. Let x_n be the J_n-component of x in this direct sum. The mapping $x \rightarrow \{x_n\}$ is an isomorphism

---

[3] To apply [10, Theorem 7] we have to know that x is in the closure of xA. This is true for any C^*-algebra A; in fact, the argument of [12, Theorem 7.2] shows that x is in the closed right ideal generated by xx^*.

that sends at least the socle of A into B, for, if x is in the socle, all but a finite number of x_n's vanish. It follows that all of A is mapped into B and by Lemma 2.1 the mapping is onto and an isometry.

The criterion (2) of Theorem 2.1 gives us an immediate corollary.

COROLLARY. *A closed self-adjoint subalgebra of a dual C*-algebra is dual.*

Before proceeding to the next theorem, we prove a lemma in somewhat greater generality than needed.

LEMMA 2.5. *Let A be an irreducible algebra of operators on a Banach space E. Then A has no ideal divisors of 0, that is, if I and J are two-sided ideals in A with $IJ=0$, then I or $J=0$.*

Proof. Let F be the closed subspace spanned by the ranges of the operators in I. Then F is a closed invariant subspace, and it is annihilated by J. If $F=0$, then $I=0$. If $F=E$, then $J=0$.

The following theorem is now proved in preparation for a later study of the Weierstrass-Stone theorem. We remark that the theorem is actually true with A any dual C^*-algebra; in fact a good deal more will be proved in Theorem 7.2. But for the present this special case will suffice.

THEOREM 2.2. *Let A be a dual C*-algebra which is either simple or the direct sum of two simple algebras. Let B be a closed self-adjoint subalgebra with the following property: for any distinct regular maximal right ideals M and N in A, B contains an element in M but not in N. Then $B=A$.*

Proof. First we consider the case where A is simple. We can then suppose A concretely given as the algebra of all completely continuous operators on a Hilbert space H. We assert that the induced representation of B is irreducible. For suppose on the contrary that $H=F\oplus G$ with F, G invariant under B. Let α, β be nonzero vectors in F, G and let M, N be the (regular maximal) right ideals in A which annihilate $\alpha+\beta$ and $\alpha-\beta$ respectively. Then any operator in $B\cap M$ or $B\cap N$ must annihilate both α and β, and so B cannot distinguish M and N. Hence the representation of B on H is irreducible. By the corollary to Theorem 2.1, B is dual. By Lemma 2.5, B has no ideal divisors of 0. It follows that B cannot be a nontrivial direct sum, and so by Lemma 2.3, B is simple. Now Neumark [15] has shown that, up to unitary equivalence, a simple dual C^*-algebra has precisely one irreducible representation, and it consists of course of all completely continuous operators on the Hilbert space in question. Hence $B=A$.

Suppose now that A is the direct sum of two ideals I and J, each of which is a simple dual C^*-algebra. Let us write $I_1=B\cap I$, $J_1=B\cap J$. The algebra B/I_1 can, in a natural way, be regarded as a subalgebra of A/I, the mapping being implemented by sending the coset $b+I_1$ into the coset $b+I$. Let M_0, N_0 be regular maximal right ideals in A/I, and let M, N be their inverse images in A; they will again be regular maximal. By hypothesis, B contains an element x in M but not in N. The homomorphic image of x mod I_1 is an element of B/I_1 lying in M_0 but not in N_0. It follows that B/I_1 satisfies the hypothesis of our theorem relative to the dual simple algebra A/I. By the preceding paragraph, B/I_1 is all of A/I_1. This can be translated to say $B+I=A$.

We look again at the ideals I_1, J_1 and observe $I_1 \cap J_1 = 0$. Next, $I_1 + J_1$ is closed [12, Lemma 8.1]. But since B/I_1 is simple, there are no closed ideals between I_1 and B. Hence either (1) $I_1 + J_1 = I_1 = J_1$, or (2) $I_1 + J_1 = B$. (1) In this case $I_1 = J_1 = 0$, and so B is simple. We return to the algebras I, J and note that they have representations as all completely continuous operators on Hilbert spaces, say H and K. In a natural way this induces a representation of A, and hence B, on $H \oplus K$. In this representation of B, it will still be true that H and K are irreducible; the proof is virtually the same as the corresponding one in the second last paragraph. By the theorem of Neumark cited above, these two irreducible representation of B on H and K are unitarily equivalent. Let γ, δ be nonzero vectors of H and K which correspond under this unitary equivalence. Let M', N' be the (regular maximal) right ideals in A which annihilate γ and δ respectively. Then B is unable to distinguish M' and N', and we have a contradiction.

We are thus led to conclude that case (2) holds: $I_1 + J_1 = B$. Combining this with $B+I=A$, we have $I_1 + J_1 + I = A$, $J_1 + I = A$. This is possible only if $J_1 = J$, $B \supset J$. Similarly $B \supset I$ and $B = A$.

In concluding this section we remark that the study of dual C^*-algebras can be pursued further, and that they appear to be nearly as well behaved as finite-dimensional algebras. In particular, the classical theorems about pairs of commutators and extensions of isomorphisms have satisfactory analogues.

3. **Partitions.** Much of our later work will rest on results obtainable by a suitable application of partitions. We state the requisite topological lemma in the form given by Bourbaki [3, p. 65].

LEMMA 3.1. *Let X be a normal Hausdorff space and $U_1, \cdots, U_n$ a covering of X by open sets. Then:* (a) *there exists a covering by open sets $V_1, \cdots, V_n$ such that the closure of V_i is in U_i,* (b) *we may write $1 = f_1 + \cdots + f_n$ where f_i is a real non-negative continuous function on X vanishing outside U_i.*

We are now going to apply partitions to the study of algebras of functions. For the purpose of later applications, it is essential to take up the case of functions whose value $f(x)$ at x lies in an algebra A_x *depending on x.* In doing this, we shall follow Godement [6], with some slight modifications.

Let then X be a compact Hausdorff space and for each $x \in X$ let there be given a Banach algebra A_x. The case of functions "vanishing at ∞" can be covered by allowing one or more A_x's to be 0 (there is actually no loss of generality in insisting that there be at most one $A_x = 0$). Write $C(X)$ for the algebra of all real continuous functions on X. We consider an algebra A of functions f on X with $f(x) \in A_x$, and for the purposes of Theorem 3.1, we impose four conditions:

(a) $\|f(x)\|$ is continuous on X,

(b) A is complete under the norm $\|f\| = \sup \|f(x)\|$,

(c) at each point x, A fills out A_x,

(d) A is closed under multiplication by $C(X)$, and moreover the result of multiplying $f \in A$ by an element of $C(X)$ lies in the closed right ideal generated by f.

We pause to assess the significance of the assumption (d). The simplest way of fulfilling it is of course to assume outright that A contains $C(X)$, or at any rate contains all continuous real functions that vanish at the points x where $A_x = 0$. As a preliminary application of partitions, we shall now show that under suitable circumstances we can weaken (d) to the mere assumption that A is closed under multiplication by $C(X)$.

LEMMA 3.2. *Let X be a compact Hausdorff space at each point of which a Banach algebra A_x is given. Suppose that each A_x satisfies the condition that every $z \in A_x$ lies in the closure of zA_x*[4]. *Let A be an algebra of functions f on X with $f(x) \in A_x$, satisfying postulates* (a), (b), *and* (c) *above, and further satisfying the portion of* (d) *that asserts closure of A under multiplication by $C(X)$. Then A satisfies all of* (d).

Proof. Let $f \in A$, $\lambda \in C(X)$, and a positive ϵ be given. For each $x \in X$ we can find an element $g \in A$ whose value $g(x)$ at A_x satisfies $\|f(x)g(x) - f(x)\| < \epsilon$. In a suitable neighborhood U of x this continues to be true. Select a finite covering of X, say by $U_1, \cdots, U_n$, and corresponding elements $g_1, \cdots, g_n$. By Lemma 3.1, select a partition of unity relative to the U's, that is, write $1 = \theta_1 + \cdots + \theta_n$ with θ_i continuous, non-negative and vanishing outside U_i. By hypothesis, $h = \lambda \sum \theta_i g_i$ is a well defined element of A. We assert

$$(1) \qquad \|fh - \lambda f\| < \epsilon.$$

The argument for this is standard, and we shall give it only on this first occasion. To prove (1), we must verify it at the general point y of X. Suppose for definiteness that y lies in $U_1, \cdots, U_r$, but not in $U_{r+1}, \cdots, U_n$. Then

$$(2) \qquad \|f(y)g_i(y) - f(y)\| < \epsilon$$

holds for $i = 1, \cdots, r$. Moreover at y, $\theta_{r+1} = \cdots = \theta_n = 0$, and $\theta_1, \cdots, \theta_r$ are non-negative and add up to 1. On taking the appropriate convex combination of the elements (2), we find that the result is still less than ϵ:

$$(3) \qquad \left\| \sum \theta_i(y)[f(y)g_i(y) - f(y)] \right\| < \epsilon.$$

---

[4] This hypothesis is of course satisfied if A_x has a unit element; it is also satisfied if A_x is a C^*-algebra (footnote 3).

A further multiplication of (3) by $\lambda(y)$ yields (1) evaluated at y. Hence (1) is proved, and it shows that λf can be approximated arbitrarily closely by right multiples of f, as desired.

We now prove a general result concerning the form of closed ideals in algebras of functions.

THEOREM 3.1. *Let X be a compact Hausdorff space at each point of which a Banach algebra A_x is given. Let A be an algebra of functions f on X with $f(x) \in A_x$, satisfying postulates* (a) *to* (d) *above. Then any closed right ideal I in A has the following form: for each x a closed right ideal I_x in A_x is given, and I consists of all $f \in A$ with $f(x) \in I_x$ for all x.*

Proof. Let H_x be the set of all values taken at x by members of I; H_x is in any event a right ideal. A priori H_x might not be closed, and we let I_x be its closure. Let $h \in A$ be an element with $h(x) \in I_x$; we must prove $h \in I$. It will suffice to approximate h in I within ϵ. For any $x \in X$ we can find $g \in I$ such that $g(x)$ is within ϵ of $h(x)$. This continues to be true in a suitable neighborhood of x. A finite number of these neighborhoods cover X, say $U_1, \cdots, U_n$. Let $\theta_1, \cdots, \theta_n$ be a corresponding partition of unity (Lemma 3.1). Then $\sum \theta_i g_i \in I$ (we are using hypothesis (d) and the fact that I is closed). The observation $\| \sum \theta_i g_i - h \| < \epsilon$ concludes the proof.

Many corollaries can be deduced from Theorem 3.1. Without attempting to be exhaustive, we shall mention three; for each of them the hypothesis of Theorem 3.1 is assumed to be in force.

COROLLARY 1. *Any closed maximal (right) ideal in A consists of all functions confined at a certain point x to a closed maximal (right) ideal in A_x.*

COROLLARY 2. *If each A_x is simple (that is, has no proper closed two-sided ideals), then a closed two-sided ideal in A consists of all functions vanishing on a closed subset of X.*

COROLLARY 3. *If in each A_x it is true that a closed right ideal is the intersection of the regular maximal right ideals containing it, then the same is true in A.*

A fourth corollary is a fairly immediate consequence of Lemma 3.2 and Theorem 3.1.

COROLLARY 4. *Let B be a Banach algebra with the property that for any $z \in B$, z is in the closure of zB. Let X be a locally compact Hausdorff space, and A the Banach algebra of all continuous functions vanishing at ∞ from X to B. Then any closed right ideal in A has the structure described in Theorem 3.1.*

REMARK. One can carry out a similar discussion of the case where each A_x is merely a suitable locally convex topological algebra, for example, the algebra of all bounded operators on a Banach space in its strong or weak topology. Thus generalized, Theorem 3.1 subsumes a result of Hing Tong

[20, Theorem 7]. It is possible further to consider the case where A_x is a locally convex topological linear space; Theorem 3.1 then describes the closed $C(X)$-submodules of A.

We are now going to specialize the discussion of algebras of functions, by requiring that each A_x be a C^*-algebra. Before doing so, several remarks on C^*-algebras are in order.

(1) We use the definition of the spectrum given in [12]: if x is an element of a Banach algebra A, the nonzero spectrum of x consists of all scalars λ such that $-\lambda^{-1}x$ does not have a quasi-inverse, and we insert 0 in the spectrum unless A has a unit element and x has an inverse with respect to it. The insertion of 0 in the spectrum is handled differently by other authors, but the ambiguity is not of great importance. However we take the definition, there are special difficulties at 0, as is illustrated by Lemma 3.3 below.

(2) Let A be a C^*-algebra and x a self-adjoint element of A. The closed subalgebra B generated by x is known to consist of all continuous complex functions vanishing at ∞ on a locally compact Hausdorff space X (in fact X is homeomorphic to the nonzero spectrum of x). Let p be any continuous real function with $p(0) = 0$. Then $p(x)$ is a well defined element of B (and is again self-adjoint). This technique of applying a scalar continuous function to a self-adjoint element will be used repeatedly.

Suppose further that I is a closed two-sided ideal in A and let x' be the homomorphic image of x mod I. Then $p(x')$ may be formed, and we assert that it is the homomorphic image of $p(x)$. This is clear if p is a polynomial (with no constant term of course). The general p is a uniform limit of such polynomials, and the result follows from the continuity of the map from A to A/I.

(3) As an application of the preceding remark, we prove a result which will be used later. Let A be a C^*-algebra, I_n a set of closed two-sided ideals with intersection 0, and x a self-adjoint element in A such that each image x_n mod I_n satisfies $x_n \geq 0$; then $x \geq 0$. To see this, define p by $p(t) = 0$ for $t \geq 0$, $p(t) = -t$ for $t \leq 0$. Then $p(x_n) = 0$, hence $p(x) = 0$, hence $x \geq 0$. (By a slight extension of this remark, one can prove that the spectrum of x is the closure of the union of the spectra of x_n.)

(4) Suppose finally that we have a self-adjoint C^*-algebra A of functions from X to C^*-algebras A_x, as described below. Then (by considering the ideals of functions vanishing at a given coordinate) it follows from the above that: (a) in applying a continuous real function p to an element in A, the result is obtained by applying p coordinate-wise, (b) if a self-adjoint element of A is not less than 0 at each coordinate, then it is not less than 0 in A.

Let now X be any topological space at each point of which a C^*-algebra A_x is given. If we wish to combine these to get a C^*-algebra A of functions, it is in the first place necessary to assume that A is self-adjoint, that is, contains f^* along with f, f^* being defined by taking the * pointwise. Besides this, we are going to analyze more closely the assumptions (a)–(d) above. We begin by observing that the condition,

(b′) $\|f(x)\|$ is bounded, and A is complete under the norm $\|f\| = \sup \|f(x)\|$, suffices to ensure that a self-adjoint algebra A is a C^*-algebra. (In other words, we are taking A to be a closed self-adjoint subalgebra of the C^*-sum of the algebras A_x.) We next single out a piece of continuity:

(a′) $\|f(x)\|$ is continuous at 0, that is, if $f(x) = 0$, then for any $\epsilon > 0$, there exists a neighborhood U of x such that $\|f(x)\| < \epsilon$ for x in U.

From this assumption we proceed to deduce (for later use) a sort of continuity of the spectrum.

LEMMA 3.3. *Let X be a topological space at each point of which a C^*-algebra A_x is given, and let A be a self-adjoint algebra of functions from X to $\{A_x\}$, with $f(x) \in A_x$, satisfying (a′) and (b′) above. Then for any self-adjoint element $f \in A$ the spectrum of $f(x)$ is a continuous function of x in the following sense: for any $x \in X$ and $\epsilon > 0$ there is a neighborhood U of x such that for all $y \in U$, the spectrum of $f(y)$ is contained in an ϵ-neighborhood of the set consisting of 0 and the spectrum of $f(x)$.*

Proof. Write V for the set consisting of 0 and the spectrum of $f(x)$, and W for an ϵ-neighborhood of it. Let p be a continuous real-valued function which vanishes on V and is equal to 1 on the complement of W. Then $g = p(f)$ vanishes at x, and so by the continuity of the norm, $\|g(y)\| < 1$ for y in a suitable neighborhood U of x. For y in U the spectrum of $g(y)$ must lie in W.

We now turn our attention to assumption (d) above. In view of Lemma 3.2, and the fact that the hypothesis on A_x in that lemma is satisfied if each A_x is a C^*-algebra, we are concerned only with the part of assumption (d) asserting closure under multiplication by $C(X)$. In the next two theorems we show that it is possible to derive this closure under $C(X)$ from weaker assumptions about what the algebra contains at pairs of points. When specialized to the commutative case, a statement of this kind coincides with the Weierstrass-Stone theorem, and so we may regard Theorems 3.2–3.4 as noncommutative generalizations of the Weierstrass-Stone theorem.

It turns out that the simplest case to handle is that where each A_x has a unit element. We denote them all ambiguously by 1, and it then becomes meaningful to formulate another piece of continuity:

(a″) $\|f(x)\|$ is continuous at 1, that is, if $f(x) = 1$ then for any $\epsilon > 0$ there exists a neighborhood U of x such that $\|f(y) - 1\| < \epsilon$ for y in U. We use this as part of the hypothesis in the following theorem.

THEOREM 3.2. *Let X be a compact Hausdorff space at each point of which a C^*-algebra A_x is given. We suppose that each A_x has a unit element, but we admit the possibility of an exceptional point u where $A_u = 0$. Let B be a self-adjoint algebra of functions f on X with $f(x) \in A_x$; we suppose that (a′), (a″), and (b′) above are satisfied, that is, $\|f(x)\|$ is bounded and continuous at 0 and 1, and B*

is complete under the sup norm. Suppose further that for any distinct points y and z ($z \neq u$), B contains a function f with $f(y) = 0$, $f(z) = 1$. Write $C'(X)$ for the algebra of real continuous functions on X, vanishing at the exceptional point u (if there is one). Then, B contains $C'(X)$.

Proof. The idea of the proof is to approximate an element of $C'(X)$ by the use of a partition. However it would be futile to select a partition in the ordinary way, since its members would not be known to be in B; a priori, B is not known to contain any elements of $C'(X)$ at all. Instead we build a partition out of whatever is available in B—a "noncommutative" partition, so to speak.

It is convenient to suppose that an exceptional point u is really present—we can adjoin an isolated point if necessary. Let C be the C*-algebra obtained by adjunction of a unit element to B; it consists of all functions obtained by adding constant scalar functions to B. If we prove that C contains all real continuous functions on X, we will know that B contains those vanishing at u. We note that continuity of $\|f(x)\|$ at 0 and 1 survives in the enlarged algebra C (as a matter of fact we shall need continuity only at 0 in C, but to get this requires continuity at both 0 and 1 in B).

In this paragraph we shall prove that C contains a function equal to 0 and 1 respectively on prescribed disjoint compact subsets of X. We know that for any distinct points y and z, C contains a self-adjoint element k with $k(y) = 0$, $k(z) = 1$. Let p be a real-valued continuous function which vanishes in a neighborhood of 0 and satisfies $p(1) = 1$. Then $k_1 = p(k)$ vanishes in a neighborhood of y and still satisfies $k_1(z) = 1$. Let K be a compact set disjoint from z. For each of its points we find an element in C vanishing in a neighborhood of the point and equal to 1 at z. A finite number of these neighborhoods cover K. Multiplication of the corresponding functions gives us one which vanishes on all of K and is equal to 1 at z. Finally, let K and L be disjoint compact sets. For any point z in L, take the above constructed k_2 which vanishes on K and satisfies $k_2(z) = 1$; we may assume k_2 self-adjoint. With the same real-valued function p as above, form $k_3 = p(1 - k_2)$; $k_3 = 1$ on K and vanishes in a neighborhood of z. We do this at every point of L, select a finite covering of L by the resulting neighborhoods, and multiply the corresponding functions k_3. The result is the desired function in C equal to 1 and 0 on K and L.

Now let h be a continuous real function on X. We propose to approximate it in C within ϵ. Decompose X into open sets $U_1, \cdots, U_n$ on which the oscillation of h is less than ϵ, and let r_i be one of the values assumed by h in U_i. Let $V_1, \cdots, V_n$ be a covering of X by open sets with the closure of V_i in U_i—we are citing part (a) of Lemma 3.1. By the preceding paragraph, C contains a function d_i which equals 1 on V_i and 0 on the complement of U_i. Write $e_i = d_i^* d_i$, $e = \sum e_i$. Then $e \geq 1$ everywhere, and it follows (see the remarks above) that $e \geq 1$ also holds in C. From this we deduce that e has an inverse in C. Moreover that inverse is again positive definite, and has a unique positive square root g, so that $g^2 = e^{-1}$. We write $f_i = ge_ig$, and observe that f_i is self-adjoint, $f_i \geqq 0$, and $\sum f_i = 1$. Then $\sum r_i f_i$ is in C, and we assert that it approximates h within ϵ. Since the convex combination here involves ring elements, this claim needs further substantiation beyond the standard argument. So let x be any point in X, lying say in $U_1, \cdots, U_m$, but not in $U_{m+1}, \cdots, U_n$. We have that for $i \leqq m$, $r_i - h(x)$ is a real number s_i with $|s_i| < \epsilon$. Now choose a fixed operator *-representation of the C^*-algebra A_x, representing $f_i(x)$ say by T_i; we have $T_i \geqq 0$ and $T_1 + \cdots + T_m =$ the identity operator. Our task is to prove that $\| \sum r_i f_i - h \| < \epsilon$ holds at the point x, and this is equivalent to $\| \sum s_i T_i \| < \epsilon$. For this purpose it suffices to show that

$$(4) \qquad \left| \left(\alpha \sum s_i T_i, \alpha \right) \right| = \left| \sum s_i (\alpha T_i, \alpha) \right| < \epsilon$$

holds for any vector α of length 1. But for $i = 1, \cdots, m$ the numbers $(\alpha T_i, \alpha)$ are real non-negative and have sum $= 1$; (4) then follows from $|s_i| < \epsilon$. This concludes the proof of Theorem 3.2.

REMARKS. 1. The assumption (a'') of continuity at 1 cannot be dropped in Theorem 3.2. An example showing this is the following: take A to be the algebra of all sequences of 2 by 2 matrices approaching a matrix of the form

$$\begin{pmatrix} a & 0 \\ 0 & 0 \end{pmatrix}.$$

A is in a natural way a C^*-algebra of functions on a space X consisting of a sequence and a limit point. The norm is continuous on X (not just at 0, but unrestrictedly). However the norm is not continuous at 1 in the sense of (a''), and though A satisfies all the other hypotheses of Theorem 3.2, it fails to contain $C(X)$. On the other hand, A does admit *multiplication* by $C(X)$, and this checks with Theorem 3.3 below.

This same example, incidentally, illustrates that the precautions taken at 0 in Lemma 3.3 were indispensable.

2. If it is the case that not only does each A_x have a unit, but A itself has a unit, then (a'') follows from (a').

3. The assumption in Theorem 3.2 of a unit element in each A_x is alarmingly restrictive for some applications. Whether a theorem like 3.3 or 3.4 can be proved without any restriction on the A's is not known to the author. However a satisfactory result is available in the case where each A_x is dual. In connection with Theorem 3.3, we note: (a) it seems to be necessary to revert to hypothesis (a) above, full-blown continuity of the norm; (b) since there are no unit elements in sight, closure under multiplication by $C(X)$ is the strongest conclusion we can hope for.

THEOREM 3.3. *Let X be a compact Hausdorff space at each point of which* a dual C^*-algebra A_x is given (some of the A_x's may be 0). Let B be a self-adjoint algebra of functions f on X with $f(x) \in A_x$, satisfying postulates (a) and (b) above, that is, $\|f(x)\|$ is continuous and B is complete under the sup norm. Suppose further that for any distinct points x, $y \in X$, B contains functions taking arbitrary pairs of values in A_x, A_y at x, y. Then B is closed under multiplication by $C(X)$, the algebra of all real continuous functions on X.

Proof. Let an element g of B, $\lambda(x) \in C(X)$, and a positive ϵ be given. Write $h = g_* g$. We pick a point y in X and examine the spectrum of $h(y)$. We shall treat explicitly the case where this spectrum is infinite; with some slight modifications the argument applies to the case where it is finite. Then the spectrum of $h(y)$ consists of 0 and a descending sequence $\alpha_1, \alpha_2, \cdots$ of positive real numbers approaching 0—this follows from Theorem 2.1 and known facts about completely continuous operators. Suppose $\alpha_r < \epsilon$, and choose β and γ with $\alpha_r > \beta > \gamma > \alpha_{r+1}$. Define $p(t)$, a real-valued function of the real variable t as follows: $p(t) = 0$ for $t \leq \gamma$, $p(t) = 1$ for $t \geq \beta$, and p is linear between γ and β. Write $e = p(h)$. At the point y we have $\|eh - h\| < \epsilon$, and (by the hypothesis of continuity of the norm) this persists in a suitable neighborhood U_1 of y. Further, Lemma 3.3 shows that for z in a neighborhood U_2 of y, the spectrum of $h(z)$ lies outside (γ, β); it follows that $e(z)$ is an idempotent for z in U_2. Let $U = U_1 \cap U_2$, and let V be an open set containing y, with closure V' in U.

In this paragraph we shall show that, within V', e can be multiplied by an arbitrary continuous real function. For this purpose we introduce I, the (closed two-sided) ideal in B consisting of all functions vanishing on V'. Let e_1 denote the image of e mod I. We consider the algebra $C = e_1(B/I)e_1$. In a natural way, C admits a functional representation on the space V', with the functional values at $x \in V'$ lying in $e(x)A_x e(x)$, and it is still true that for any distinct points x, w in V', C contains a function taking any prescribed pair of values in $e(x)A_x e(x)$ and $e(w)A_w e(w)$. Moreover the continuity and completeness conditions—labelled (a) and (b) above—are still satisfied. All the more so are the hypotheses (a') and (b') of Theorem 3.2 fulfilled. Finally, C has e_1 as unit element, and from this (remark 2 above) we deduce that C satisfies the hypothesis (a'') of continuity of the norm at 1. All the hypotheses of Theorem 3.2 have now been verified and we are entitled to conclude that C contains all continuous real functions on V', or more precisely, all multiples of e_1 by continuous real functions on V'. By taking the inverse image in B of the elements just constructed, we achieve the desired elements which are, within V', multiples of e by arbitrary real continuous functions on V'.

While the functions just constructed behave well on V', we have as yet no control over them outside V'. We shall now adjust this situation, by an argument like that used in Theorem 3.2. Let W be an open set with closure W' in V, $y \in W$. We are going to construct an element f in B such that $f = e$ on W', $f = 0$ on the complement Z of V. Reference to the construction of e as $p(h)$ shows that for any z in Z, the spectrum of $e(z)$ is finite. Hence there exists a self-adjoint element d in B such that $d(z)$ is an idempotent and a unit element for $e(z)$, and we can further suppose that d vanishes at a selected point w of W'. Apply to d a (real-valued continuous) function which vanishes in a neighborhood of 0 and takes the value 1 at 1; this will not change d at z and will make it vanish in a neighborhood of w. A covering of W' reduced to a finite covering, plus a multiplication of the resulting functions, gives us an element vanishing on W' and satisfying $d_1(z) = d(z)$. Now consider $e - d_1 e$; this coincides with e on W' and vanishes at z. It should be clear how another application of the same device gives us the desired element f—vanishing on Z and equal to e on W'.

We next put our two results together. Let θ be a real continuous function on X which vanishes outside W. By our first result we know that B contains an element which on V' coincides with θe. Multiply this element by f and we have one which coincides everywhere with θe. In short, e admits multiplication by real continuous functions vanishing on the complement of W.

We are now ready for the final construction. The neighborhood W and its associated element e have been constructed relative to the point y. Do this at every point of X and extract a finite covering, say by sets $W_1, \cdots, W_r$ with associated elements $e_1, \cdots, e_r$. Split 1 by a partition: $1 = \theta_1 + \cdots + \theta_r$, where θ_i is real continuous, non-negative on X and vanishes outside W_i. Then, by the preceding paragraph, $\lambda \theta_i e_i$ is a well defined element of B. Write $k = \lambda g (\theta_1 e_1 + \cdots + \theta_r e_r)$. Then

$$(k - \lambda g)^*(k - \lambda g) = \lambda^2 \sum_{i,j} \theta_i \theta_j (e_i - 1) h (e_j - 1).$$

In view of the fact that $\|e_i h - h\| < \epsilon$ in W_i this gives

$$\|k - \lambda g\|^2 \leq [\sup |\lambda|]^2 \epsilon,$$

and shows that we can approximate λg arbitrarily closely in B. This concludes the proof of Theorem 3.3.

We note (for later use) a corollary of Theorem 3.3. We first cite [12, Theorem 8.4] and deduce from it that every closed right ideal in a simple dual C^*-algebra is an intersection of regular maximal right ideals. Next, this result extends to any dual C^*-algebra, that is, to a $C^*(\infty)$-sum of simple dual C^*-algebras; this can be seen directly, or we may regard a $C^*(\infty)$-sum as an algebra of functions and cite Corollary 3 of Theorem 3.1. Finally, we put together Theorem 3.3, Lemma 3.2, and Corollary 3 of Theorem 3.1, and we have proved the following result.

COROLLARY. *Let B be an algebra satisfying the hypotheses of Theorem 3.3; then any closed right ideal in B is an intersection of regular maximal right ideals.*

We conclude this section with a more conventional version of the Stone-Weierstrass theorem, in which we speak of continuous functions to a fixed algebra, and show that a suitable subalgebra contains all continuous functions. Theorem 3.4 is readily deduced from the two preceding theorems; one need only remark that by these theorems B admits multiplication by continuous real functions, and it then becomes possible to approximate an arbitrary element of A by a partition.

THEOREM 3.4. *Let X be a locally compact Hausdorff space, D a C^*-algebra which is either dual or has a unit element, and A the C^*-algebra of all continuous functions vanishing at ∞ from X to D. Let B be a closed self-adjoint subalgebra of A, which contains functions taking arbitrary prescribed pairs of values in D at pairs of points in X. Then $B = A$.*

4. Hausdorff structure space. In the remainder of the paper we shall make extensive use of the structure space introduced by Jacobson [8]: the space X of primitive ideals, topologized by making the closure of $\{P_i\}$ the set of primitive ideals containing $\cap P_i$. This definition always makes X at least a T_0-space. We quote the following useful result from [13, Theorem 3.1].

LEMMA 4.1. *Let A be any ring and B either a two-sided ideal in A, a subring of the form eAe, or a subring of the form $(1-e)A(1-e)$, e being an idempotent. Then there is a one-one correspondence between the primitive ideals of B and those primitive ideals of A not containing B. The mapping is implemented by $P \to P \cap B$, P primitive in A, and it is a homeomorphism in the topologies of the structure spaces of A and B.*

In the context of C^*-algebras, the use of the Jacobson structure space is perhaps open to suspicion on two grounds. (1) A primitive ideal is the kernel of a purely algebraic irreducible representation. For C^*-algebras one naturally prefers to use irreducible *-representations. Since in the latter case irreducibility means the absence of *closed* invariant subspaces, the connection between the two concepts is not clear. However, in one direction we can clear up the ambiguity: any primitive ideal P is also the kernel of an irreducible *-representation. To see this, we note that there is a regular maximal right ideal M such that P is the kernel of the natural representation on A/M. Now, in the terminology of Segal [19], there exists a state vanishing on M [17, p. 390]. By an appropriate application of the Krein-Milman theorem we can further get a pure state ω vanishing on M. The *-representation attached to ω is irreducible and has P as kernel. Whether it is true conversely that the kernel of an irreducible *-representation is primitive is an open question; we shall settle a special case in the affirmative in Theorem 7.3. But in any event the structure space of Jacobson is, if anything, the smaller of the two spaces, and for many purposes this more or less justifies its use.

(2) For commutative Banach algebras, it has been found that the structure space has unsatisfactory properties, and that the right topology is the weak topology introduced by Gelfand. Of course for commutative C^*-algebras, the two are known to coincide. The next three results will indicate that

for arbitrary C^*-algebras, the Jacobson structure space is reasonably well behaved.

Having selected the structure X of primitive ideals $\{P_i\}$ in A, we may represent an arbitrary element a of A by the set $\{a_i\}$ of its images in the C^*-algebras $\{A/P_i\}$. We shall write $a(P)$ for the value of a at the point P of X. As observed on page 411 of [12], this functional representation preserves norm, that is, we have $\|a\| = \sup \|a(P)\|$, taken over P in X. But in this context where X is the structure space, even more is true: the sup is attained for some P in X. Because of the identity $\|a^*a\| = \|a\|^2$, one needs to prove this only in case $a \geqq 0$, and we may assume $\|a\| = 1$. To say that $\|a(P)\|$ is less than 1 is to say that the spectrum of $a(P)$ does not contain 1, that is, that $-a$ has a quasi-inverse modulo P. If this is true for every P, then $-a$ has a quasi-inverse modulo every primitive ideal. It is known that this implies that $-a$ has a quasi-inverse in A itself, contradicting $\|a\| = 1$.

The following lemma will be used on several later occasions.

LEMMA 4.2. *Let a be a self-adjoint element of a C^*-algebra with structure space X. Let E be a closed set of real numbers containing 0. Then the set Z of $P \in X$, such that the spectrum of $a(P)$ is contained in E, is a closed subset of X.*

Proof. Suppose that Q is in the closure of Z, and $a(Q)$ has α in its spectrum, $\alpha \notin E$. Let p be a continuous real-valued function vanishing on E but not at α. Then $p(a)$ vanishes on Z but not at Q, contradicting the definition of the topology of X.

The next result indicates that it is quite generally true that the functions on the structure space "vanish at infinity."

LEMMA 4.3. *Let a be any element of a C^*-algebra with structure space X, and ϵ a positive real number. Then the set K of $P \in X$ for which $\|a(P)\| \geqq \epsilon$ is a compact subset of X.*

REMARK. Since X is not necessarily Hausdorff, we are not asserting that K is closed; indeed simple examples show that K need not be closed.

Proof. Because of the identity $\|a^*a\| = \|a\|^2$, we need consider only the case where a is self-adjoint. Let $\{F_j\}$ be a family of (relatively) closed subsets of K having void intersection. We must prove that a finite subset of the F's already have void intersection. Let I_j be the intersection of the primitive ideals comprising F_j, H_0 the ideal generated by the I_j's (in the purely algebraic sense), and H the closure of H_0. We observe that H is not contained in any of the primitive ideals comprising K. For if $H \subset R$, $R \in K$, then R contains each I_j. By the definition of the topology of the structure space, we see that R lies in each F_j, a contradiction. Next we remark that A/H is semisimple (indeed a C^*-algebra). So H is the intersection of the primitive ideals containing it. We write L for this set of primitive ideals, and observe (as just shown) that L is disjoint from K. Hence for every Q in L, we have $\|a(Q)\| < \epsilon$.

Write $r = \sup \|a(Q)\|$, taken over Q in L. If we write a_1 for the image of a mod H, we see that r is precisely the sup of the norms of all the images of a_1 at primitive ideals of A/H. As we remarked above, such a sup is actually attained; in other words we have $r = \|a(Q_0)\|$ for a suitable Q_0 in L. Hence r is itself less than ϵ. Let $p(t)$ be a continuous real-valued function of the real variable t, which vanishes for $|t| \leq r$, equals 2 for $|t| \geq \epsilon$, and is linear between; write $b = p(a)$. Then b is in H since it vanishes on L, and $\|b(P)\| = 2$ for P in K. Since H_0 is dense in H, there exists an element c in H_0 with $\|c - b\| < 1$. The element c must already lie in the union of a finite number of I's, say $I_1, \cdots, I_r$. Then $F_1 \cap \cdots \cap F_r$ must be void; for at any primitive ideal in this intersection c would vanish, whereas $\|c(P)\| \geq 1$ throughout K.

The question as to when the functional representation on the structure space gives functions with continuous norm is completely answered in the following theorem.

THEOREM 4.1. *For any C*-algebra A with structure space X, the following statements are equivalent*: (1) X *is Hausdorff*, (2) *for any $a \in A$, the function $\|a(P)\|$ is continuous on X.*

Proof. That (2) implies (1) is immediate. If Q, $R \in X$, $Q \neq R$, then there exists $a \in A$ vanishing say at Q, but not at R. The continuous real-valued function $\|a(P)\|$ yields disjoint neighborhoods of Q, R.

To prove that (1) implies (2), we begin by investigating continuity of $\|a(P)\|$ at 0. Take a fixed $Q \in X$, and let I be the closure of the set of all $a \in A$ such that $a(P)$ vanishes in a neighborhood of Q; I is a closed two-sided ideal in A. We claim that $I = Q$. If not, since I is the intersection of the primitive ideals containing it, I will also be contained in a different primitive ideal Q_0. By the Hausdorff separation property, there exists a neighborhood U of Q whose closure does not contain Q_0. By the definition of the topology of X, A contains an element vanishing on U but not at Q_0, and this is a contradiction. At this point we know that anything vanishing at Q is a limit of elements vanishing in a neighborhood of Q; this proves continuity of $\|a(P)\|$ at 0.

We pass to the general proof of continuity. Because of the equation $\|a^*a\| = \|a\|^2$, it is enough to do this for self-adjoint a. Let $Q \in X$ and $\epsilon > 0$ be given, and write $r = \|a(Q)\|$. It follows from Lemma 3.3 that in a suitable neighborhood of Q, $\|a(P)\| < r + \epsilon$. To complete the proof it will suffice to show that the set of P with $\|a(P)\| > r - \epsilon$ is open; or alternatively, that the set of P with $\|a(P)\| \leq r - \epsilon$ is closed. This is a consequence of Lemma 4.2.

REMARK. Putting together Lemma 4.3 and Theorem 4.1, we see that when the structure space is Hausdorff it is also locally compact. It is perhaps worth mentioning however that the structure space of a C^*-algebra is not always locally compact. A counter-example is the following: let H be a Hilbert space of dimension $\aleph_\omega$ and take the algebra of bounded operators on H whose range has dimension less than $\aleph_\omega$. It must be admitted however that in this example the structure space is not T_1; the author does not have an example which is T_1 but not locally compact.

In the following theorem we treat the fundamental case in which we are able to prove that the structure space is Hausdorff. The result indicates, roughly speaking, that the structure space being Hausdorff is related to the algebra being "homogeneous" in a suitable sense.

THEOREM 4.2. *Let A be a C^*-algebra in which for every primitive ideal P, A/P is finite-dimensional and of order independent of P. Then the structure space of A is Hausdorff.*

Proof. We shall not prove this by a direct assault on the structure space; instead, following the idea of [12, Theorem 9.2], we provisionally introduce another space. Let M be the finite-dimensional C^*-algebra (a full matrix algebra) to which each A/P is isomorphic. Let Y be the space of all *-homomorphisms of A into M, including the 0 homomorphism. In the weak topology, Y is a compact Hausdorff space, and the elements of A are represented by continuous functions from Y to M—these details go through more or less as usual, and we leave them to the reader. Each primitive ideal in A gives rise to an orbit of points in Y, the orbit being in fact induced by the group G of *-automorphisms of M. Now G is compact in its natural topology, and the mapping from G onto an orbit is readily seen to be continuous. Hence the orbits are closed, and we may form a well defined quotient space X relative to this decomposition of Y. The points of X are of course in one-one correspondence with the set consisting of the structure space of A and a point at infinity. Being a continuous image of Y, X is again compact. We can no longer speak of elements of A as being represented by continuous functions $a(x)$ on X; but the function $\|a(y)\|$ is constant on orbits, and so gives us a uniquely defined function on X, which is again continuous. Moreover these functions $\|a(x)\|$ exist in sufficient abundance to separate points of X, for given two distinct points of X, we can find an element $a \in A$ vanishing at one but not at the other. From this it follows that X is Hausdorff. A final remark is that the functions $\|a(x)\|$ are continuous at 1 in the sense of hypothesis (a'') of Theorem 3.2. We are now ready to cite Theorem 3.2, and we arrive at the main fact we are after: A contains all the real continuous functions that vanish at the 0 homomorphism. From this it would not be difficult to see directly that X (with 0 deleted) is homeomorphic to the structure space; however it is faster to observe that we now know A to be "central" in the sense of [12], and quote [12, Theorem 9.1].

REMARKS. 1. Another way of saying what we have just proved is that, in the homogeneous case, the hypothesis that A is central in [12, Theorem 9.2] is redundant. We can, if we like, cite that theorem to see that the representation of A on Y gives rise to all continuous functions satisfying a suitable "covariance" condition. However, for the purposes of the present paper the representation on X is decidedly more convenient, since on X we get *all* continuous scalar functions.

2. We shall say a few words concerning how the above considerations fit into a more general framework. We stick to the homogeneous case: a Banach algebra A such that all A/P are isomorphic to a fixed finite-dimensional matrix algebra M. Without the aid of a well behaved *-operation, it seems to be difficult to construct a satisfactory theory; in general terms one may trace the trouble to the fact that the group of automorphisms of M is not compact. So let us assume that A has a continuous *-operation which is symmetric (x^*x has a quasi-inverse). Then the construction of the above space Y, and its reduction to X, go through; this appears to be a satisfactory beginning for the theory. Still, one is troubled by the fact that there are no fewer than three further plausible candidates for a space of primitive ideals: (1) the structure space, (2) primitive ideals with the weak topology induced by traces of elements, (3) the space of maximal ideals (in Gelfand's sense) of the center Z of A. If A is a C^*-algebra, all four versions can be seen to coincide, but the general situation is not clear. In particular the choice (3) is in jeopardy, since perhaps $Z = 0$. The case of degree two is an honorable exception, since then $(xy - yx)^2$ is always in the center.

From the homogeneous case treated in Theorem 4.2, we pass on to the case where A is a C^*-algebra such that each A/P is finite-dimensional with a fixed upper bound on the order. This hypothesis can be described more briefly by saying that A satisfies a polynomial identity. It will clarify our future considerations to make a few general remarks at this point. Consider the algebra of n by n matrices over a field. In the notation of $[11]$, this satisfies the identity

$$(5) \qquad\qquad [x_1, \cdots, x_{r(n)}] = 0$$

where $r(n)$ is a certain function of n[5], and matrix algebras of higher order do not satisfy this identity. Now let A be any Banach algebra; let C_n denote the set of all primitive ideals P such that A/P is a k by k matrix algebra with $k \leq n$, and let I_n be the intersection of these ideals. Then A/I_n satisfies the identity (5), and so of course does every primitive image of A/I_n. It follows that C_n is a *closed* subset of the structure space[6]. Moreover it follows from Lemma 4.1 that I_{r-1}/I_r is homogeneous: each of its primitive images is an r by r matrix algebra.

Let A be any C^*-algebra satisfying a polynomial identity. The above defined chain of ideals I_n reaches 0 in a finite number of steps. We have thus

---

(5) In a paper to appear in the Proceedings of the American Mathematical Society, Amitsur and Levitzki prove that $r(n) = 2n$. For our purposes, all that matters is that some identity shall exist that characterizes n by n matrices.

(6) By way of justifying the somewhat artificial introduction of polynomial identities into the discussion, it seems appropriate to remark that I see no other way to prove that C_n is closed.

constructed a finite composition series for A, with the property that every factor algebra has a Hausdorff structure space. This is the embryonic version of the more general theorems to follow in §6.

5. **A category theorem.** From a certain point of view, the study of C^*-algebras may be divided into two parts: the determination of the primitive ones, and the study of how the latter combine to form a general C^*-algebra. We now pick out a class of C^*-algebras for which the first problem evaporates, and so attention is concentrated on the second.

DEFINITION. A CCR-algebra is a C^*-algebra for which every primitive homomorphic image A/P is isomorphic to the algebra of all completely continuous operators on a Hilbert space; more briefly, each A/P is dual.

We noted in §4 that any primitive ideal in a C^*-algebra is the kernel of an irreducible *-representation. Hence if we state that every irreducible *-representation of A consists of completely continuous operators, we are uttering a hypothesis which is formally stronger than the assertion that A is CCR. However we shall later see (Theorem 7.3) that the two hypotheses are equivalent.

In connection with Theorem 5.1, a remark is in order. It seems to be somewhat unorthodox to make use of the category concept in connection with non-Hausdorff spaces. It should be noted that compactness no longer implies second category: a countable space, in which the closed sets are finite or the whole space, is a compact T_1-space of the first category. So Theorem 5.1 tells us roughly that the structure space X of a CCR-algebra cannot be as badly non-Hausdorff as this last example (and in fact one can deduce from Theorem 6.2 that every open subset of X contains an open Hausdorff subset).

We begin with a preliminary lemma.

LEMMA 5.1. *Let $T_1, \cdots, T_n$ be self-adjoint completely continuous operators on a Hilbert space H, r a non-negative real number, and α a vector of unit length in H such that*

$$(I + T_1)^2 + \cdots + (I + T_n)^2 - r^2 I$$

annihilates α, I being the identity operator. Then for each i there exists a self-adjoint completely continuous operator U_i such that $\alpha(I+U_i)=0$ and $\|U_i - T_i\| \leqq 3r$.

Proof. Let E be the projection on α, and define

$$V_i = E(I + T_i) + (I + T_i)E - E(I + T_i)E.$$

Then V_i is self-adjoint and completely continuous. We have

$$\|V_i\| \leqq 3\|E(I + T_i)\| = 3\|\alpha(I + T_i)\|,$$
$$\|V_i\|^2 \leqq 9\sum (\alpha(I + T_i)^2, \alpha) = 9r^2,$$

and so $\|V_i\| \leqq 3r$. We note that $\alpha V_i = \alpha(I+T_i)$. Hence the choice $U_i = T_i - V_i$

satisfies the requirements of the lemma.

THEOREM 5.1. *The structure space X of a CCR-algebra A is of the second category.*

Proof. Suppose on the contrary that X is the union of an increasing sequence C_n of closed nowhere dense sets. We proceed to define stepwise an array of objects.

(a) An increasing sequence of integers $k(1)$, $k(2)$, $\cdots$,

(b) For $j = 2, 3, \cdots$, a point (equal to a primitive ideal) P_j; we realize the algebra A/P_j as the algebra of all completely continuous operators on a Hilbert space H_j,

(c) A vector α_j of unit norm in H_j,

(d) Self-adjoint elements $a(i, j) \in A$, defined for $j = 2, 3, \cdots$ and $1 \leq i \leq j$.

To begin the process, we set $k(1) = 1$, $k(2) = 2$ and take $a(1, 2)$ to be a self-adjoint element vanishing on C_1 and having -1 in its spectrum at a certain point P_2 not in C_2; and we pick α_2 to be a vector of unit norm annihilated by the homomorphic image of $1 + a(1, 2)$ mod P_2. (This formal use of 1 is legal in all the ensuing manipulations even if A lacks a unit element.) Then take $a(2, 2)$ to be a self-adjoint element vanishing on C_2 and such that α_2 is also annihilated by $1 + a(2, 2)$ at P_2.

Now suppose the selection of $k(j)$, P_j, α_j, $a(i, j)$ has been made for $i \leq j \leq n$ so as to satisfy

(1) P_{j-1} is not in $C_{k(j-1)}$ but is in $C_{k(j)}$,

(2) α_j is annihilated by the homomorphic image of $1 + a(i, j)$ mod P_j,

(3) $a(i, j)$ vanishes on $C_{k(i)}$,

(4) $a(i, j-1)$ and $a(i, j)$ coincide on $C_{k(j)}$,

(5) $\|a(i, j-1) - a(i, j)\| \leq 2^{-i}$.

Our selections in the preceding paragraph satisfied these assumptions, insofar as they were as yet meaningful. We show how to push on to $n+1$. Take $k(n+1)$ large enough so that P_n is in $C_{k(n+1)}$. We write

$$b_n = [1 + a(1, n)]^2 + \cdots + [1 + a(n, n)]^2 - n,$$

observing that b_n is an actual ring element. It follows from (2) that b_n has $-n$ in its spectrum at P_n. We claim that there exists a point P_{n+1} outside $C_{k(n+1)}$, where the spectrum of b_n contains a number arbitrarily close to $-n$; concretely, we specify that the spectrum of $b_n(P_{n+1})$ shall contain a number lying within $m = 2^{-2n-8}$ of $-n$. For if not, let $p(t)$ be a continuous real-valued function which is nonzero at $-n$ and vanishes outside the open interval from $-n-m$ to $-n+m$. Then $p(b_n)$ vanishes in the complement of $C_{k(n+1)}$ but is different from 0 at P_n, and this contradicts the assumption that $C_{k(n+1)}$ is nowhere dense (whence its complement is dense). Hence we may choose $P_{n+1} \notin C_{k(n+1)}$ in such a way that $b_n(P_{n+1})$ has in its spectrum a real number s

with $|s+n| <m$. In accordance with the above notation we realize A/P_{n+1} as the algebra of all completely continuous operators on a Hilbert space H_{n+1}. We note that s is nonzero, and consequently we may find in H_{n+1} a unit vector α_{n+1} which is a characteristic vector for the operator $b_n(P_{n+1})$, corresponding to the characteristic root s. Let us write T_i for the operator representing $a(i, n)(P_{n+1})$. Then, in view of the definition of b_n, we find that the operator

$$(I + T_1)^2 + \cdots + (I + T_n)^2 - (s + n)I$$

annihilates α_{n+1}. We are now in a position to apply Lemma 5.1, and we deduce the existence of self-adjoint completely continuous operators $U_1, \cdots, U_n$ on H_{n+1} with $\alpha_{n+1}(I+U_i) = 0$ and $\|T_i - U_i\| \leq 3m^{1/2} < 2^{-n-2}$.

Let us write J for the intersection of the ideals comprising $C_{k(n+1)}$, and $K = J \cap P_{n+1}$. We note that J is not contained in P_{n+1}, for $C_{k(n+1)}$ is a closed set not containing P_{n+1}. Since there are no closed ideals between P_{n+1} and A, $J+P_{n+1}$ is dense in A. Hence [12, Lemma 8.1], A/K is the direct sum of A/J and A/P_{n+1}, in the C^* as well as in the algebraic sense. It is therefore possible to find in A/K a (unique) element g_i which vanishes mod J, and maps on $U_i - T_i$ mod P_{n+1}. We shall have $\|g_i\| = \|U_i - T_i\| < 2^{-n-2}$. Going back to A, we can find a self-adjoint element h_i, mapping on g_i mod K, and with norm arbitrarily close to $\|g_i\|$; we do this so as to get $\|h_i\| < 2^{-n-1}$.

We are ready to define $a(i, n+1)$ as $a(i, n) + h_i$. In view of the estimate for $\|h_i\|$, this satisfies condition (5) above. If we further note that $a(i, n+1)$ was constructed so as to agree with $a(i, n)$ on $C_{k(n+1)}$, and map on U_i at P_{n+1}, we verify conditions (2)–(4). P_{n+1} was selected so as to verify (1).

To complete this step of the induction, it now only remains to choose $a(n+1, n+1)$. For this purpose we need only choose it to be a self-adjoint element vanishing on $C_{k(n+1)}$, and such that its image mod P_{n+1} annihilates α_{n+1}. That such a selection is possible follows from the fact that (in the notation above) A/K is the direct sum of A/J and A/P_{n+1}.

Having completed the construction, we proceed as follows. From (5) we have that for each fixed i, the sequence $a(i, j)$ converges, say to c_i. We note two properties of the c's. Since property (3) is evidently preserved on passage to the limit, we have:

(6) c_i vanishes on $C_{k(i)}$.

Next we deduce from (4), and the fact that the C's are an increasing sequence of sets, that $a(i, j-1)$ coincides with $a(i, s)$ on $C_{k(j)}$ for all $s \geq j$. Now $P_{j-1} \in C_{k(j)}$ by (1); it then follows from (2) that α_{j-1} is annihilated by the homomorphic image at P_{j-1} of every $a(i, s)$ with $s \geq j$. This is preserved in the limit, and we state:

(7) α_j is annihilated by the homomorphic image of $1+c_i \bmod P_j$.

We are now ready to exhibit the contradiction. Consider the right ideal defined by

$$I = (1 + c_1)A + \cdots + (1 + c_n)A + \cdots.$$

Let Q be any primitive ideal in A, say $Q \in C_{k(n)}$. Since, by (6), c_n vanishes on $C_{k(n)}$, we have $c_n \in Q$. The equation $x = (1+c_n)x - c_n x$ shows that $I + Q = A$. Next, I is regular, since for example $-c_1$ is a left unit for I. If $I \neq A$, then I can be embedded in a regular maximal right ideal M. There is a primitive ideal Q contained in M (Q is the kernel of the natural representation of A on A/M), and this contradicts $I + Q = A$. Hence $I = A$, and for suitable elements d_i:

$$(1 + c_1)d_1 + \cdots + (1 + c_n)d_n = - c_1$$

or

$$(1 + c_1)(1 + d_1) + (1 + c_2)d_2 + \cdots + (1 + c_n)d_n = 1.$$

This contradicts the assertion in (7) that the homomorphic images mod P_n of $1+c_1, \cdots, 1+c_n$ all anihilate α_n. This completes the proof of Theorem 5.1.

6. Composition series. By a composition series of a C^*-algebra A, we mean a well-ordered ascending series of closed two-sided ideals I_ρ, beginning with 0 and ending with A, and such that for any limit ordinal λ, I_λ is the closure of the union of the preceding I's. The use of an ascending series here, as opposed to a descending series, is typical of ring theory, and is analogous to the enormous superiority of minimal over maximal ideals.

THEOREM 6.1. *Let A be a C^*-algebra such that every primitive image A/P is finite-dimensional. Then A has a composition series I_ρ such that each $I_{\rho+1}/I_\rho$ satisfies a polynomial identity.*

Proof. In the structure space X of A, let C_n be the set of primitive ideals P such that A/P has degree not greater than n. Then X is the union of the countable family of closed sets C_n. Since X is of the second category (Theorem 5.1), one of them, say C_r, must have a nonvoid interior U. Let I denote the intersection of the primitive ideals comprising the complement of U. Since the latter is closed, these are precisely the primitive ideals containing I; in particular we see that I is nonzero. By Lemma 4.1, the primitive ideals in I itself are in one-one correspondence with the members of U. It follows that the primitive images of I are all of degree not greater than r. Hence I satisfies a polynomial identity: to be precise, the identity for r by r matrices. This is the beginning of our composition series. The algebra A/I again satisfies the hypothesis of our theorem (any primitive image of A/I is a primitive image of A), and we continue by transfinite induction.

As we observed at the end of §4, a C^*-algebra B with a polynomial identity has a (finite) composition series such that all factor algebras possess a Hausdorff structure space. In particular the first nonzero ideal in this series has a Hausdorff structure space, and by Lemma 4.1 the latter is homeomorphic to an open subset of the structure space of B. If we combine this with Theorem 6.1, and another application of Lemma 4.1, we obtain the following corollary.

COROLLARY. *If A is a C^*-algebra such that every primitive image A/P is finite-dimensional, then the structure space of A has a nonvoid open Hausdorff subset.*

We are now able to prove our main theorem.

THEOREM 6.2. *A CCR-algebra has a composition series I_ρ such that each $I_{\rho+1}/I_\rho$ has a Hausdorff structure space.*

Proof. Let A be the algebra and X its structure space. Select a self-adjoint element a in A whose spectrum lies in $(0, 1)$ and actually contains 1. At any $P \in X$, the spectrum of $a(P)$ is a finite or countable set with at most 0 as a limit point. Let $p(t)$ be a continuous real-valued function, vanishing in a neighborhood of 0, satisfying $p(1) = 1$, and, say, linear between. We pass to $b = p(a)$, and observe that every $b(P)$ has a finite spectrum lying between 0 and 1. Let B denote the intersection of the primitive ideals containing b, or in other words, the set of P with $b(P) = 0$; let Y be the structure space of B. Lemma 4.1 shows that B is again CCR, and that Y is in a natural way an open subset of X. Our next efforts will be devoted to proving that Y has an open Hausdorff subset.

For this purpose we first note that $b(Q) \neq 0$ for any $Q \in Y$. For $n = 2, 3, \cdots$, let C_n denote the set of $Q \in Y$ for which the spectrum of $b(Q)$ lies in the closed set consisting of 0 and the closed interval from $1/n$ to 1. By Lemma 4.2, C_n is closed. Also, since the spectrum of each $b(Q)$ is finite, $Y = \cup C_n$. Hence (Theorem 5.1) one of the C's, say C_r, has a nonvoid interior U. Let $q(t)$ be a continuous real-valued function satisfying $q(0) = 0$, $q(t) = 1$ for $t \geq 1/r$, and write $c = q(b)$. Then at every point of C_r, c maps into a non-zero self-adjoint idempotent, and this is a fortiori true at all the points of the closure V of U. Let J be the intersection of the primitive ideals in B which comprise V. Let $D = B/J$; the structure space of D can be identified with V. Let e denote the homomorphic image of c mod J; then e is a self-adjoint idempotent not vanishing at any point of V. We consider finally the algebra eDe; by Lemma 4.1 its structure space is again homeomorphic to V. Now it follows from Lemma 4.1 again that the primitive ideals in eDe are of the form $R \cap eDe = eRe$, where R is primitive in D. Thus the primitive homomorphic images of eDe are of the form $e_1(D/R)e_1$, e_1 being the image of e mod R. We know that D/R is the algebra of all completely continuous operators on a Hilbert space. It follows that $e_1(D/R)e_1$ is finite-dimensional. In short, all the primitive images of eDe are finite-dimensional. The corollary of Theorem 6.1 is therefore applicable, and tells us that V has a nonvoid open Hausdorff subset, say Z. The intersection T of Z and U is a nonvoid open Hausdorff subset of Y.

The same set T is open in X (since Y is open in X). Let I be the intersection of the primitive ideals comprising the complement of T. Then I is a nonzero closed two-sided ideal in A whose structure space is homeomorphic to the Hausdorff space T (Lemma 4.1 is being used again). This is the beginning of our composition series; we continue with a similar treatment of A/I, and so on by transfinite induction.

7. Ideals and subrings. Our study of CCR-algebras now stands as follows. We have established the existence of a composition series such that all the factor algebras have a Hausdorff structure space. Next, if A is a CCR-algebra with a Hausdorff structure space X, then Lemma 4.3 and Theorem 4.1 show that X is locally compact, and that the representation of A on X gives us functions with continuous norm vanishing at ∞. This is precisely the setup to which Theorem 3.3 is applicable, and we are entitled to make use of that theorem and its corollary.

While this cannot be considered a complete structure theory, it does provide us with sufficient information to settle certain problems. We have selected two problems for treatment: the structure of closed one-sided ideals, and the Weierstrass-Stone theorem. The mechanism for climbing up a composition series, for the first of these problems, is provided by Lemmas 7.2 and 7.3.

LEMMA 7.1. *Let A be a ring, R a right ideal in A, I a two-sided ideal in A, and N a regular maximal right ideal in I containing $R \cap I$. Then there exists a regular maximal right ideal M in A, containing R, and such that $M \cap I = N$.*

Proof. The proof is nearly identical with a portion of that of [13, Theorem 3.1], but we repeat it for completeness. Let $e \in I$ be a left unit modulo N. It cannot be the case that $A = N + NA + (1-e)A$, for then

$$n + \sum n_i a_i + (1 - e)a = e$$

for suitable n, $n_i \in N$, a, $a_i \in A$. From this equation we first deduce $a \in I$ (since all other terms are in I). Then $(1-e)a \in N$. A right multiplication by e yields $e^2 \in N$, a contradiction. Hence $N + NA + (1-e)A$ is proper and may be expanded to a regular maximal right ideal M in A with e as left unit. We must verify: (1) $M \cap I = N$, (2) $M \supset R$. (1) Certainly $M \cap I$ contains N. If equality fails, then (since N is maximal) $M \cap I = I$ and M contains e, a contradiction. (2) If not, $M + R = A$, $m + r = e$ with $m \in M$, $r \in R$. We have $mi + ri = ei$ for any i in I, and mi and ri lie in $M \cap I$ and $R \cap I$ which are both contained in N. Hence $ei \in N$, $I \subset N$, a contradiction.

LEMMA 7.2. *Let A be a topological ring with a closed two-sided ideal I. If in I and A/I it is true that every closed right ideal is an intersection of regular maximal right ideals, then the same is true in A.*

Proof. Let R be a closed right ideal in A, and x an element lying in every regular maximal right ideal containing R; our problem is to prove $x \in R$. Denote images mod I by primes. Then x' lies in every regular maximal right ideal containing R', and therefore by hypothesis $x' \in R'$, $x \in R + I$. Write $x = r + y$, $r \in R$, $y \in I$. Our task has been reduced to proving that y is in $R \cap I$. This will be established as soon as we know that y lies in every regular maximal right ideal N in I containing $R \cap I$. But y (like x) lies in the ideal M given by Lemma 7.1, and so y lies in $M \cap I = N$.

LEMMA 7.3. *Let A be a topological ring such that every $x \in A$ lies in the closure of xA. Suppose further that A is the closure of the union of closed two-sided ideals I_ρ, in each of which it is true that every closed right ideal is an intersection of regular maximal right ideals. Then the same is true in A.*

Proof. Let x be an element lying in every regular maximal right ideal M containing the closed right ideal R. Then $xI_\rho \subset M \cap I_\rho$. It follows from our hypothesis and Lemma 7.1 that $xI_\rho \subset R \cap I_\rho \subset R$. Hence $xA \subset R$, $x \in R$.

Before proceeding to use these lemmas, we remark that they allow us to handle a composition series, provided we can manage each factor algebra. It is therefore appropriate to generalize further the concept of CCR-algebra.

DEFINITION. A GCR-algebra is a C^*-algebra having a composition series I_ρ such that each $I_{\rho+1}/I_\rho$ is CCR.

In view of Theorem 6.2, it amounts to the same thing to assume a composition series such that each factor algebra is CCR with a Hausdorff structure space. That this is really a wider class of algebras is shown by the example of the algebra of all operators of the form $\lambda I + U$, where λ is a scalar, I is the identity operator, and U is completely continuous.

THEOREM 7.1. *Any closed right ideal in a GCR-algebra is an intersection of regular maximal right ideals.*

Proof. We remarked at the beginning of this section that for a CCR-algebra with a Hausdorff structure space, the corollary of Theorem 3.3 is available. Hence Theorem 7.1 is known in that case. Lemmas 7.2 and 7.3 extend the result to an arbitrary GCR-algebra (concerning the hypothesis of Lemma 7.3, see footnote 3).

In generalizing the Weierstrass-Stone theorem we shall confine ourselves to CCR-algebras. In connection with Theorem 7.2, it should perhaps be remarked that this is not the only possible form for the Weierstrass-Stone theorem to take—one might for example have the closed subalgebra separate suitable two-sided ideals, or pure states. However, it can be seen that for CCR-algebras these alternative versions come to essentially the same thing.

THEOREM 7.2. *Let A be a CCR-algebra and B a closed self-adjoint subalgebra with the following property: for any distinct regular maximal right ideals in A, B contains an element in one but not in the other. Then $B = A$.*

Proof. Let P, Q be any distinct primitive ideals in A. It follows from [12, Lemma 8.1] that the algebra $A_1 = A/(P \cap Q)$ is the direct sum of A/P and A/Q, that is, A_1 is the direct sum of two simple dual C^*-algebras. The algebra $B_1 = B/(B \cap P \cap Q)$ is in a natural way embedded in A_1, and B_1 separates the regular maximal right ideals of A_1 in the same sense that B does so in A (the simple argument for this was provided in the proof of Theorem 2.2). It follows from Theorem 2.2 that $B_1 = A_1$; in other words, B contains elements taking arbitrary pairs of values at pairs of primitive ideals in A.

Now let $\{I_\rho\}$ be a composition series for A such that each $I_{\rho+1}/I_\rho$ is CCR with a Hausdorff structure space. Let us for brevity write I for the first nonzero ideal of the series; I is thus itself a CCR-algebra with Hausdorff structure space. Let R be any primitive ideal in I, and e any self-adjoint idempotent in I/R. The next few paragraphs will be devoted to proving that $B \cap I$ contains an element mapping on e mod R.

We first consider an arbitrary I_ρ and observe (by Lemma 4.1) that there exists a unique primitive ideal R_ρ in I_ρ such that $R_\rho \cap I = R$. In the present context where all factor algebras modulo a primitive ideal are simple, the two algebras I/R and I_ρ/R_ρ are necessarily isomorphic. There will be no ambiguity if we use the same letter e to denote the element in I_ρ/R_ρ which corresponds to e in I/R.

Suppose we have obtained a self-adjoint element x in $B \cap I_\lambda$ such that $x(R_\lambda) = e$. We shall show how to achieve this with a smaller ordinal than λ.

Case I. λ a limit ordinal. We note that I_λ is the closure of the union of I_μ with $\mu < \lambda$. Hence for a suitable I_μ we have an element $y \in I_\mu$ with $\|y - x\| < 1/2$. In other words, the image of x in I_λ/I_μ is an element of norm less than $1/2$. Let $p(t)$ be a continuous real-valued function vanishing for $|t| \leq 1/2$, and satisfying $p(1) = 1$. Then $p(x)$ is in $B \cap I_\mu$, and $p(x)$ is again equal to e at R_μ.

Case II. λ not a limit ordinal. The algebra $I_\lambda/I_{\lambda-1}$ has a structure space Y which is locally compact Hausdorff, and the elements of $I_\lambda/I_{\lambda-1}$ are represented by functions with continuous norm vanishing at ∞. It will be convenient to think of Y as consisting of the primitive ideals of I_λ which contain $I_{\lambda-1}$. Among these we do not find R_λ (unless $I_\lambda = I$, in which case the present paragraph is superfluous). Let S be the general point of Y, and let us write S' for the unique primitive ideal in A with $S' \cap I_\lambda = S$; similarly write R' for the unique expansion of R to a primitive ideal in A. Then by the opening paragraph of the proof, we know that B contains an element y with $y(S') = 0$, $y(R') = e$. We pass to the element $z = y^* x^* x y$ which is self-adjoint, lies in $B \cap I_\lambda$, and enjoys the properties $z(S) = 0$, $z(R_\lambda) = e$. Let q be a continuous real-valued function vanishing in a neighborhood of 0 and satisfying $q(1) = 1$. Then when we take $w = q(z)$, we achieve the following: $w \in B \cap I_\lambda$, $w(R_\lambda) = e$, w vanishes in a neighborhood of S, and further vanishes outside a compact set $K \subset Y$. For any point in K we can similarly get hold of an element in $B \cap I_\lambda$ which vanishes in a neighborhood of that point, and takes the value e

at R_λ. By a finite number of such elements we can cover all of K. Let v be the result of multiplying w by these elements; then v vanishes on Y and so lies in $I_{\lambda-1}$. If we finally take v^*v we have found the desired self-adjoint element in $B \cap I_{\lambda-1}$, mapping on e at $R_{\lambda-1}$.

Now to start the process going we pick any self-adjoint element x_0 in B with $x_0(R') = e$. We apply the result just obtained, first with $I_\lambda = A$, then with I_λ the next ideal we reach, and so forth. This gives us a chain of elements $x_0, x_1, \cdots$ which must terminate in a finite number of steps, for we are descending a well ordered set. The last element x_k of the chain lies in $B \cap I$ and satisfies $x_k(R) = e$, as desired.

We can now complete the proof of the theorem. Let P, Q be distinct primitive ideals in I. Let e, f be self-adjoint idempotents in I/P, I/Q. Following the same notation as above, we write P', Q' for the extensions of P and Q to primitive ideals in A, and we continue to write e and f for the corresponding elements in A/P', A/Q'. By the first paragraph of the proof, we know that B contains elements y_1, y_2 with $y_1(P) = e$, $y_2(Q) = f$, $y_1(Q) = y_2(P) = 0$. We have further just proved that $B \cap I$ contains y_3, y_4 with $y_3(P) = e$, $y_4(Q) = f$. Then $y_1y_3 + y_2y_4$ lies in $B \cap I$ and maps on e, f at P, Q. Translated, this says that the algebra $B' = (B \cap I)/(B \cap P \cap Q)$ contains all the self-adjoint idempotents of $I/(P \cap Q)$ (observe that the former algebra is naturally embedded in the latter). But the algebra $I/(P \cap Q)$ is dual, and the self-adjoint idempotents of any dual C^*-algebra generate a dense subalgebra. By Lemma 2.1, B' is all of $I/(P \cap Q)$. In other words, $B \cap I$ contains elements taking arbitrary pairs of values at P and Q. Since I is CCR with a Hausdorff structure space, we are in precisely the situation to which Theorem 3.3 is applicable (compare the opening remarks of this section). We deduce that $B \cap I$ is closed under multiplication by any continuous real-valued function on the structure space of I. A partition argument now allows us to approximate elements of I arbitrarily closely by elements of $B \cap I$; this is done exactly as in Theorem 3.1, and we leave the details to the reader. Hence $B \cap I = I$, that is B contains I.

Suppose we have reached the statement $B \supset I_\rho$ for a certain ordinal ρ. The algebra $B/(B \cap I_\rho)$ satisfies the hypothesis of our theorem relative to A/I_ρ, that is, $B/(B \cap I_\rho)$ separates the regular maximal right ideals of A/I_ρ. We may therefore apply the result just proved, and we have $B \supset I_{\rho+1}$. By a transfinite induction, $B = A$. This completes the proof of Theorem 7.2.

REMARK. It seems to be a plausible conjecture that Theorems 7.1 and 7.2 are actually valid for any C^*-algebra.

We shall next briefly discuss the *-representations of a GCR-algebra, confining our discussion to factor representations. A *-representation of a C^*-algebra is said to be a *factor representation* if the weak closure of the set of representing operators is a factor in the sense of Murray and von Neumann [14]. (The Murray-Neumann theory is formally available only in the separable case. However, the sole facts we shall need are easily supplied in the nonseparable case, and are as follows: (1) a factor has no ideal divisors of 0, in the sense of Lemma 2.5. The same condition is called quasi-transitivity by Rickart [18]. (2) If a factor has a minimal right ideal, then it is isomorphic to the algebra of all bounded operators on a Hilbert space. One might call this "type $I_\aleph$," where $\aleph$ is the dimension of the Hilbert space, but we shall simply refer to it as type I.)

Lemma 7.4. *A GCR-algebra with no ideal divisors of* 0 *is primitive with a minimal right ideal.*

Proof. Suppose first that A is a *CCR*-algebra with a Hausdorff structure space X. If X contains more than a single point, take two disjoint open sets, and form the ideals I, J of all functions vanishing on their respective complements. Then I, J are nonzero but $IJ = 0$. Hence A must be primitive.

In the general case where A is *GCR*, there will exist a closed ideal I_1 (the beginning of a composition series for A) such that I_1 is a *CCR*-algebra with a Hausdorff structure space. As a ring on its own merits, I_1 has no ideal divisors of 0 (any closed ideal in I_1 is an ideal in A). Hence I_1 is primitive. Let e be a primitive idempotent in I_1. Then $eI_1 = eA$ is a minimal right ideal in A, and the fact that there are no ideal divisors of 0 implies that the natural representation of A on eA is faithful. Hence A is primitive with a minimal right ideal.

An immediate corollary of this lemma, in view of the remarks above, is the following theorem.

Theorem 7.3. *Any factor representation of a GCR-algebra is of type* I, *and the kernel of the representation is a primitive ideal.*

Theorem 7.3 is in particular applicable to irreducible *-representations, since they are factor representations. In the case of *CCR*-algebras, this tells us that all irreducible *-representations of a *CCR*-algebra consist of completely continuous operators, and we have thus cleared up the question discussed at the beginning of §5. We use this fact in proving the next theorem.

Theorem 7.4. *If A is a CCR-algebra (resp. GCR-algebra), so is any homomorphic image or closed self-adjoint subalgebra of A.*

Proof. (1) If A is *CCR*, then so is any homomorphic image B of A (by a homomorphic image we mean the C^*-algebra obtained in canonical fashion by reducing A modulo a closed two-sided ideal); this is immediate from the definition of *CCR* and the fact that the primitive images of B are also primitive images of A.

(2) Let A be *CCR* and B a closed self-adjoint subalgebra. To prove that B is *CCR* it will suffice (in fact by Theorem 7.3 it is equivalent) to prove that any irreducible *-representation of B consists of completely continuous

operators. Suppose then we have an irreducible *-representation θ of B on a Hilbert space H. It is known [15] that θ can be extended to A in the following sense: there is an irreducible *-representation ϕ of A on a Hilbert space K, and when ϕ is specialized to B there is a closed invariant subspace on which ϕ is unitarily equivalent to θ. Since ϕ sends A into completely continuous operators, θ does the same to B.

(3) Let A be GCR and I_ρ a composition series with each $I_{\rho+1}/I_\rho$ CCR. Suppose that for all ordinals ρ less than λ, we know that the homomorphic images of I_ρ are GCR, and let J be a closed ideal contained in I_λ. Write $K_\rho = (J+I_{\rho+1})/(J+I_\rho)$; K_ρ is isomorphic to $I_{\rho+1}/[I_{\rho+1}\cap(J+I_\rho)]$. For $\rho+1 < \lambda$, K_ρ is GCR by our inductive assumption, while if λ is not a limit ordinal, $K_{\lambda-1} = I_\lambda/(J+I_{\lambda-1})$ is CCR, being a homomorphic image of the CCR-algebra $I_\lambda/I_{\lambda-1}$. We have thus verified that $\{J+I_\rho; \rho \leq \lambda\}$ is a composition series running from J to I_λ, with all quotient algebras GCR. We interpolate further composition series into each of these GCR-algebras, and we have shown that I_λ/J is GCR.

(4) Again let A be GCR with composition series I_ρ, and let B be a closed self-adjoint subalgebra. The algebra $(B\cap I_{\rho+1})/(B\cap I_\rho)$ is CCR, being in a natural way a subalgebra of $I_{\rho+1}/I_\rho$. We shall know that B is GCR as soon as we verify that $\{B\cap I_\rho\}$ is indeed a composition series, that is, for a limit ordinal, we must show that $B\cap I_\lambda$ is the closure of the union of the preceding $B\cap I_\rho$. So let x be a self-adjoint element in $B\cap I_\lambda$. For any $\epsilon > 0$ there exists $\mu < \lambda$ and a self-adjoint y in I_μ, such that $\|x-y\| < \epsilon/2$. Let $f(t)$ be a continuous real-valued function with $f(t) = 0$ for $|t| \leq \epsilon/2, f(t) = t$ for $|t| \geq \epsilon$, and linear between. Then $f(x) \in B\cap I_\mu$ and $\|f(x)-x\| < \epsilon$. This completes the proof of Theorem 7.4.

The next theorem indicates, roughly speaking, the limits of our theory when applied to a general C^*-algebra; we have extensive information on K, but can say nothing about A/K.

THEOREM 7.5. *In any C^*-algebra A there exists a unique maximal GCR-ideal K, and A/K has no nonzero GCR-ideals.*

Proof. (By a GCR-ideal we mean a closed two-sided ideal which is itself a GCR-algebra.) Only two things need verification: (1) the union of two GCR-ideals is a GCR-ideal, (2) the union of a chain of GCR-ideals is a GCR-ideal. (1) If I, J are GCR, then $(I+J)/J \cong I/(I\cap J)$ is GCR by Theorem 7.4. We may then run a CCR composition series from 0 to J and thence to $I+J$. (2) We may assume that the chain $\{I_\rho\}$ is well ordered. By Theorem 7.4 each $I_{\rho+1}/I_\rho$ is GCR, and by interpolation we can build a composition series from 0 to $\cup I_\rho$.

As further results one might mention that the algebra of continuous functions from a compact Hausdorff space to a GCR-algebra is GCR, and that the $C^*(\infty)$-sum of GCR-algebras is GCR. In this last result one cannot replace the $C^*(\infty)$-sum by the C^*-sum (the C^*-sum is the algebra of all bounded sequences). In fact we prove a counter-theorem.

THEOREM 7.6. *If the C^*-sum of C^*-algebras $\{A_i\}$ is GCR, then all but a finite number of the A's satisfy a fixed polynomial identity.*

Proof. Suppose the contrary. Then we can find a sequence of A's, say $B_1, B_2, \cdots$, such that B_n does not satisfy the polynomial identity for n by n matrices. For at least one primitive ideal P_n in B_n, it will likewise be the case that B_n/P_n fails to satisfy this identity. We write $C_n = B_n/P_n$, C for the C^*-sum of $\{C_n\}$, and observe that there is a natural homomorphism of the C^*-sum of $\{A_i\}$ onto C. By Theorem 7.4, C is GCR. Next we note that C_n is primitive, and that a primitive ring has no ideal divisors of 0 [7, Lemma 4]. It follows from Lemma 7.4 that C_n has a minimal right ideal. Now there are two possibilities: C_n may be finite-dimensional, in which case it has degree greater than n. Otherwise, we cite [12, Theorem 7.3] to deduce that C_n contains all the completely continuous operators on an infinite-dimensional Hilbert space. The latter algebra contains subalgebras isomorphic to total matrix algebras of arbitrarily high degree. So in either case we have shown that C_n contains a subalgebra isomorphic to M_n, the n by n total matrix algebra. The C^*-sum of the algebras $\{M_n\}$ is thus a closed self-adjoint subalgebra of C, and by Theorem 7.4 is GCR. The proof will therefore be complete when we establish the following lemma.

LEMMA 7.5. *Let M_n ($n=1, 2, \cdots$) be the n by n total matrix C^*-algebra, and A the C^*-sum of $\{M_n\}$. Then A is not a GCR-algebra.*

Proof. Let I be any maximal ideal in A. If A is GCR, so is A/I (Theorem 7.4). Also A/I is simple with a unit element (since A has a unit), certainly has no ideal divisors of 0, and so by Lemma 7.4 has a minimal right ideal. Theorem 7.3 of [12] is applicable, and shows that A/I must be finite-dimensional, for otherwise it would have a proper two-sided ideal (the ideal of completely continuous operators, for example).

On the other hand, let T be the $C^*(\infty)$-sum of the M's, that is, the set of sequences vanishing at ∞; T is a closed ideal in A. Since A has a unit, T can be enlarged to at least one maximal ideal. In contradiction to the preceding paragraph, we shall show that A/I is infinite-dimensional for any maximal ideal I containing T.

For suppose on the contrary that A/I is an r by r total matrix algebra. Consider an idempotent e in A, $e = \{e_n\}$, where e_n is a projection on a k_n-dimensional subspace. We assert the following: if for sufficiently large n (say $n \geq N$) we have $k_n(r+1) \leq n$, then $e \in I$. Since the unit element of A is evidently the sum of a finite number of such idempotents, this will complete the proof. For $n \geq N$, it is possible to find in M_n, r further idempotents to go along with e_n, all of them projections on k_n-dimensional subspaces, and any two of them orthogonal. Say we pick $e_n(i)$, $i=1, \cdots, r+1$, where $e_n(1)=e_n$. Define $e(i)$ to be the element in A whose nth coordinate is $e_n(i)$; it does not matter what we do with the coordinates before N. Then $e(1), \cdots, e(r+1)$ map into orthogonal idempotents in A/I, and one of them must be 0, say $e(i)$. But $e(1)$ can be expressed in the form $e(1)=fe(i)g$, and so $e(1)=e$ is in I.

8. **Regular ideals.** Ideal theory in rings without a unit element is complicated by the possibility that maximal right ideals, or even closed maximal right ideals, may be nonregular. We shall now indicate that, to some extent, this does not happen in the C^*-algebras under discussion.

LEMMA 8.1. *Let A be a ring with a two-sided ideal I such that $AI=I$. Suppose that all maximal right ideals in I are regular, and that M is a maximal right ideal in A not containing I. Then M is regular.*

Proof. We assert that $N=M\cap I$ is maximal in I. For let $x\in I$, $x\notin N$. For any a in A there exist $b\in A$, $m\in M$, and an integer p such that $x(b+p)+m=a$. A right multiplication by I shows that $xI+N\supset AI=I$, as desired. So N is maximal and hence regular, say with left unit $e\in I$. By Lemma 7.1 there exists a regular maximal right ideal M_1 in A with $M_1\cap I=N$. If $M_1\neq M$, then $M+M_1=A$, and we may write $m'+m_1=e$, $m'\in M$, $m_1\in M_1$. A right multiplication by I shows $eI\subset N$, whence $I\subset N$, a contradiction. Hence $M=M_1$ is regular[7].

LEMMA 8.2. *Let A be a ring with a two-sided ideal I such that $AI=I$. If in I and in A/I all maximal right ideals are regular, then the same is true in A.*

Proof. Let M be a maximal right ideal in A. If $M\supset I$ the result is clear, and otherwise we cite Lemma 8.1.

LEMMA 8.3. *Let A be a Banach algebra with $A^2=A$ which is the closure of the union of two-sided ideals $\{I_r\}$ satisfying $AI_r=I_r$. Suppose that every maximal right ideal in each I_r is regular. Then the same is true in A.*

Proof. Let M be a maximal right ideal in A. If M fails to contain some I_r, we cite Lemma 8.1. So assume $M\supset\cup I_r$. Let P be the primitive ideal attached to M: the set of all x in A with $Ax\subset M$. Then P contains $\cup I_r$. In a Banach algebra any primitive ideal is closed, and so $P=A$, $A^2\subset M$, a contradiction.

LEMMA 8.4. *Let A be a ring with center Z. Suppose that every element of A is a multiple of some element of Z, and that for any primitive ideal P, A/P has a unit element. Then any maximal right ideal M in A is regular.*

Proof. Let P be the primitive ideal attached to M: the set of all x in A with $Ax\subset M$. It will suffice to prove $P\subset M$. Let $x\in P$ and write $x=yz$, $z\in Z$. If $y\in M$, we are finished. Otherwise any a in A can be written a

---

[7] This argument in effect proves that the ideal M of Lemma 7.1 is unique.

$=y(b+n)+m$ for $b \in A$, $m \in M$, and n an integer. A multiplication by zz_1, where z_1 is a general element of Z, shows that $Az \subset M$. Now suppose $z \notin M$. Then (with a similar notation) we have $a_1 = z(b_1 + n_1) + m_1$, and multiplication by the general element of Z gives $M \supset AZ = A$, a contradiction. Hence $z \in M$ and $x \in M$, as desired.

This sequence of lemmas can be applied as follows. First let A be a C^*-algebra for which each primitive image is finite-dimensional of a fixed degree, and let Z be its center. Then we may deduce from Theorem 4.2 that the hypothesis of Lemma 8.4 is satisfied; in fact, if $a \in A$ has the functional representation $f(x)$ on the structure space of A, then $z = \|f(x)\|^{1/2}$ is in Z and a is a multiple of z([8]). So we know that the maximal right ideals of A are regular. Then by applying Theorem 6.1 and Lemmas 8.2 and 8.3, we obtain the following theorem. (It is to be observed that the hypotheses of these lemmas are fulfilled since any C^*-algebra is equal to its square.)

THEOREM 8.1. *If A is a C^*-algebra for which every primitive image is finite-dimensional, then the maximal right ideals of A are regular.*

We conclude with two similar results for which we omit the proofs. (1) For two-sided ideals an analogue of Lemma 8.3 seems not to be available. However the analogues of Lemmas 8.2 and 8.4 hold and we find: *in a C^*-algebra with a polynomial identity all maximal two-sided ideals are regular.* (2) When we push on to GCR-algebras, we are impeded by a lack of information about maximal ideals in the algebra of completely continuous operators on Hilbert space. However *closed* ideals are easily treated and we have: *the closed maximal right ideals in a GCR-algebra are regular.*

9. **Weakly closed algebras.** There is virtually no intersection between the Murray-von Neumann theory of rings of operators and the results of the present paper. In fact, the GCR hypothesis is so restrictive in the context of weakly closed operator algebras that the algebras become trivial.

THEOREM 9.1. *A weakly closed self-adjoint algebra of operators on Hilbert space is GCR if and only if it satisfies a polynomial identity, and it is then the direct sum of a finite number of algebras, each of which is a full matrix algebra over a commutative algebra.*

Theorem 9.1 can (at least in the separable case) be deduced from von Neumann's reduction theory [16]; it is also possible to give a brief direct proof. We omit the details.

10. **Algebraic algebras.** There appears to be a rather far-reaching analogy between algebraic algebras and C^*-algebras. In the commutative case this analogy was explored in [2]. In the more general case of a polynomial identity, there is a striking similarity between Theorem 4.2 of this paper and [13,

---

([8]) That A does actually contain an element b with $a = bz$ can be seen by approximating b locally, and combining the parts by a partition.

Theorem 4.1], each of these theorems asserting that for a homogeneous algebra the structure space is Hausdorff. In this section we shall show that the main device of the present paper—the category argument—also has a perfect counterpart in algebraic algebras, and we shall derive some corollaries concerning Kurosch's problem.

We shall make free use of the results in [2] and [13], but for the reader's convenience we collect the relevant definitions. A ring is biregular if every principal two-sided ideal is generated by a central idempotent, strongly regular if for every a there exists an x with $a^2x = a$, π-regular if for every a there exists an x and an integer n with $a^n x a^n = a^n$. An algebra is algebraic if every element satisfies a polynomial equation, locally finite if every finitely generated subalgebra is finite-dimensional. Any algebraic algebra is π-regular; any strongly regular ring is both π-regular and biregular. Kurosch's problem is to determine whether every algebraic algebra is locally finite.

We first exhibit the category argument in a context which is simplified by the fact that we already know the structure space is Hausdorff. We use the term "composition series" as above; I_ρ is a composition series for A if it is a well ordered ascending chain of ideals running from 0 to A, such that for any limit ordinal λ, I_λ is the union of the preceding I's.

THEOREM 10.1. *Let A be a biregular ring such that every primitive image A/P satisfies a polynomial identity (or equivalently is finite-dimensional over its center). Then A has a composition series I_ρ such that each $I_{\rho+1}/I_\rho$ satisfies a polynomial identity.*

Proof. Let C_n be the set of all primitive ideals P such that A/P satisfies the identity for n by n matrices. Then C_n is a closed subset of the structure space X, and $X = \cup C_n$. Since X is locally compact Hausdorff [2, Theorem 2.2], it is of the second category. Hence some C_r contains a nonvoid open set U. (We could take U open and closed and get a direct summand if we wished.) Let I be the intersection of the ideals comprising the complement of U. Then I satisfies the identity for r by r matrices; it is the beginning of our composition series, we continue on A/I, and so on by transfinite induction.

We now prove the analogue of Theorem 5.1. It is typical that the proof here is simpler: the elaborate construction of Theorem 5.1 is replaced by some simple juggling of idempotents.

THEOREM 10.2. *The structure space of any π-regular ring is of the second category.*

Proof. Suppose on the contrary that the structure space X is the union of the ascending sequence C_n of closed nowhere dense sets. We proceed to construct a sequence $e_1, e_2, \cdots$ of nonzero idempotents satisfying: (1) e_n vanishes on C_n, (2) $e_n \in e_{n-1} A e_{n-1}$. Suppose this has been done as far as e_r. We note that the complement of C_{r+1} is dense (since C_{r+1} is nowhere dense).

It cannot be the case that the nonzero element e_r vanishes everywhere on the complement of C_{r+1}. Hence there exists $P \notin C_{r+1}$ such that $e_r(P) \neq 0$. Write J for the intersection of the ideals comprising C_{r+1}. We observe that J is not contained in P, for C_{r+1} is closed and does not contain P. If we can find a nonzero idempotent in $e_r J e_r$, this will be a suitable choice for e_{r+1}. Let x be any element in $e_r J e_r$. We have $x^s y x^s = x^s$ for suitable s and y. Then $x^s y e_r$ is an idempotent in $e_r J e_r$; if it is zero, so is $x^s = (x^s y e_r) x^s$. Thus the absence of non-zero idempotents in $e_r J e_r$ means that it is a nil ring. Then $e_r J$ is a nil right ideal in A, hence contained in the radical of A, hence contained in P. Likewise $(A e_r A) J (A e_r A)$ is contained in P. By [7, Lemma 4] either J or $A e_r A$ is in P, and both possibilities are excluded. Thus the construction of the e's can be successfully continued.

Let I be the right ideal in A defined by

$$I = (1 - e_1)A + (1 - e_2)A + \cdots + (1 - e_n)A + \cdots .$$

The proof that $I = A$ is identical with the corresponding portion of the proof of Theorem 5.1. Hence for suitable n and elements a_i:

$$(1 - e_1)a_1 + (1 - e_2)a_2 + \cdots + (1 - e_n)a_n = e_1.$$

We observe that $e_n e_i = e_n$ for $i \leq n$, and hence a left-multiplication by e_n yields $e_n = 0$, a contradiction.

Theorem 10.2 could be used for a study of ideals and subrings, along the lines of §7. However we shall here confine ourselves to deriving some consequences concerning Kurosch's problem.

THEOREM 10.3. *Let A be a semi-simple algebraic algebra with the property that every primitive image satisfies a polynomial identity. Then A is locally finite.*

Proof. The structure space X of A is of the second category by Theorem 10.2. Exactly as in the proof of Theorem 10.1 we find in X an open set U such that for $P \in U$, A/P satisfies a fixed polynomial identity, say the one for r by r matrices; and we let I denote the intersection of the ideals comprising the complement of U. Since A is semi-simple, so is I. Moreover every primitive image of I corresponds to an A/P with $P \in U$; hence I itself satisfies the identity for r by r matrices. The algebra A/I again satisfies the hypotheses of our theorem, and the process can be continued. In this way we build a composition series for A, say I_ρ, with each $I_{\rho+1}/I_\rho$ satisfying a polynomial identity. Suppose that for all ρ less than a certain ordinal λ we have proved that I_ρ is locally finite. If λ is a limit ordinal, it is clear that I_λ is again locally finite; indeed any finitely generated subalgebra of I already lies in some I_ρ. If λ is not a limit ordinal, we quote [13, Theorem 6.1] which asserts that any algebraic algebra satisfying a polynomial identity is locally finite. Thus $I_\lambda/I_{\lambda-1}$ is locally finite, $I_{\lambda-1}$ is locally finite by induction, and hence I_λ is locally finite by [9, Theorem 15].

THEOREM 10.4. *If Kurosch's problem has an affirmative answer for nil algebras and for primitive algebraic algebras, then it has an affirmative answer for
all algebraic algebras.*

Proof. Let A be a finitely generated algebraic algebra. Our task is to
prove A finite-dimensional, granted the above hypotheses. The radical R of
A is a nil algebra, and so by hypothesis locally finite. Next we observe that
every primitive image of A/R is (like A) finitely generated, and hence finite-
dimensional by hypothesis. This makes Theorem 10.3 applicable, and we
deduce that A/R is locally finite (and indeed finite-dimensional since it is
finitely generated). Now that we know that both R and A/R are locally
finite, we quote [9, Theorem 15] to conclude the proof. (We remark that
the proof has actually used a slightly weaker hypothesis; we needed only
that a finitely generated primitive algebra is finite-dimensional).

Our final theorem gives another conditional result of this kind.

THEOREM 10.5. *If Kurosch's problem has an affirmative answer for all algebraic division algebras, then it has an affirmative answer for all algebraic algebras
of bounded index.*

Proof. It was observed after [13, Theorem 6.1] that to solve Kurosch's
problem for algebraic algebras of bounded index, it would suffice to treat
the case of index 0 (that is, no nilpotent elements). Such an algebra is
strongly regular, and so is any subalgebra. Thus we have reduced the problem to the case of a finitely generated algebraic algebra A without nilpotent
elements. The primitive images of A are division algebras, are again finitely
generated, and consequently are finite-dimensional by hypothesis. Quotation
of Theorem 10.3 completes the proof (actually the more elementary Theorem
10.1 would suffice).

BIBLIOGRAPHY

1. A. A. Albert, *Structure of algebras*, Amer. Math. Soc. Colloquium Publications, vol. 24,
New York, 1939.

2. R. Arens and I. Kaplansky, *Topological representation of algebras*, Trans. Amer. Math.
Soc. vol. 63 (1948) pp. 457–481.

3. N. Bourbaki, *Élements de mathématique*, vol. 3, *Topologie générale*, Actualités Scientifiques et Industrielles, no. 1045, Paris, 1948.

4. I. Gelfand and M. Neumark, *On the imbedding of normed rings into the ring of operators
in Hilbert space*, Rec. Math. (Mat. Sbornik) N. S. vol. 12 (1943) pp. 197–213.

5. ———, *Unitary representations of the Lorentz group*, Izvestia Akademii Nauk SSSR
Ser. Mat. vol. 11 (1947) pp. 411–504 (Russian).

6. R. Godement, *Théorie générale des sommes continues d'espaces de Banach*, C. R. Acad.
Sci. Paris vol. 228 (1949) pp. 1321–1323.

7. N. Jacobson, *The radical and semi-simplicity for arbitrary rings*, Amer. J. Math. vol. 67
(1945) pp. 300–320.

8. ———, *A topology for the set of primitive ideals in an arbitrary ring*, Proc. Nat. Acad.
Sci. U.S.A. vol. 31 (1945) pp. 333–338.

9. ——, *Structure theory for algebraic algebras of bounded degree*, Ann. of Math. vol. 46 (1945) pp. 695–707.

10. I. Kaplansky, *Dual rings*, Ann. of Math. vol. 49 (1948) pp. 689–701.

11. ——, *Groups with representations of bounded degree*, Canadian Journal of Mathematics vol. 1 (1949) pp. 105–112.

12. ——, *Normed algebras*, Duke Math. J. vol. 16 (1949) pp. 399–418.

13. ——, *Topological representation of algebras*, II, Trans. Amer. Math. Soc. vol. 68 (1950) pp. 62–75.

14. F. Murray and J. von Neumann, *On rings of operators*, Ann. of Math. vol. 37 (1936) pp. 116–229.

15. M. Neumark, *Rings with involutions*, Uspehi Matematičeskih Nauk (N. S.) vol. 3 (1948) pp. 52–145.

16. J. von Neumann, *On rings of operators. Reduction theory*, Ann. of Math. vol. 50 (1949) pp. 401–485.

17. D. Raikov, *To the theory of normed rings with involutions*, C. R. (Doklady) Acad. Sci. URSS. vol. 54 (1946) pp. 387–390.

18. C. Rickart, *Banach algebras with an adjoint operation*, Ann. of Math. vol. 47 (1946) pp. 528–550.

19. I. Segal, *Irreducible representations of operator algebras*, Bull. Amer. Math. Soc. vol. 53 (1947) pp. 73–88.

20. Hing Tong, *On ideals of certain topologized rings of continuous mappings associated with topological spaces*, Ann. of Math. vol. 50 (1949) pp. 329–340.

INSTITUTE FOR ADVANCED STUDY,
    PRINCETON, N. J.
UNIVERSITY OF CHICAGO,
    CHICAGO, ILL.

Afterthought

My afterthoughts concerning this paper are presented as a series of eight remarks.

1. First and foremost: In [2] Glimm advanced the theory by an order of magnitude. Among numerous other things he proved that a separable C^*-algebra is *GCR* if and only if it is of type I (i.e., all its representations are of type I).

2. Page 220. The projected follow-up paper never happened. My interests shifted.

3. Page 236. The advent of central polynomials makes it possible to streamline the proof of Theorem 4.2.

4. Page 233. Kadison [3] abolished the difficulty discussed on this page by showing that for representations of C^*-algebras topological irreducibility implies algebraic irreducibility.

5. Page 244, Theorem 7.1. In [3] Kadison extended this theorem to all C^*-algebras with unit element.

6. Page 244, Theorem 7.2. However, the Weierstrass-Stone theorem, as formulated here, is still holding out.

7. Page 239, Theorem 5.1. Dixmier [1, Theorem 1, p. 97] proved that the structure space of any C^*-algebra is of the second category.

8. It took forty years, but a purely algebraic analogue of *CCR*-algebras finally saw the light of day [4].

References

1. Jacques Dixmier, Sur les C^*-algèbres, *Bull. Soc. Math. France* 88(1960), 95–112.
2. James Glimm, Type I C^*-algebras, *Ann. of Math.* 73(1961), 572–612.
3. Richard V. Kadison, Irreducible operator algebras, *Proc. Natl. Acad. Sci. USA* 43(1957), 273–76.
4. Irving Kaplansky, *CCR*-rings, *Pac. J. of Math.* 137(1989), 155–57.

A GENERALIZATION OF ULM'S THEOREM (*)

By Irving Kaplansky and George W. Mackey

Ulm's thecrem [3], [5] gives a complete classification of countable torsion groups (countable abelian groups with all elements of finite order). Now an abelian group may be regarded as a module over the ring of integers. As a suitable generalization of abelian groups, one may therefore consider modules over a principal ideal ring (integral domain in which every ideal is principal) (¹). It is a fact that the proof of Ulm's theorem goes through without change for countably generated torsion modules over a principal ideal ring R. Ulm himself [4] observes this in the important special case of linear transformations, where R is the ring of polynomials in one variable over a field.

Still another point of view is possible. An abelian group is said to be primary if every element has order a power of a prime, and every torsion group is a direct sum of primary groups. The study of torsion groups is thereby reduced to the primary case. For primary groups the ring of operators can be enlarged beyond the ring of integers to the ring of rational numbers with denominator prime to p, or better yet, to the p-adic integers. As far as primary groups are concerned, this change in the ring of operators simply makes no difference. But it is another matter when we leave torsion groups. For example, countable torsion-free abelian groups have yet to be classified. But countably generated torsion-free modules over the p-adic integers have a trivial structure, being merely direct sums of replicas of the p-adic numbers or p-adic integers (²).

Let us now turn to the mixed case (neither torsion nor torsion--free), where very little is known in the case of ordinary abelian

(*) Manuscrito recebido a 20 de Junho de 1951.
(1) All rings are supposed to have a unit element, which acts as unit operator on any module.
(2) This fact is a consequence of the results in Prüfer's paper quoted in the Bibliography·

1 (Sum. Bras. Math. II — Fasc. 13, 195 - 1951)

groups. It therefore seems to be appropiate to make a flank attack
on the problem by first studying modules over the p-adic integers
(or more generally one can replace the p-adic integers by any complete discrete valuation ring) ([3]). The purpose of this paper is to
make an initial contribution by classifying countably generated
modules over a complete discrete valuation ring, in the case where
the torsion-free rank is one. We find that the Ulm invariants
have to be supplemented by a further invariant: a certain equivalence class of sequences of ordinals. With some fairly obvious
simplifications (Lemma 2 in particular becoming much easier), this
paper can also be read as a proof of the original Ulm theorem, and
it is somewhat shorter than the proofs in the literature.

Before stating the theorem precisely, we discuss some matters
of notation. Let R be a complete discrete valuation ring with
prime p; we write K for the residue class field $R/(p)$. Let M be an
R-module. The torsion submodule T of M consists of all elements
annihilated by some power of p, and M/T is a torsion-free R-
module. The rank of M/T (the maximum number of linearly independent elements) will be referred to as the *torsion-free rank* of M.

A submodule N of M is said to be *divisible* if $pN = N$; any divisible
submodule is known to be a direct summand of M([4]). There exists
a unique largest divisible submodule B in M, and $M = B \oplus C$ where
C has no divisible submodules. The structure of B is completely
known. There is thus no loss of generality in confining our attention to modules with no divisible submodules; we shall call such
a module *reduced*.

Let then R be a reduced R-module. We define

$$M_0 = M, \; M_{\alpha+1} = pM_\alpha$$

for any ordinal α, and

$$M_\alpha = \bigcap M_\beta \; (\beta < \alpha)$$

---

(3) A discrete valuation ring is a principal ideal ring with a unique prime p; completeness
is meant relative to the topology given by the powers of p.

(4) Cf. the paper by Baer quoted in the Bibliography for a complete study of this question.

2 (SUM. BRAS. MATH. II — FASC. 13, 196 - 1951)

for α a limit ordinal. Since M has no divisible submodules, this descending chain must terminate at 0. Let λ be the first ordinal such that $M_\lambda = 0$. For any non-zero x in M there is a last ordinal α such that $x \in M_\alpha$; we call α the *height* of x and write it $h(x)$. It is technically convenient to assign to 0 the height $\lambda + 1$. If S is a submodule of M and x is not in S, we shall say that x is *proper* with respect to S if $h(x)$ is maximal in the coset $x + S$. In a given coset such a proper element may or may not exist, and Lemma 2 below is devoted to a special case where it does exist.

We shall make repeated use of the following easily verified remarks.

1. If $h(y) = h(x)$, then $h(x+y) \geqq h(x)$.
2. If $h(y) > h(x)$, then $h(x+y) = h(x)$.
3. If $x \neq 0$, then $h(px) > h(x)$.

4. If x is proper with respect to S, then for any s in S, $h(x+s)$ is determined by $h(x)$ and $h(s)$, being in fact the smaller of the two.

For any submodule S of M, we write $S_\alpha = S \cap M_\alpha$. In particular we do this for the submodule P consisting of all elements of M annihilated by p. Then $P_\alpha / P_{\alpha+1}$ can be regarded as a vector space over K; its dimension is a cardinal number for which we use the notation $f(\alpha)$. The set of cardinals $f(\alpha)$ we shall call the *Ulm invariants* of M. (These invariants are defined in a slightly different, but essentially equivalent way in [3] and [5]).

Now let us further suppose that M has torsion-free rank 1. Let x be any element of infinite order in M. We introduce the sequence of ordinals $g(i) = h(p^i x)$. Suppose x' is another element of infinite order, giving rise to the sequence $g'(i)$. There exists a relation of the form $x = p^m x' + y$, where y has finite order, say $p^n y = 0$. It follows that the sequences $g(i)$ and $g'(i)$ become identical after the first n, respectively $m+n$ terms are deleted. We are led to call two sequences of ordinals equivalent if they become identical after the omission of suitable initial segments from each. In terms of this equivalence relation, *the equivalence class of $g(i)$ is an invariant of M*.

THEOREM. *Let R be a complete discrete valuation ring, and M a countably generated reduced R-module with torsion-free rank one.*

3 (SUM; BRAS. MATH. II — FASC. 13, 197 - 1951)

Then a complete set of invariants for M is given by the Ulm invariants $f(\alpha)$, together with the equivalence class of $g(i)$, as defined above.

We shall break the proof into several lemmas. For the first of these lemmas, R can be any discrete valuation ring and M any reduced R-module. Before stating the lemma we introduce a certain natural isomorphism. We follow the notation of the preceding discussion; further for any submodule S of M we write

$$S_\alpha{}^* = S_\alpha \cap p^{-1} M_{\alpha+2} \, .$$

In words: $S_\alpha{}^*$ is the set of all x in S_α with px in $M_{\alpha+2}$. For any x in $S_\alpha{}^*$ we may write $px = py$ with $y \in M_{\alpha+1}$. This y is not unique, but it is unique modulo $P_{\alpha+1}$. Thus the mapping $x \to x - y$, followed by the natural homomorphism from P_α to $P_\alpha/P_{\alpha+1}$, gives rise to a homomorphism from $S_\alpha{}^*$ into $P_\alpha/P_{\alpha+1}$. The kernel is precisely $S_{\alpha+1}$. We now have an isomorphism, which we shall call ρ, from $S_\alpha{}^*/S_{\alpha+1}$ into $P_\alpha/P_{\alpha+1}$. In terms of these concepts, we state our first lemma.

LEMMA 1. *The following two statements are equivalent: (a) the isomorphism ρ is not onto, (b) there exists in P_α an element of height α proper with respect to S.*

Proof. $(a) \to (b)$. Let u be an element of P_α whose coset is not in the range of ρ. Then u is surely not in $P_{\alpha+1}$, so its height is α. Suppose u is not proper with respect to S. This means that there exists s in S with $h(s-u) > \alpha$. We write $s - u = pt$ with t in M_α. Since $pu = 0$, we have $ps = p^2 t \in M_{\alpha+2}$. It follows that s is in $S_\alpha{}^*$. Now by definition, in getting the image of s under ρ we send s into $s - pt = u$. This is a contradiction.

$(b) \to (a)$. Let u be the element in question. Then the coset containing u is not in the range of ρ. For if it were we would have

$$u \equiv x - y \quad (\text{mod } P_{\alpha+1})$$

for suitable elements x in S_α, y in $M_{\alpha+1}$. Then $h(u-x) > \alpha$, contradicting the fact that u is proper with respect to S.

LEMMA 2. *Let R be a complete discrete valuation ring, M a reduced R-module, S a finitely generated submodule with torsion-free* *rank one, and x an element not in S. Then the coset $x+S$ contains an element proper with respect to S.*

Proof. It is easy to see that at most a countable number of different heights can occur in $x+S$. Consequently it will suffice for us to examine the following situation: suppose we have a sequence s_i in S such that $h(x+s_i)=\alpha_i$ is monotone increasing. We must produce $s \in S$ with $h(x+s) \geq$ each α_i.

The module S is the direct sum of cyclic modules, generated say by $y, z_1, \ldots, z_r$ where y has infinite order and the $z's$ have finite order. We are going to argue by an induction on r, and for this purpose we first take up the case $r=0$. Write

$$s_n = a_n y, \quad b_n = a_{n+1} - a_n .$$

We claim that b_{n-1} divides b_n properly $(n=2, 3, \ldots)$. First we exclude the possibility that on the contrary b_n divides b_{n-1} properly; for then $b_n/(b_n+b_{n-1})$ is a unit and the identity

$$[b_n/(b_n + b_{n-1})] (x + s_{n-1}) = (x + s_n) - [b_{n-1}/(b_n + b_{n-1})] (x + s_{n+1})$$

shows that $x+s_{n-1}$ has height $\geq \alpha_n$, a contradiction. Hence b_{n-1} at least divides b_n. If this division is not proper, then b_n/b_{n-1} is a unit, and we similarly get a contradiction from the identity

$$(b_n/b_{n-1}) (x + s_{n-1}) = (1 + b_n/b_{n-1}) (x + s_n) - (x + s_{n+1}) .$$

We have proved that b_{n-1} divides b_n properly, and hence a_n is a Cauchy sequence in R, say with limit a. We observe that $a-a_n$ is an associate of $a_{n+1}-a_n$, so that $(a-a_n)y$ has the same height as $(a_{n+1}-a_n)y$; this latter height is precisely α_n, as follows from

$$(a_{n+1}-a_n) y = (x + s_{n+1}) - (x + s_n) .$$

The equation

$$x + ay = (x + s_n) + (a-a_n)y$$

shows that $h(x+ay) \geq \alpha_n$, and so the choice $s=ay$ meets our requirements. This concludes the discussion of the case $r=0$.

We pass to the general case. Suppose $p^m z_1=0$, and let c_n be the coefficient of z_1 in s_n. We can suppose that no c_n is divisible by p^m. Then for a suitable k, there will be an infinite number of c's which are divisible by precisely p^k. After a change of notation (dropping down to a subsequence, and replacing z_1 by $p^k z_1$), we may suppose that each c_n is a unit. Since we are prepared to drop down to a subsequence again if necessary, we need only consider the following two cases.

Case I. All the c's are incongruent mod p. In particular, $c_{n+1} - c_n$ is a unit. The expression

$$c_{n+1}\,(x + s_n) - c_n\,(x + s_{n+1})$$

has height α_n, and it has no term in z_1. We set

$$t_n = (c_{n+1} - c_n)^{-1}\ (c_{n+1} s_n - c_n s_{n+1})\,.$$

and observe that t_n is free of z_1, and that $h(x + t_n) = \alpha_n$. We apply induction on r.

Case II. All the c's are congruent mod p. Write

$$x' = x + c_1 z_1,\quad u_n = s_n - c_1 z_1$$

We have $x' + u_n = x + s_n$ and so $h(x' + u_n) = \alpha_n$. Moreover, in each u_n the coefficient of z_1 is divisible by p. We may replace z_1 by $p z_1$, thus reducing the order from p^m to p^{m-1}. An induction on m is applicable, and this completes the proof of Lemma 2.

LEMMA 3. *Let R be a complete valuation ring, M and N reduced R-modules with the same Ulm invariants, S and T finitely generated submodules of M and N with torsion-free rank one, σ a height-preserving isomorphism of S onto T, and x an element of M with $px \in S$. Then σ can be extended to a height-preserving isomorphism between the submodule generated by S and x and a suitable submodule of N containing T.*

Proof. By Lemma 2, we may assume that x is proper with respect to S. Write $h(x) = \alpha$, and let $z \in S$ be the image of px under σ. Then our problem comes down to this: find an element

w in N which satisfies $pw=z$, has height α, and is proper with respect to T; for once w has been found, we may extend σ by mapping x on w. The resulting mapping will still be an isomorphism and height-preserving. We distinguish two cases.

Case I. $h(px)>\alpha+1$. Then px may be written pv with v in $M_{\alpha+1}$. The element $x-v$ is then in P_α; like x it has height α, and so it is likewise proper with respect to S (since the element v does not interfere in computations of heights $\leq\alpha$). We are now entitled to apply Lemma 1. Since $S_\alpha^*/S_{\alpha+1}$ is a finite-dimensional vector space over K, Lemma 1 tells us that its dimension is less than the Ulm invariant $f(\alpha)$. Since σ is height-preserving, we have that $T_\alpha^*/T_{\alpha+1}$ has the same dimension as $S_\alpha^*/S_{\alpha+1}$. So the dimension of $T_\alpha^*/T_{\alpha+1}$ is likewise less than $f(\alpha)$, which by hypothesis is also the α'th Ulm invariant of N. By Lemma 1 again, there exists in N an element w_1 which satisfies $pw_1=0$, has height α, and is proper with respect to T. Next we note that

$$h(z)=h(px)>\alpha+1$$

and so z may be written pw_2 with w_2 in $N_{\alpha+1}$. We are now ready to take w to be w_1+w_2. It has the desired properties: $pw=z$, $h(w)=\alpha$, and w is proper with respect to T.

Case II. $h(px)=\alpha+1$. If there exists an element s in S with

$$h(x+s)=\alpha, \quad h(px+ps)>\alpha+1,$$

we can replace x by $x+s$ and revert to Case I. So we can even assume the following: for any $s \in S$ with $h(x+s)=\alpha$, we have $h(px+ps)=\alpha+1$. It is to be noted that by our convention about the height of 0, px (and hence z) must be non-zero. We now simply choose for w any element in N_α with $pw=z$ (such an element certainly exists since $h(z)=\alpha+1$). The height of w must be precisely α, for if it exceeded α, the height of $pw=z$ would exceed $\alpha+1$. (Whenever w is non-zero, the height of pw exceeds the height of w). Next, it cannot be the case that w is in T. For suppose it corresponds under σ to y in S. Then $px=py$; also $x-y$ is not in S for otherwise x would be in S; and finally $x-y$ has height α, since each has height α and x is proper with respect to S. But

$$h(px-py)=h(0)>\alpha+1,$$

and this contradicts the (strengthened) hypothesis of Case II. To complete the proof, it only remains to see that w is proper with respect to T. Suppose on the contrary that $h(w+t) \geqq \alpha+1$, where t is an element of T corresponding under σ to s in S. Since $w+t \neq 0$, we have $h(pw+pt) \geqq \alpha+2$. Likewise $h(px+ps) \geqq \alpha+2$. Now t necessarily has height at least α, hence so has s, and $h(x+s) = \alpha$. We have again contradicted the hypothesis of Case II.

With Lemma 3 at hand, it is easy to complete the proof of the main theorem. Suppose that M and N are countably generated reduced R-modules of torsion-free rank one, that they have the same Ulm invariants, and the same equivalence class $g(i)$. Take any elements u und v of infinite order in M and N respectively. By hypothesis, the sequences $h(p^i u)$ and $h(p^i v)$ coincide after the deletion of say m and n terms. Let S and T be the submodules generated by $p^m u$ and $p^n v$; then the mapping $p^m u \rightarrow p^n v$ implements a height-preserving isomorphism of S and T. This isomorphism is now extended stepwise to an isomorphism of M and N. To make sure that we catch all of M and N, we take fixed countable sets of generators for each, and alternate between adjoining an element of M and adjoining an element of N. Since the elements of M and N have finite order modulo S and T respectively, we can suppose that at each step we are adjoining an element x such that px lies in the preceding submodule. This is precisely the situation to which Lemma 3 applies.

Bibliography

1. R. BAER, *Abelian groups that are direct summands of every containing abelian group*, Bulletin of the American Mathematical Society, vol. 46 (1940) pp. 800-806.
2. H. PRÜFER, *Theorie der Abelschen Gruppen*. II, Mathematische Zeitschrift, vol. 22 (1925) pp. 222-249.
3. H. ULM, *Zur Theorie der abzählbar-unendlichen Abelschen Gruppen*, Mathematische Annalen, vol. 107 (1933), pp. 774-803.
4. H. ULM, *Elementarteilertheorie unendlicher Matrizen*, Mathematische Annalen, vol. 114 (1937) pp. 493-505.
5. L. ZIPPIN, *Countable torsion groups*, Annals of Mathematics, vol. 36 (1935) pp. 86-99.

UNIVERSITY OF CHICAGO
HARVARD UNIVERSITY

8 (SUM. BRAS. MATH. II — Fasc. 13, 202 - 1951)

Afterthought

I drafted an afterthought on this paper and submitted it to my co-author, George Mackey, for comments. The following is an extract (slightly edited) from a letter from him, dated December 4, 1993.

"I did not contemplate a systematic study of linear transformations in infinite dimensional spaces. It was rather that I was intrigued by the Jordan canonical form and the fact that it is so easy to reduce to the case of nilpotent linear transformations, and wondered to what extent this argument could be extended to the infinite dimensional case. I was of course delighted to find that I could manage in the countable case and that the extension was far from trivial. As I recall I worked quite hard for several weeks at least. I also seem to recall that I became aware of Zippin's simplification of Ulm's work while I was still struggling to prove my conjecture, and was encouraged to proceed by noting the analogy. However, I did not try to follow Zippin's rather complicated arguments and continued to work independently. When my proof turned out to be distinctly simpler than Zippin's, I attributed this to the difference between vector spaces and Abelian groups and did not even try to prove Ulm's theorem by my methods. Also, at this time I had not yet absorbed the module concept and saw only a vague analogy rather than two different cases of the same concept. I planned to write up my results as a paper about operators in vector spaces of countable dimension and then was crushed to discover that a module version of Ulm's theorem had been explicitly stated in the Russian literature and of course included my theorem as a special case. The linear transformation case may have been an explicit corollary. I found this Russian paper purely by accident while browsing in the relevant journal. I may have found Zippin's paper in the same way, but here my memory fails me. In any event I am reasonably sure that I heard about Ulm from Zippin's paper.

"All of this happend in early 1944 shortly before I took leave to do war work in England just after Easter. In the interval you made a visit to Cambridge and I told you my sad story. You asked to see my manuscript and reported back that my arguments extended without difficulty to the module case and provided a proof of Ulm's theorem that was distinctly simpler than Zippin's. You advised that I should rewrite my paper accordingly and submit it for publication. I was delighted but felt (mistakenly, I now realize) that you should be a joint author. I now see that I should have done as you suggested, giving you due credit for calling my attention to the possibility. However, I was stubborn and you quite rightly refused to accept joint authorship. There the matter rested for many years. Then you had your idea for invading mixed modules and wrote a paper with both our names signed to it. By this time I thought your contribution outweighted mine but I read the paper carefully and gave my approval."

I have been unable to identify the Russian paper in question. However, Ulm himself published a second paper [6], proving his theorem for the case of appropriate modules over a polynomial ring in one variable over a field, that is, linear transformations. He used the infinite matrix methods of his groundbreaking paper [5].

I am able to have the last word, and I record my dissent from the "outweighed" judgment in George's last sentence. At any rate, I am glad that his beautiful proof got published. The history of Abelian groups would surely have been quite different otherwise. In fact it promptly got published three times:

(a) in the paper under discussion.
(b) as exercises in Bourbaki. In the latest version [1] it is still there as Exercises 13 and 14 on pages AVII 56–57. (I believe this came about as follows: I showed the proof to my colleague André Weil at the University of Chicago and he sent it on to the late Jean Dieudonné.)
(c) When I wrote my monograph [3], this is of course the proof I used.

Abelian group theory has developed mightily and, in particular, Ulm's theorem has been greatly generalized. This is no place to try to survey it. Let me just mention Fuchs' definitive treatise [2] and my account [4] of five advances that I considered particularly striking in 1978.

References

1. N. Bourbaki, *Algèbre*, Chs. 4–7, Masson, Paris, 1981.
2. L. Fuchs, *Infinite Abelian Groups*, Academic Press, New York, vol. 1, 1970; vol. 2, 1973.
3. I. Kaplansky, Infinite Abelian Groups, Univ. of Michigan Press, Ann Arbor, 1st ed. 1954; rev. ed. 1969.
4. _____, Five theorems on Abelian groups, pp. 47–51 in *Topics in Algebra*, Proc. of the 18th Summer Research Institute of the Australian Math. Soc., Springer Lecture Notes no. 697, 1978.
5. H. Ulm, Zur Theorie der abzählbar-unendlichen abelschen Gruppen, *Math. Ann.* 107(1933), 774–803.
6. _____, Elementarteilertheorie unendlicher Matrixen, *Math. Ann.* 114(1937), 493–505.

MODULES OVER DEDEKIND RINGS AND VALUATION RINGS

BY

IRVING KAPLANSKY

1. Introduction. This paper has two purposes. In §4 our objective is to push forward the theory of modules over Dedekind rings to approximately the same point as the known theory of modules over principal ideal rings. In §5 we prove several results concerning modules over valuation rings. In Theorem 12 we give a complete structure theorem for torsion-free modules of countable rank over a maximal valuation ring. In Theorem 14 we show that any finitely generated module over an almost-maximal valuation ring is a direct sum of cyclic modules; this provides a simpler proof and generalization of Theorem 11.1 in [6](1).

2. Basic notions. Let R be an integral domain (commutative ring with unit and no divisors of 0). In saying that M is an R-module, we shall always mean that the unit element of R acts as unit operator on M. If x is any element in M, the set of α in R with $\alpha x = 0$ is called the *order ideal* of x. If all elements of M have a nonzero order ideal, we call M a *torsion module*. If all nonzero elements of M have 0 as their order ideal, we say that M is *torsion-free*. In general, the elements of M with a nonzero order ideal form a submodule T of M called the *torsion submodule*, and the quotient module M/T is torsion-free.

In case M is a torsion-free R-module, we may speak unambiguously of the maximum number of linearly independent elements in M, and we call this cardinal number the *rank* of M. There is another point of view that is useful. We may extend the coefficient domain of M from R to its quotient field K; that is, we may form the Kronecker product $K \times M$ of the R-modules K and M. Then $K \times M$ may be regarded as a vector space over K, and its dimension coincides with the rank of M.

We shall not attempt the delicate task of assigning a meaning to "rank" in the case of torsion modules.

If M is a torsion-free R-module of rank one, it is easy to see that M is isomorphic to an (integral or fractional) ideal I in R. (In this connection, no demand should be made concerning the boundedness of the denominators occurring in I; such boundedness is equivalent to the possibility of selecting I to be an integral ideal.)

We next introduce Prüfer's [10] concept of "Servanzuntergruppe," but we prefer to use Braconnier's terminology [3]. The submodule S of the

Received by the editors July 30, 1951.

(1) Numbers in brackets refer to the bibliography at the end of the paper.

327

R-module M is said to be *pure* if $\alpha S = S \cap \alpha M$ for every α in R.

The reader may find the following remarks helpful. They are easily proved and will be used without further comment. (a) Any direct summand is pure. (b) A pure submodule of a pure submodule is pure. (c) The torsion submodule is pure. (d) If M/S is torsion-free, then S is pure. (e) If M is torsion-free, then the intersection of any set of pure submodules of M is again pure. Hence any subset of M is contained in a unique smallest pure submodule.

Numerous examples show that a pure submodule need not be a direct summand. However in various special cases we shall be able to prove that this converse does hold.

Finally we define an R-module M to be *divisible* if $\alpha M = M$ for all non-zero α in R. We note that if M is divisible, then a submodule of M is divisible if and only if it is pure; also if M is any R-module, then any divisible submodule of M is pure (by a divisible submodule we mean a submodule which is divisible as a module in its own right). A torsion-free divisible R-module is evidently the same thing as a vector space over the quotient field of R.

3. Some lemmas. Before proceeding to special classes of rings, we shall prove three lemmas that are valid for an arbitrary integral domain. The first concerns the question of the uniqueness of the representation of a torsion-free module as a direct sum of modules of rank one. If I and J are (integral or fractional) ideals in R, one knows that they are isomorphic R-modules if and only if there exists a nonzero α in the quotient field of R, with $I = \alpha J$. Following the usual terminology, we then say that I and J are in the same *class*. We now show that for the direct sum of a finite number of ideals, the class of the product is an invariant. In general there are other invariants, but in a particular case (Theorem 2) we shall find it to be the sole invariant.

LEMMA 1. *Let R be an integral domain, and let $I_1, \cdots, I_m, J_1, \cdots, J_m$ be (integral or fractional) ideals in R such that $I_1 \oplus \cdots \oplus I_m$ and $J_1 \oplus \cdots \oplus J_m$ are isomorphic R-modules. Then $I_1 I_2 \cdots I_m$ and $J_1 J_2 \cdots J_m$ are in the same class.*

Proof. It is convenient to assume (as we can without loss of generality) that each of the ideals contains R. In the isomorphism between the two direct sums, suppose the element 1 in I_r corresponds to $(\alpha_{r1}, \cdots, \alpha_{rm})$, where $\alpha_{rs} \in J_s$. We then find

$$(1) \qquad\qquad J_s = \alpha_{1s} I_1 + \cdots + \alpha_{ms} I_m \qquad\qquad (s = 1, \cdots, m).$$

Let γ be the determinant $|\alpha_{rs}|$, and let δ be a typical term in the expansion of this determinant. On multiplying the equations (1), we express $J_1 \cdots J_m$ as a sum of ideals, among which we find all $\delta I_1 \cdots I_m$. Hence

$$J_1 \cdots J_m \supset \gamma I_1 \cdots I_m.$$

We now invert the procedure. Suppose the element 1 in J_r corresponds to $(\beta_{r1}, \cdots, \beta_{rm})$, where $\beta_{rs} \in J_s$. Then the matrix (β_{rs}) is the inverse of the matrix (α_{rs}). We thus obtain the inclusion

$$I_1 \cdots I_m \supset \gamma^{-1} J_1 \cdots J_m,$$

and this proves the lemma.

The next lemma could easily be stated and proved in more general contexts. (By the term "direct sum" we mean the "weak" direct sum, where only a finite number of nonzero components are allowed in each element.)

LEMMA 2. *Let M be any module, S a submodule, M/S a direct sum of modules U_i, and T_i the inverse image in M of U_i. Suppose S is a direct summand of each T_i. Then S is a direct summand of M.*

Proof. Let W_i be a complementary summand of S in T_i, and W the union of $\{W_i\}$. Then W maps onto all of M/S, whence $S + W = M$. Again suppose $x = \sum y_i$ is in $S \cap W$, where $y_i \in W_i$. Then on passing to M/S we see that each y_i is in S, $y_i = 0$, $x = 0$. Thus $S \cap W = 0$ and $M = S \oplus W$.

Let R be any integral domain with quotient field K, and I an (integral or fractional) ideal in R. We recall that I^{-1} is defined as the set of all α in K with $\alpha I \subset R$; I is said to be *invertible* if $II^{-1} = R$. Any invertible ideal is finitely generated.

LEMMA 3. *Let R be an integral domain, M an R-module, S a submodule such that M/S is a direct sum of modules which are isomorphic to invertible ideals in R. Then S is a direct summand of M.*

Proof. Lemma 2 reduces the problem to the case where M/S is itself isomorphic to an invertible ideal I. We therefore make this assumption, and we furthermore make the provisional assumption that M is torsion-free. Since $II^{-1} = R$, there exist elements α_i, β_i in I and I^{-1} such that $\sum \alpha_i \beta_i = 1$. Let y_i be the elements in M/S which correspond to α_i, let γ be any nonzero element in I, let x_i be any elements in M mapping on y_i, and set $z = \gamma \sum \beta_i x_i$. (Notice that $\gamma \beta_i \in R$, so that z is a well defined element of M.) Let T be the pure submodule generated by z; since M is torsion-free, this is a well defined submodule of rank one. We claim that T is the desired complementary summand of S. That $S \cap T = 0$ is clear; we need only show $S + T = M$, that is, we must show that T maps on all of M/S. For this it suffices to prove that y_i is in the image of T (note that the elements α_i generate I, and so the elements y_i generate M/S). Now $\alpha_i \beta_j \in R$. Hence $t = \alpha_i \gamma^{-1} z = \alpha_i \sum \beta_j x_j$ is in M, and consequently t is in T. The image of t is $\alpha_i \sum \beta_j y_j$. In the isomorphism of M/S and I, $\sum \beta_j y_j$ corresponds to $\sum \beta_j \alpha_j = 1$. Hence $\alpha_i \sum \beta_j y_j = y_i$, as desired.

We pass to the general case where M is no longer assumed to be torsion-free. Select a homomorphism ϕ of a free R-module F onto M. This gives rise

149

to a homomorphism from F onto M/S, say with kernel G (G is also the inverse image of S under ϕ). By what we have just proved, G is a direct summand of F: say $F = G \oplus H$. Let $V = \phi(H)$. Manifestly $S + V = M$. Again, since the kernel of ϕ is contained in G, $\phi(G)$ and $\phi(H)$ have only 0 in common. Thus $S \cap V = 0$ and $M = S \oplus V$. This concludes the proof of Lemma 3.

4. Dedekind rings. A Dedekind ring is an integral domain in which classical ideal theory holds: that is, every ideal is uniquely a product of prime ideals or alternatively, every ideal is invertible. For our first two theorems we find it possible to consider the more general class of integral domains in which it is assumed only that every finitely generated ideal is invertible.

THEOREM 1. *Let R be an integral domain in which every finitely generated ideal is invertible, and let M be a finitely generated R-module with torsion submodule T. Then T is is a direct summand of M, and M/T is a direct sum of modules of rank one (necessarily isomorphic to invertible ideals).*

Proof. That T is a direct summand of M will follow from Lemma 3, as soon as we have proved the last statement of the theorem. So let us suppose that M is actually a torsion-free finitely generated R-module. Take any nonzero element in M and form the pure submodule S of rank one that it generates. Then M/S is again torsion-free and finitely generated. Its rank is one less than the rank of M. By induction on the rank, we may assume that M/S is a direct sum of modules of rank one. Lemma 3 now applies to show that M is the direct sum of S and M/S, and this proves the theorem.

We turn now to the uniqueness question associated with Theorem 1. It is a classical result of Steinitz [13] that, over a Dedekind ring, the rank n and the class of the product $I_1 I_2 \cdots I_n$ are a complete set of invariants for a torsion-free module $I_1 \oplus I_2 \oplus \cdots \oplus I_n$. In Theorem 2 we are able to generalize this result, provided we supplement the hypothesis that finitely generated ideals are invertible with another hypothesis known to be satisfied in a Dedekind ring([2]).

We also take up this uniqueness question in the case of infinite rank. The result is typical of infinite algebra: the invariant evaporates.

THEOREM 2. *Let R be an integral domain which satisfies the following two conditions: every finitely generated ideal in R is invertible, and if I is a nonzero finitely generated ideal in R, R/I is a ring in which every finitely generated ideal is principal. Then:*

(a) If M is a finitely generated torsion-free R-module, a complete set of invariants for M is the rank n together with the class of the product $I_1 I_2 \cdots I_n$ in a representation of M as a direct sum of ideals $I_1 \oplus \cdots \oplus I_n$.

([2]) A suitable infinite algebraic extension of a Dedekind ring gives an example of a ring (other than a Dedekind ring) satisfying the hypotheses of Theorem 2. For an explicit example we may take the ring of all algebraic integers.

(b) *If N is an R-module which is a direct sum of an infinite number of invertible ideals, then N is free.*

Proof. (a) That the class of $I_1 \cdots I_n$ is an invariant was shown in Lemma 1. We have to show that this class, together with the integer n, forms a complete set of invariants. To do this, it suffices by an evident induction to treat the case $n = 2$, that is to say, we wish to show that $M = I_1 \oplus I_2$ is isomorphic to $R \oplus I_1 I_2$. Suppose we have shown that M contains a pure submodule S isomorphic to R. Then by Lemma 3, M is the direct sum of S and M/S, and M/S will necessarily be isomorphic to $I_1 I_2$. So our task is reduced to finding the submodule S. We can of course suppose that I_2 is an integral ideal. Let α be any nonzero element in I_1. Then $\alpha I_1^{-1} I_2$ is a nonzero integral ideal contained in I_2. By hypothesis, I_2 maps onto a principal ideal in $R/\alpha I_1^{-1} I_2$, that is, there exists an element β in I_2 such that $I_2 = (\beta) + \alpha I_1^{-1} I_2$, whence

$$(2) \qquad \alpha I_1^{-1} + \beta I_2^{-1} = R.$$

Now consider the element $x = (\alpha, \beta)$ of $M = I_1 \oplus I_2$. The pure submodule generated by x is isomorphic to the ideal K consisting of all γ (in the quotient field of R) with $\gamma x \in M$; that is, all γ with $\gamma \alpha \in I_1$ and $\gamma \beta \in I_2$. Thus $K = \alpha^{-1} I_1 \cap \beta^{-1} I_2$ and is the inverse of the ideal appearing on the left of (2). Hence $K = R$, as desired.

(b) If N is of uncountable rank, we may break it into a direct sum of pieces of countable rank. Since it suffices to treat each of these pieces separately, we may assume that N itself is of countable rank, whence it is countably generated. Let $x_1, x_2, \cdots$ be a countable set of generators for N.

We proceed to build a sequence of submodules $\{K_j\}$ of N with the properties:

(1) $K_1 \subset K_2 \subset K_3 \subset \cdots$,

(2) each K_j is free,

(3) K_{j+1} is the direct sum of K_j and a free module,

(4) N is the direct sum of K_j and a module L_j which is an infinite direct sum of finitely generated modules of rank one,

(5) the union of all the K's is N.

It is evident that the existence of such a sequence implies that N is free. We take $K_1 = 0$, and suppose K_r has been selected. Let y denote the first of the x's which is not in K_r, and z the L_r-component of y in the decomposition $N = K_r \oplus L_r$. By hypothesis, L_r is a direct sum of modules $\{P_t\}$ of rank one where P_t is isomorphic to the invertible ideal I_t in R. Suppose z is in $P_1 \oplus \cdots \oplus P_s$, and write $I = I_1 I_2 \cdots I_s$. By part (a) of the present theorem, $P_{s+1} \oplus P_{s+2}$ can be rewritten as $G \oplus H$ where G and H are isomorphic to I^{-1} and $I I_{s+1} I_{s+2}$ respectively. Then the module $P_1 \oplus \cdots \oplus P_s \oplus G$ is free, and we take K_{r+1} to be the direct sum of it and K_r, while $L_{r+1} = H \oplus P_{s+3} \oplus P_{s+4} \oplus \cdots$. With this choice of K_{r+1} and L_{r+1} we continue to satisfy properties (1)–(4) above. Thus the sequence $\{K_j\}$ can be selected, and this concludes the proof of Theorem 2.

From now on, R will be an actual Dedekind ring, and we shall study R-modules. We begin by disposing of torsion modules. Call a torsion R-module *primary* (for the prime ideal p) if each of its elements is annihilated by a suitable power of p. It can be proved (in virtually the same way as for principal ideal rings) that every torsion R-module is uniquely a direct sum of primary modules.

Next let R_p denote the quotient ring of R relative to p, that is, the set of all elements α/β in the quotient field with β prime to p. Then any p-primary R-module can as well be regarded as an R_p-module. Since R_p is a principal ideal ring (indeed a discrete valuation ring), this makes the known theory of modules over principal ideal rings fully available. Consequently we have virtually nothing to say about torsion modules over a Dedekind ring; our interest is confined to the torsion-free and mixed cases.

At the risk of possible ambiguity, we shall call an R-module *decomposable* if it is a direct sum of cyclic modules and finitely generated torsion-free modules of rank one. It is to be noted that when R is a principal ideal ring, a decomposable module is precisely a direct sum of cyclic modules. A fundamental theorem of Steinitz [13] asserts that every finitely generated module over a Dedekind ring is decomposable[3]. In Theorem 1, together with the remarks above concerning torsion modules, we have in effect given a new proof of this theorem.

We shall proceed to derive some results concerning decomposable modules.

LEMMA 4. *Let R be a Dedekind ring, M an R-module, S a pure submodule, x_0 an element of M/S. Then there exists an element x in M, mapping on x_0 mod S, and having the same order ideal as x_0*[4].

Proof. The lemma is known if R is a principal ideal ring, and the proof will consist of a reduction to that case. Let I be the order ideal of x_0. If $I=0$, any choice of x will do. So we suppose $I\neq0$, let α be a nonzero element of I, and pick any y in M mapping on x_0. Then αy is in S and, by the purity of S, there exists s in S with $\alpha y = \alpha s$. If $z = y - s$, then $\alpha z = 0$, and z (like y) maps on x_0 mod S.

Now let T be the torsion submodule of M. Then $S \cap T$ is pure in T. Moreover $T/(S \cap T)$ is in a natural way a submodule of M/S, and (in this sense) it contains x_0. The problem of lifting up x_0 has thus been reduced to the cor-

---

[3] A brief proof of this theorem is given by Chevalley in [4, Appendix 2].

[4] It is conversely true for a module M over an arbitrary ring that purity of a submodule S is implied by the ability to lift elements of M/S with preservation of the order ideal. This stronger property should perhaps be used as the definition of purity when working over general rings.

responding one for torsion modules, and it then reduces further to the primary case, since we can work separately in each primary case. This reduces the problem to the case of a principal ideal ring, and proves the lemma.

The next three theorems were proved for ordinary abelian groups by Kulikoff in [7, Theorem 1], [8, Theorem 2], and [7, Theorem 5]. His proofs extend without change to any principal ideal ring. We now provide the extension to Dedekind rings.

THEOREM 3. *Let R be a Dedekind ring, M an R-module, and S a pure submodule such that M/S is decomposable. Then S is a direct summand of M.*

Proof. In the light of Lemma 2, it suffices to prove this theorem in the cases where M/S is itself cyclic or torsion-free of rank one. In the latter case the result follows from Lemma 3, and in the former case it is an immediate consequence of Lemma 4.

THEOREM 4. *Let R be a Dedekind ring and M a decomposable R-module. Then any submodule of M is decomposable.*

Proof. Case I. M is a torsion module. We may work separately in each primary summand, and this reduces the problem to the case of modules over a principal ideal ring, where the theorem was proved by Kulikoff [8, Theorem 2].

Case II. M is torsion-free. We may regard M as the set of all well-ordered vectors $\{\alpha_\lambda\}$, all but a finite number of coordinates zero, with α_λ ranging over an ideal I_λ in R. Let N be a submodule of M, and N_λ the submodule of N consisting of all elements whose coordinates all vanish after the λth. By mapping the general element of N_λ onto its λ-coordinate we define a homomorphism of N onto an ideal J_λ contained in I_λ. (It may be that $J_\lambda = 0$; this makes no difference.) By Lemma 3, the kernel of this homomorphism is a direct summand of N_λ. Let S_λ be a complementary summand. Then each S_λ is of rank one (or 0), and it is easy to see that N is the direct sum of $\{S_\lambda\}$. Hence N is decomposable.

Case III. M is any decomposable module. Suppose N is any submodule of M, and let T be the torsion submodule of M. Then $N \cap T$ is the torsion submodule of N and, by Case I, it is a direct sum of cyclic modules. Again $N/(N \cap T)$ may be regarded as a submodule of M/T and is therefore decomposable by Case I. By Theorem 3 (or Lemma 3), $N \cap T$ is a direct summand of N. Hence N is decomposable.

THEOREM 5. *Let R be a Dedekind ring, M an R-module, and S a pure submodule of bounded order (that is, $\gamma S = 0$ for some nonzero γ in R). Then S is a direct summand of M.*

Proof. We begin by noting that the theorem is essentially known when M is finitely generated. For then the torsion submodule T of M is a direct summand (Theorem 1 or Steinitz's classical theorem). Moreover S is a direct summand of T; for this may be checked separately in each primary component, which means we are over a principal ideal ring where the theorem is known [7, Theorem 5].

Another result we shall need is the classical theorem of Prüfer [10] which asserts that a module of bounded order is a direct sum of cyclic modules; this extends at once from the principal ideal to the Dedekind case, by passing as usual to the primary components.

We turn to the proof of Theorem 5. We note (by the purity of S) that $S \cap \gamma M = 0$. Let $P = S + \gamma M$, and write $M^* = M/\gamma M$, $P^* = P/\gamma M$. We propose to prove that P^* is pure in M^*. For this purpose, suppose that $x^* = \alpha y^*$, where x^* is in P^*, y^* in M^*. Choose representatives x, y in M for x^*, y^*; we may take x to be in S. We have an equation

$$(3) \qquad\qquad x = \alpha y + \gamma z.$$

Since S is a direct sum of cyclic modules, x may be embedded in a finitely generated direct summand Q of S; and since S is pure in M, Q is likewise pure in M. Let Z be the submodule generated by Q, y, and z. Then (as observed above for the finitely generated case) Q is a direct summand of Z. Let y_1, z_1 be the Q-components of y, z in this decomposition. Then (3) yields $x = \alpha y_1 + \gamma z_1$. But $\gamma z_1 = 0$, since z_1 is in S. Hence $x = \alpha y_1$. Passing to M^*, we obtain $x^* = \alpha y_1^*$, where y_1^* is in P^*. We have thus proved that P^* is pure in M^*.

Next, by a second application of Prüfer's theorem, M^*/P^* is a direct sum of cyclic modules. By Theorem 3, P^* is a direct summand of M^*, say $M^* = P^* + L^*$. Let L be the inverse image in M of L^*. Then $S \cap L = S \cap P \cap L = S \cap \gamma M = 0$; and $S + L = S + \gamma M + L = P + L = M$. Hence $M = S \oplus L$, as desired.

There is a corollary of Theorem 5 which is worth stating.

COROLLARY. *Let M be a module over a Dedekind ring. If the torsion submodule of M is of bounded order, then it is a direct summand of M.*

We next take up the question of divisible modules, as defined in §2.

THEOREM 6. *Let R be a Dedekind ring, and D a divisible R-module. (a) If M is an R-module, S a submodule of M, and f a homomorphism of S into D, then f can be extended to all of M. (b) If D is a divisible submodule of an R-module E, then D is a direct summand of E[5].*

Proof. Let us begin by observing that (b) follows from (a). For let D be a divisible submodule of E. We let f be the identity map of D onto itself, and, on the authority of (a), we extend f to E. If K is the kernel of the

---

[5] Baer [2] has given a necessary and sufficient condition, applicable to any ring R, for an R-module to be a direct summand of every larger module. What we are doing in effect is to show that, for Dedekind rings, his condition is equivalent to the simpler one which we are using.

extended map, we have $E = D \oplus K$.

We proceed to prove (a), and we take up the torsion and torsion-free cases separately.

Case I. D is torsion-free. By a transfinite induction, it suffices to demonstrate the possibility of extending f to a single element x. Let I denote the set of α in R with αx in S. If $I = 0$, it is evident that $f(x)$ may be chosen arbitrarily in D. Otherwise take any $\gamma \neq 0$ in I, and choose y in D with $\gamma y = f(\gamma x)$. We define f on $S + Rx$ by $f(s + \alpha x) = f(s) + \alpha y$. To see that this is a consistent definition, we have to check that $f(\beta x) = \beta y$ holds for all β in I. This follows since D is torsion-free and

$$\gamma[f(\beta x) - \beta y] = \gamma f(\beta x) - \beta f(\gamma x) = 0.$$

(It is to be observed that Case I works when R is any integral domain.)

Case II. D is a torsion module. We define I as above and again assume $I \neq 0$. Define further J to be the subset of I consisting of all α with $f(\alpha x) = 0$. Since D is a torsion module, $J \neq 0$. Modulo J, I is therefore a principal ideal: that is, $I = J + (\gamma)$. We again choose y in D so as to satisfy $\gamma y = f(\gamma x)$, and extend f as in Case I. The fact that $f(\beta x) = \beta y$ for all β in I now follows since β is a multiple of γ mod J.

Case III. D is any divisible R-module. The torsion submodule T of D is again divisible. Now for torsion modules we have proved part (a) of the theorem in Case II. Moreover, as we observed in the first paragraph of the proof, part (b) follows from part (a). In other words, we know that T is a direct summand of D. We may now perform the extension of f separately in each summand, by citing Cases I and II. This completes the proof of Theorem 6.

It is now easy to give a complete structure theory for a divisible module M over a Dedekind ring R. For by Theorem 6, M is a direct sum of a torsion module and a torsion-free module. The latter is merely a vector space over the quotient field K of R. The structure of the former is known from the principal ideal ring case. In order to state the result, we define a module of type p^∞: take the torsion R-module K/R and split it into its primary parts; the summand for the prime ideal p is called the module of type p^∞. The following theorem summarizes the facts.

THEOREM 7. *Let R be a Dedekind ring with quotient field K. Then any divisible R-module is the direct sum of a vector space over K and modules of type p^∞ for various prime ideals p.*

The following theorem is also an immediate consequence of Theorem 6. It shows that in the study of Dedekind modules we can stick to the case where there are no divisible submodules.

THEOREM 8. *Any module M over a Dedekind ring possesses a unique largest divisible submodule D; $M = D \oplus E$ where E has no divisible submodules.*

Finally we take up the question of indecomposable[6] modules over a Dedekind ring, and we generalize [7, Theorem 7].

THEOREM 9. *Let R be a Dedekind ring and M an R-module which is not torsion-free. Then M possesses a direct summand which is either of type p^∞ or isomorphic to R/p^n for some prime ideal p.*

Proof. Take a primary summand P of the torsion submodule of T. It is proved in [7] that P has a direct summand Z which is either of type p^∞ or isomorphic to R/p^n. In the first case Z is divisible and so is a direct summand of M by Theorem 6. In the second case Z is pure in M and of bounded order, and so is a direct summand of M by Theorem 5.

We state explicitly the following consequence of Theorem 9.

THEOREM 10. *An indecomposable module over a Dedekind ring R cannot be mixed, i.e. it is either torsion-free or torsion. In the latter case it is either of type p^∞ or isomorphic to R/p^n for some prime ideal p.*

5. **Valuation rings.** In this paper we are not attempting to handle anything more general than integral domains. So a valuation ring is an integral domain R such that for any α and β in R, either α divides β or β divides α.

Essential to our purpose is the concept of maximality of a valuation ring. However we do not need either the original definition in [9], or the characterization by pseudo-convergence in [5]. We shall instead use the characterization by systems of congruences in [12][7]. The property we need may as well be taken as the definition. We also introduce at this point a weakened form of maximality that we shall need; it is identical with that used in the hypothesis of [6, Theorem 11.1].

DEFINITION. Let R be a valuation ring with quotient field K. We say that R is *maximal* if the following is true: whenever $\alpha_r \in K$ and (integral or fractional) ideals I_r are such that the congruences[8]

$$x \equiv \alpha_r \ (\text{mod } I_r)$$

can be solved in pairs, then there exists in K a simultaneous solution of all the congruences. If this condition holds whenever the intersection of the ideals I_r is nonzero, we shall say that R is *almost-maximal*.

REMARK. It should be noted that when R is a discrete valuation ring, maximality coincides with completeness, and almost-maximality is vacuous.

It is convenient also to define maximality of modules[9].

---

[6] A module is indecomposable if it cannot be written as a direct sum, except in the trivial way. This is not the negation of "decomposable" in the sense used above.

[7] The fact that the definition about to be given is equivalent to maximality can be seen by combining remarks 2 and 4 on pages 48 and 50 of [12].

[8] Here, and in (4), the subscript r has range in an arbitrary set.

[9] There should be no confusion between this meaning of "maximal" and the possible connotation of "largest."

DEFINITION. Let R be any ring, and M an R-module. We say that M is *maximal* if the following is true: whenever ideals $I_r \subset R$ and elements $s_r \in M$ are such that

$$(4) \qquad\qquad x \equiv s_r \pmod{I_r M}$$

can be solved in pairs in M, then there exists in M a simultaneous solution of all the congruences.

LEMMA 5. *If R is a maximal valuation ring and J an (integral or fractional) ideal in R, then J is a maximal R-module. If R is an almost-maximal valuation ring and J a nonzero integral ideal in R, then R/J is a maximal R-module. The direct sum of a finite number of maximal modules is maximal.*

Proof. We shall prove the first statement, leaving the others to the reader. We have then elements s_r in J and ideals $I_r \subset R$ such that the congruences (4), with M replaced by J, can be solved in pairs within J. By hypothesis there exists in the quotient field of R a simultaneous solution x of these congruences, and the fact that x satisfies a single one of the congruences suffices to show that x itself is in J.

Our main theorems will be proved by a suitable process of "lifting up" elements, like that employed in Lemmas 3 and 4. The first such theorem is the following.

THEOREM 11. *Let R be a valuation ring, M a torsion-free R-module, and S a pure maximal submodule of M such that M/S is a direct sum of modules of rank one. Then S is a direct summand of M.*

Proof. As usual, Lemma 2 reduces the problem to the case where M/S is itself of rank one. Let x_0 be any nonzero element in M/S. We must find an element x in M which maps on x_0 and has the following property: whenever x_0 is a multiple of $\alpha \in R$ in M/S (that is, whenever $x_0 = \alpha y_0$ with y_0 in M/S), then x is likewise a multiple of α in M; for when such an element x is found, we need only take the pure submodule it generates to get the desired complementary summand to S.

We first pick any element x^* in M mapping on x. If $x_0 = \alpha_r t_r$ ($\alpha_r \in R$, $t_r \in M/S$), pick u_r in M mapping on t_r; then $s_r = x^* - \alpha_r u_r \in S$. We do this for every possible choice of α_r and consider the set of congruences

$$(5) \qquad\qquad y \equiv s_r \pmod{\alpha_r S}$$

thus obtained. We note that any two of these congruences have a simultaneous solution. In fact, if α_j divides α_i, then s_i works for both i and j; for

$$s_i - s_j = \alpha_i u_i - \alpha_j u_j = \alpha_j(\alpha_j^{-1}\alpha_i u_i - u_j),$$

and the quantity in parentheses lies in S since S is pure. Let y denote a simultaneous solution in S of the congruences (5). Then the element $x = x^*$ $- y$ meets our requirements.

THEOREM 12. *Let R be a maximal valuation ring and M a torsion-free R-module of countable rank. Then M is a direct sum of modules of rank one.*

Proof. We may exhibit M as the union of an ascending sequence of pure submodules $S_1 \subset S_2 \subset \cdots$, where S_i has rank i. Suppose by induction that we know that S_i is a direct sum of modules of rank one. Then by Lemma 5, S_i is maximal. Also S_{i+1}/S_i has rank one. Hence Theorem 11 is applicable and shows that S_i is a direct summand of S_{i+1}. This in turn shows us that S_{i+1} is a direct sum of modules of rank one. Hence M is likewise a direct sum of modules of rank one.

The following consequence of Theorems 11 and 12 is worth stating.

THEOREM 13. *Let R be a maximal valuation ring, M a torsion-free R-module of countable rank, S a pure submodule of finite rank. Then S is a direct summand of M.*

Proof. The module M/S is again torsion-free and again has countable rank; by Theorem 12, it is a direct sum of modules of rank one. Again, by Theorem 12, S is the direct sum of a finite number of modules of rank one, hence maximal by Lemma 5, hence a direct summand of M by Theorem 11.

REMARKS. 1. In the special case where R is a complete discrete valuation ring, Theorem 12 may be stated as follows: any countably generated torsion-free R-module is the direct sum of a divisible module and a free module. This Theorem is implicit in the work of Prüfer [11].

2. The assumption of countable rank in Theorem 12 cannot be dropped. A counter-example is given by Baer in [1, Theorem 12.4]. Indeed let R be any principal ideal ring (not a field) and M the *complete* direct sum of an infinite number of copies of R. Then if M (as an R-module) were the direct sum of modules of rank one, it would have to be free. But Baer proves that M is not free.

3. In Theorem 13 we cannot allow S to be of infinite rank. For example, let R be any principal ideal ring (not a field), and M a free R-module of infinite rank. Then we know that M can be mapped homomorphically onto the quotient field of R, say with kernel S. Then although S is pure it cannot possibly be a direct summand of M.

4. On the other hand, I have been unable to determine whether Theorem 13 still holds if M is of uncountable rank. Indeed it seems to be unknown whether a module of uncountable rank over a maximal valuation ring (say the p-adic integers) might even be indecomposable.

5. Theorem 12 may be supplemented by a uniqueness theorem which is easily proved by the methods used by Baer in proving [1, Corollary 2.9]: let R be any valuation ring and M an R-module which is a direct sum of

modules of rank one; then the expression is unique up to isomorphism.

Our final theorem will deal with torsion modules, and in preparation for it we prove a lemma.

LEMMA 6. *Let R be an almost-maximal valuation ring, M an R-module, S a pure cyclic submodule with a nonzero order ideal, and suppose that M/S is a direct sum of cyclic modules. Then S is a direct summand of M.*

Proof. It suffices (Lemma 2) to handle the case where M/S is cyclic. Let t_0 be a generator of M/S, and H the order ideal. Our problem is to find in M an element t mapping on t_0 and again having order ideal H, for then the cyclic submodule generated by t will be a complementary summand to S. First pick any element y in M mapping on t_0. For any $\alpha_r \in H$ we have $\alpha_r y \in S$. Since S is pure, $\alpha_r y = \alpha_r s_r$ with s_r in S. Let J be the order ideal of S, and let K_r denote the ideal in R consisting of all β with $\beta\alpha_r \in J$; note that $K_r S$ is precisely the submodule of S annihilated by α_r. We consider the system of congruences thus obtained for all choices of α_r in H:

$$(6) \qquad\qquad x \equiv s_r \ (\mathrm{mod}\ K_r S),$$

and observe that any two have a simultaneous solution; for if α_j divides α_i, then

$$\alpha_i(s_i - s_j) = \alpha_i s_i - (\alpha_i \alpha_j^{-1})\alpha_j s_j$$
$$= \alpha_i y - (\alpha_i \alpha_j^{-1})\alpha_j y \equiv 0,$$

whence

$$s_j \equiv s_i \ (\mathrm{mod}\ K_i S),$$

that is to say, s_j works for both the ith and jth congruences. Now by Lemma 5, $S = R/J$ is a maximal module. Hence the congruences (6) have a simultaneous solution x in S. The element $t = y - x$ then has the desired properties: it maps on t_0 in M/S, and it has the order ideal H.

THEOREM 14. *Any finitely generated module over an almost-maximal valuation ring is a direct sum of cyclic modules.*

Proof. Let M be the module. We first apply Theorem 1, and learn that the torsion submodule T of M is a direct summand, and that M/T is free[10]. The problem is consequently reduced to the torsion case, and we accordingly assume that M is a torsion module.

Let $z_1, \cdots, z_n$ be a set of generators for M. Suppose that of these n elements, z_1 has the smallest order ideal. This implies that the cyclic submodule S generated by z_1 is pure. By induction on n, we may assume that

---

[10] One should of course observe that in a valuation ring all finitely generated ideals are principal, and hence invertible; thus the theorem is applicable.

M/S is a direct sum of cyclic modules. By Lemma 6, S is a direct summand of M. Hence M is a direct sum of cyclic modules.

Bibliography

1. R. Baer, *Abelian groups without elements of finite order*, Duke Math. J. vol. 3 (1937) pp. 68–122.

2. ———, *Abelian groups that are direct summands of every containing abelian group*, Bull. Amer. Math. Soc. vol. 46 (1940) pp. 800–806.

3. J. Braconnier, *Sur les groupes topologiques localements compacts*, J. Math. Pures Appl. vol. 27 (1948) pp. 1–85.

4. C. Chevalley, *L'arithmétique dans les algèbres de matrices*, Actualités Scientifiques et Industrielles, no. 323, Paris, 1936.

5. I. Kaplansky, *Maximal fields with valuations*, Duke Math. J. vol. 9 (1942) pp. 303–321.

6. ———, *Elementary divisors and modules*, Trans. Amer. Math. Soc. vol. 66 (1949) pp. 464–491.

7. L. Kulikoff, *Zur Theorie der abelschen Gruppen von beliebiger Mächtigkeit*, Mat. Sbornik vol. 9 (1941) pp. 165–181 (Russian with Germany summary).

8. ———, *On the theory of abelian groups of arbitrary power*, Mat. Sbornik vol. 16 (1945) pp. 129–162 (Russian with English summary).

9. W. Krull, *Allgemeine Bewertungstheorie*, Journal für Mathematik vol. 167 (1932) pp. 160–196.

10. H. Prüfer, *Untersuchungen über die Zerlegbarkeit der abzählbaren primären abelschen Gruppen*, Math. Zeit. vol. 17 (1923) pp. 35–61.

11. ———, *Theorie der abelschen Gruppen. II. Ideale Gruppen*, Math. Zeit. vol. 22 (1925) pp. 222–249.

12. O. F. G. Schilling, *Valuation theory*, Mathematical Surveys, no. 4, New York, American Mathematical Society, 1950.

13. E. Steinitz, *Rechtickige Systeme und Moduln in algebraischen Zahlkörpern*, Math. Ann. vol. 71 (1912) pp. 328–354.

University of Chicago,
    Chicago, Ill.

Afterthought

I imagine that for some time no one has read this paper for its mathematical content, because it has been largely superseded. In particular, the understanding of projectivity and injectivity that came just a little after 1952 made both Lemma 3 and Theorem 6 obsolete.

There are, however, two afterthoughts that I wish to record: a better proof and an admission of error.

The improved proof applies to part (b) of Theorem 2. There is a device that applies perfectly: the Eilenberg-Mazur trick. At the time this paper was written the device was not yet known (at least to me).

We have the infinite direct sum

$$A = I_1 \oplus I_2 \oplus I_3 + \cdots$$

of ideals, and we wish to prove that A is a free module. Pair off the summands. Replace each pair $I_{2n-1} \oplus I_{2n}$ by $R \oplus I_{2n-1}I_{2n}$. This creates inside A an infinite direct sum of copies of R. Now we are ready for the coup de grace: the observation that

$$I \oplus R \oplus R \oplus \cdots$$

is free. This is seen by writing it as

$$I \oplus (I^{-1} \oplus I) \oplus (I^{-1} \oplus I) \cdots$$

and regrouping.

The idea easily yields more general theorems. I leave it to the reader to enjoy working on this. Bass's paper [1] is relevant.

As for the error, the offending passage occurs at the bottom of page 339 and reads "This implies that the cyclic submodule S generated by z_1 is pure." Eighteen years later Warfield detected the error and supplied a correct proof [6, Theorem 1]. In the comprehensive monograph [2] the matter at hand is treated on pages 29–31.

A second identification and rectification of my error was made in [5]; there was no reference to Warfield. The papers [3] and [4] are also pertinent to this area of investigation. [3], [4].

References

1. H. Bass, Big projective modules are free, *Illinois J. of Math.* 7(1963), 24–31.
2. W. Brandal, Commutative rings whose finitely generated modules decompose, *Springer Lecture Notes* 723, 1979.
3. I. Fleischer, Modules of finite rank over Prüfer rings, *Ann. of Math.* 65(1957), 250–54.
4. E. Matlis, Injective modules over Prüfer rings, *Nagoya Math. J.* 15(1959), 57–69.
5. L. Salce and P. Zanardo, On a paper of I. Fleischer, pp. 76–86 in *Abelian Group Theory*, *Springer Lecture Notes* 874, 1981.
6. R. Warfield, Decomposability of finitely presented modules, *Proc. Amer. Math. Soc.* 25(1970), 167–72.

PRODUCTS OF NORMAL OPERATORS

By Irving Kaplansky

1. **Introduction.** In [5] Wiegmann proved the following interesting theorem: *if A and B are matrices such that A, B, and AB are all normal, then BA is also normal.* In [6] he extended this to completely continuous operators.

In the present note we shall look into the validity of this result for general operators on a Hilbert space. We hasten to inform the reader of the fact (surprising to the author) that the result may be false; an example is given in §5. On the positive side of the ledger we contribute the following: a reduction of the problem, a trace argument, and a generalization of the completely continuous case.

2. **A reduction.** The following theorem accomplishes a reduction of the problem from one of the fourth degree (the normality of BA) to one of the third degree.

THEOREM 1. *Let A and B be operators on Hilbert space such that A and AB are normal. Then the following statements are equivalent:* (1) *B commutes with* A^*A, (2) BA *is normal.*

Proof. (1) $\rightarrow$ (2). Form the polar decomposition $A = UR$. Since A is normal, U is unitary, and U and R commute. Also B commutes with R, the positive square root of A^*A. We have

$$U^*ABU = U^*URBU = BRU = BUR = BA.$$

Thus BA is unitarily equivalent to a normal operator and so is itself normal.

(2) $\rightarrow$ (1). The theorem of Fuglede [2], as generalized by Putnam [3], states the following: *if P and Q are normal and PA = AQ, then also P*A = AQ*.* We apply this with $P = AB$, $Q = BA$. The conclusion is $B^*A^*A = AA^*B^*$. In view of the normality of A, this says that B^* commutes with A^*A, and hence so does B.

3. **A trace argument.** With the aid of Theorem 1 we are able to give a trace argument for the normality of BA. It seems that such a trace argument will not work if one assaults the normality of BA directly, instead of proving that B commutes with A^*A.

In order not to tie ourselves down to any particular notion of trace, we formulate the theorem in terms of commutators (a commutator is an expression of the form $PQ - QP$).

THEOREM 2. *Suppose A, B, and AB are normal operators on Hilbert space,*

Received August 16, 1952.

257

and let $C = BA^*A - A^*AB$. *Then* C^*C *is a sum of commutators in the *-ring generated by* A *and* B.

Proof. We have .

$$C^*C = A^*AB^*BA^*A - A^*AB^*A^*AB - B^*A^*ABA^*A + B^*A^*AA^*AB.$$

The normality of AB allows us to switch the second term to

$$A^*A(ABB^*A^*).$$

In view of the normality of A and B, this differs from the fourth term by a commutator, namely $PQ - QP$, where $P = B^*A^*$, $Q = AA^*AB$. Similarly, the third term differs from the first by a commutator. Hence C^*C is a sum of commutators.

In the finite-dimensional case we can deduce that the trace of C^*C is 0, whence $C = 0$, B commutes with A^*A, BA is normal. We can argue similarly if A and B lie in a factor of type II_1, or more generally in a ring of finite type in the sense of Dixmier [1].

4. A generalization of the completely continuous case.

The following lemma embodies the essential device employed by Wiegmann.

LEMMA. *Suppose* A, B, *and* AB *are normal operators on Hilbert space. Let* λ *be the maximum of the spectrum of* A^*A, *and let* K *be the characteristic subspace of* A^*A *for* λ. *Suppose* K *is spanned by characteristic vectors of* A^*. *Then* K *is invariant under* B. *The same is true if* λ *is replaced by the minimum of the spectrum of* A^*A.

Proof. Let x be a characteristic vector of A^* in K; say $A^*x = \mu x$ with $|\mu|^2 = \lambda$. We have to prove that $y = Bx$ is in K. The normality of AB tells us that $\|ABx\| = \|B^*A^*x\|$. By a similar application of the normality of A and B^*, we change this to $\|A^*Bx\| = \|BA^*x\|$. Hence $\|A^*y\| = \|\mu y\|$, or

$$(A^*Ay,y) = (\lambda y,y).$$

If λ is either the maximum or the minimum of the spectrum of A^*A, we deduce that $A^*Ay = \lambda y$ and y is in K.

To put this lemma to work, we have to know that the orthogonal complement of K is also invariant under B (or alternatively that K is invariant under B^*). This is known to be true if K is finite-dimensional. In that case the subspace K can be factored out, and we can proceed to the next highest or lowest characteristic root of A^*A, if there is one. The process can be pursued by transfinite induction; and if we take account of the fact that we can tolerate an infinite-dimensional characteristic subspace of A^*A at the very last step, we arrive at the following theorem.

THEOREM 3. *Let* A, B *and* AB *be normal operators on Hilbert space. Suppose that* (1) A *has pure point spectrum (that is, the Hilbert space is spanned by char-*

acteristic vectors of A), (2) *the characteristic subspaces of A*A are finite-dimensional, with the possible exception of the subspace for one characteristic root* α, (3) *any set of characteristic roots of A*A, not just consisting of* α*, has a maximum or a minimum other than* α*. Then: BA is normal.*

The details of the derivation of Theorem 3 from the lemma are left to the reader. But we note that the hypothesis of Theorem 3 is satisfied if A is completely continuous. So: if A and B are normal and *one* of the two is completely continuous, then normality of AB implies that of BA.

We can obtain a result similar to Theorem 3, without the finite-dimensionality of the characteristic subspaces of A^*A, if we assume explicitly that any closed subspace invariant under B is also invariant under B^*. These are the operators of type (P) discussed by Wermer [4]. We note the trivial fact that any self-adjoint operator is of type (P).

THEOREM 4. *Let A, B and AB be normal operators on Hilbert space. Suppose that* (1) *A has pure point spectrum,* (2) *B is of type (P),* (3) *any set of characteristic roots of A*A has either a maximum or a minimum. Then: BA is normal.*

These two theorems appear to have squeezed everything possible out of the lemma. But this may not be the last word on the subject. The following conjecture might stand up: *if A is normal with pure point spectrum, B is self-adjoint, and AB is normal, then BA is normal.*

5. **An example.** It is convenient to use a (presumably self-explanatory) matrix notation for this example. Let P be the diagonal matrix with the triple 2, 1, $\frac{1}{2}$ repeated infinitely often down the diagonal. Write

$$Q = \begin{pmatrix} P & 0 \\ 0 & 0 \end{pmatrix}, \quad R = \begin{pmatrix} 0 & 0 \\ 0 & I \end{pmatrix},$$

and observe that

(1) $$QR = RQ = 0.$$

It is plain that $4Q^2 + 4R^2$ is unitarily equivalent to $4Q^2 + R^2$; thus we can find a unitary U with

(2) $$U(4Q^2 + 4R^2)U^* = 4Q^2 + R^2.$$

Similarly there exists a unitary W with

(3) $$W(Q^2 + R^2)W^* = Q^2 + 4R^2.$$

Define

$$A = \begin{pmatrix} 2U & 0 \\ 0 & W \end{pmatrix}, \quad B = \begin{pmatrix} Q & R \\ R & Q \end{pmatrix}.$$

Then A is normal, and B is even self-adjoint. By appropriate use of (1), (2) and (3) we compute

$$(AB)(BA^*) = (BA^*)(AB) = \begin{pmatrix} 4Q^2 + R^2 & 0 \\ 0 & Q^2 + 4R^2 \end{pmatrix},$$

and AB is normal. That BA is not normal may be verified conveniently with the aid of Theorem 1, for B does not commute with A^*A.

It seems noteworthy that in this example A^*A has a two-point spectrum, and B is self-adjoint with a finite spectrum. However A has a continuous spectrum and lacks property (P), so neither Theorem 3 nor Theorem 4 is applicable.

References

1. J. Dixmier, *Les anneaux d'opérateurs de classe finie*, Annales Scientifiques de l'Ecole Normale Supérieure, vol. 66(1949), pp. 209–261.
2. Bent Fuglede, *A commutativity theorem for normal operators*, Proceedings of the National Academy of Sciences, U. S. A., vol. 36(1950), pp. 35–40.
3. C. R. Putnam, *On normal operators in Hilbert space*, American Journal of Mathematics, vol. 73(1951), pp. 357–362.
4. John Wermer, *On invariant subspaces of normal operators*, Proceedings of the American Mathematical Society, vol. 3(1952), pp. 270–277.
5. N. A. Wiegmann, *Normal products of matrices*, this Journal, vol. 15(1948), pp. 633–638.
6. N. A. Wiegmann, *A note on infinite normal matrices*, this Journal, vol. 16(1949), pp. 535–538.

University of Chicago
AND
University of California at Los Angeles.

Afterthought

The passage of years has not dimmed my enthusiasm for Wiegmann's theorem. I salute him for daring to think that it might be true.

When I learned about the theorem I immediately became curious about the infinite-dimensional case. Here a little daring was again needed, but once I decided to look for a counterexample it did not take long to find one.

As far as I know, the subject died at this point. Well, actually, not quite. Around 1980 I made an unpublished investigation, which I shall report here.

We consider again the finite-dimensional case but work over a general field K. No involution is assumed, so normality of a matrix means that it commutes with its ordinary transpose. If K is formally real (i.e., -1 is not a sum of squares), then Wiegmann's theorem holds, for K can be enlarged to a real closed field where we know the theorem to be valid. The converse is true.

THEOREM. *Let K be a field with the following property: For any matrices A, B with entries in K, normality of A, B, and AB implies normality of BA. Then K is formally real.*

Proof. Let V be an n-dimensional vector space over K. Pick a basis and endow V with the inner product obtained by declaring the basis to be orthonormal. Assume, by way of contradiction, that -1 is a sum of squares. Then V is isotropic for large n and thus contains a hyperbolic plane. The validity of Wiegmann's theorem is inherited by orthogonal direct summands, where normality is taken relative to the adjoint operation in the direct summand. With the usual basis of a hyperbolic plane, the adjoint is given by

$$\begin{pmatrix} a & b \\ c & d \end{pmatrix}^* = \begin{pmatrix} d & b \\ c & a \end{pmatrix}.$$

We assume characteristic $\neq 2$; the case of characteristic 2 will be treated at the end of the proof. Then normality is equivalent to the following: Either $a = d$ or $b = c = 0$. We now simply exhibit a case of failure:

$$A = \begin{pmatrix} 1 & 0 \\ 1 & 1 \end{pmatrix}, \quad B = \begin{pmatrix} 0 & 0 \\ 0 & 1 \end{pmatrix};$$

here A, B, and AB are normal but BA is not.

In characteristic 2, 3×3 matrices are needed and I display a suitable pair:

$$C = \begin{bmatrix} 0 & 1 & 0 \\ 0 & 1 & 0 \\ 1 & 1 & 1 \end{bmatrix}, \quad D = \begin{bmatrix} 0 & 0 & 0 \\ 0 & 0 & 0 \\ 0 & 0 & 1 \end{bmatrix};$$

C, D and CD are normal, but DC is not.

ANNALS OF MATHEMATICS
Vol. 61, No. 3, May, 1955
Printed in U.S.A.

ANY ORTHOCOMPLEMENTED COMPLETE MODULAR LATTICE IS A CONTINUOUS GEOMETRY

By Irving Kaplansky

(Received September 30, 1954)

1. Introduction

A continuous geometry is a complete complemented modular lattice in which it is further assumed that the lattice operations satisfy certain continuity assumptions. It is known that one cannot drop these continuity assumptions, the pertinent example being the lattice L of all subspaces of an infinite-dimensional vector space. In fact L satisfies one of the two continuity assumptions but fails to satisfy the dual. Of course by taking the direct product of L with its dual we get a lattice satisfying neither continuity axiom. However, this lattice is not orthocomplemented, that is, there does not exist a mapping of period two assigning to each element a canonical complement. This suggests that in an orthocomplemented complete modular lattice it might be possible to prove the continuity axioms.

A more immediate motivation arises from the fact that the conjecture is easy to prove for those orthocomplemented lattices which arise from rings of operators.

THEOREM: *If the lattice L of projections in an $AW*$-algebra A is modular, then A is finite, whence by Theorem 6.5 of [3] L is a continuous geometry.*

The proof goes as follows: If A is not finite it contains an infinite set of matrix units. Then by [4, Lemma 15] there is an $AW*$-subalgebra C isomorphic to the algebra of all bounded operators on a Hilbert space H. The lattice of projections in C coincides with the lattice of closed subspaces in H, and Mackey [5] has proved that the latter is not modular.

This paper is devoted to proving the theorem stated in the title. The method is to verify first that the lattice in question can be coordinatized by a regular ring, with the usual exception of things like non-Desarguian projective planes. (However in the actual exposition we have preferred to place the introduction of coordinates last). The ring will possess an involution that makes it resemble closely a ring of operators. The arguments now needed are slight modifications of known ones—except for the proof of finiteness. In order to prove finiteness we argue indirectly that otherwise there will be an infinite set of matrix units giving rise to a ring of infinite matrices. It turns out that the rows and columns of these matrices can be arbitrarily prescribed. This bizarre state of affairs leads us ultimately to a contradiction.

2. *-Regular rings

A ring A is *regular* if for every a there exists an element x with $axa = a$. We assume that A has a unit element. Every principal right ideal in A is generated by an idempotent. The principal right ideals form a complemented modular lattice.

We say that two idempotents e and f are *equivalent*, written $e \sim f$, if there exist elements $x \in eAf$, $y \in fAe$ such that $xy = e$, $yx = f$. It is straightforward to verify that equivalence is reflexive, symmetric, and transitive. If $axa = a$, the idempotents ax and xa are equivalent (note that these idempotents generate aA and Aa respectively). If e and f generate the same right ideal, then $ef = f$, $fe = e$, whence $e \sim f$.

We shall call A a *-regular* ring if it possesses an involution $*$ such that $x*x = 0$ implies $x = 0$. An element x is self-adjoint if $x* = x$, and self-adjoint idempotents will be called *projections*. In a *-regular ring the principal right ideal aA is generated by a unique projection e [7, Theorem 4.5]; we call e the *left projection* of a. Because of this one-to-one correspondence between principal right ideals and projections, we can (and usually shall) replace the lattice of principal right ideals by the lattice of projections. The principal left ideal Aa is similarly generated by a unique projection f which we call the *right projection* of a. We note the following facts: $ea = af = a$, the left annihilator of a is $A(1 - e)$, the right annihilator of a is $(1 - f)A$. The remarks in the preceding paragraph show that the *left and right projections of any element are equivalent*. Knowing this, we can repeat the proofs of Lemma 5.3 and Theorem 6.6b in [3]; we state the results in the next two lemmas.

LEMMA 1. *For any projections e and f in a *-regular ring,*

$$e - (e \cap f) \sim (e \cup f) - f.$$

LEMMA 2. *If two projections in a *-regular ring have a common complement they are equivalent.*

If A is a regular ring and e an idempotent in A we know by [7, Theorem 2.11] that eAe is regular. From this we deduce:

LEMMA 3. *If e is a projection in a *-regular ring A, then eAe is a *-regular ring.*

LEMMA 4. *Let a be an element in a *-regular ring A, and suppose a has left projection e and right projection f. Then there exists a unique element y satisfying $fy = y$, $ay = e$. It has the following further properties: $ya = f$, the left projection of y is f, and the right projection of y is e.*

PROOF. Since $aA = eA$ there exists an element x such that $ax = e$. Define $y = fx$. Since $af = a$, we find $ay = e$. Of course $fy = y$. Suppose z shares these properties: $az = e$, $fz = z$. Then $a(y - z) = 0$ implies $f(y - z) = 0$, whence $y = z$. Next, f is a left multiple of a, say $f = ta$. Then $ya = fya = taya = tea = ta = f$. Further, $ye = yax = fx = y$. We now see that $Ay = Ae$, $yA = fA$. Hence f and e are the left and right projections of y.

Following the terminology of Rickart [9] we call the element y of the preceding lemma the *relative inverse* of a. We note that conversely a is the relative inverse of y. Since $aya = a$ we have:

LEMMA 5. *Let A be a *-regular ring and B a subring of A which contains along with every element its relative inverse. Then B is regular.*

LEMMA 6. *Let g be a projection in a *-regular ring A, and suppose that a commutes with g. Then the relative inverse of a commutes with g.*

PROOF. If B is the subring of elements commuting with g, then

$$B = gAg \oplus (1 - g)A(1 - g).$$

Since each summand is *-regular (Lemma 3), B is itself a *-regular ring. In B the element a has a relative inverse y, and left and right projections e and f. It will suffice to convince ourselves that e and f are likewise the projections of a in the ring A. Since $aB = eB$, a and e are mutual right multiples, and hence $aA = eA$. Thus e is the left projection of a in A, and similarly f is the right projection of a in A.

On combining Lemmas 5 and 6 we obtain:

LEMMA 7. *Let A be a *-regular ring and S a set of projections in A. Then the subring of elements commuting with A is a *-regular ring.*

We say that a *-regular ring is *complete* if its lattice of projections is complete. As in [3] we use the abbreviation LUB for the least upper bound of a set of projections.

LEMMA 8. *Let e be the LUB of a set of $\{e_i\}$ of projections in a complete *-regular ring. Suppose $ae_i = 0$ for all i. Then $ae = 0$.*

PROOF. Let f be the right projection of a. Then $fe_i = 0$, $e_i \leq 1 - f$ for all i, whence $e \leq 1 - f$, $fe = 0$, $ae = 0$.

LEMMA 9. *Let e be the LUB of a set $\{e_i\}$ of projections in a complete *-regular ring. Suppose $ae_i = e_i a$ for all i. Then $ae = ea$.*

PROOF. We recast slightly the proof of Lemma 2.2b in [3]. Since a commutes with e_i and $ee_i = e_i$, we have $(a - ea)e_i = 0$. By Lemma 8, $(a - ea)e = 0$. Similarly $e(a - ae) = 0$. Hence $ea = ae$.

LEMMA 10. *Let A be a complete *-regular ring and S a set of projections in A. Then the subring B of elements commuting with S is a complete *-regular ring.*

PROOF. By Lemma 7, B is *-regular. For any set of projections in B, the LUB (as computed in A) is again in B by Lemma 9. This suffices to show that the lattice of projections in B is complete.

3. Central decomposition

Let A be a complete *-regular ring, and $\{g_i\}$ a set of orthogonal central projections in A with LUB 1. By sending the general element a into $\{ag_i\}$ we evidently get a *-homomorphism (i. e. a homomorphism commuting with *) of A into the complete direct sum of the rings Ag_i. The mapping is faithful, for if each $ag_i = 0$ then $a = 0$ by Lemma 8. Ordinarily we cannot expect the mapping to be onto. A typical counterexample is afforded by the ring of all sequences of complex numbers for which all but a finite number of entries are real. However we do get all the projections of the complete direct sum. Let a projection e_i in Ag_i be given for every i, and let e be the LUB of $\{e_i\}$. Denote by f the LUB of the e_j's with $j \neq i$. Then $e = e_i + f$; also $fg_i = 0$ by Lemma 8. Hence $eg_i = e_i g_i = e_i$. Thus e is the desired projection which combines the components e_i.

Another instance in which we can combine prescribed components is that where each has square 0.

LEMMA 11. *Let a be an element of a *-regular ring with $a^2 = 0$. Let e be the left projection, f the right projection, and b the relative inverse of a. Let h be the left projection of $e + b$. (a) Then $f \leq e \cup h$. (b) If $f = ex + hy$ then $-ex = a$.*

PROOF. (a) Write $k = e \cup h$. Then $k(e + b) = e + b$, since $k \geq h$. Also $ke = e$. Hence $kb = b$. Since f is the left projection of b, $k \geq f$.

(b) Since $a^2 = 0$, e and f are orthogonal. It follows that $b^2 = 0$, $bf = 0$. Write $h = (e + b)z$, so that

$$(1) \qquad\qquad f = ex + (e + b)zy.$$

Multiply (1) on the left by b and f respectively. The resulting equations are $bex + bzy = 0$, $f = bzy$. Hence $b(-ex) = f$. This, together with $e(-ex) = -ex$, identifies $-ex$ as the relative inverse of b (Lemma 4). Hence $-ex = a$.

LEMMA 12. *Let A be a complete *-regular ring, and $\{g_i\}$ a set of orthogonal central projections in A with LUB 1. Suppose that for each i we are given an element a_i in Ag_i with $a_i^2 = 0$. Then there exists in A an element a with $ag_i = a_i$ for all i.*

PROOF. Let e_i, f_i, b_i be the left projection, right projection, and relative inverse of a_i; let h_i be the left projection of $e_i + b_i$ (these are the same whether computed in A or in Ag_i). Let e, f, h be the LUB's of $\{e_i\}$, $\{f_i\}$, and $\{h_i\}$ respectively. By part (a) of Lemma 11, $f_i \leq e_i \cup h_i$. Hence $f_i \leq e \cup h$, and since this is true for all i, $f \leq e \cup h$. Consequently we may write $f = ex + hy$. Applying g_i to this equation we get $f_i = e_i x + h_i y$. By part (b) of Lemma 11, $-e_i x = a_i$. Hence $a = -ex$ is the element we are seeking; it has the property $g_i a = -g_i ex = -e_i x = a_i$ for all i.

Using Lemma 10 we can extend Lemma 12 so as to apply to combinations built out of projections that are not necessarily central.

LEMMA 13. *Let A be a complete *-regular ring. Let $\{e_i\}$, $i = 1, 2, \cdots$ be a sequence of orthogonal projections in A with LUB e. Let elements a_k in $e_{2k-1}Ae_{2k}$ be given $(k = 1, 2, \cdots)$. Then there exists in eAe an element a satisfying $e_{2k-1}ae_{2k} = a_k$ for all k and $e_i ae_j = 0$ for all other i, j.*

PROOF. Define $g_k = e_{2k-1} + e_{2k}$. Let B be the subring of eAe consisting of all elements commuting with the g's. By Lemma 10, B is a complete *-regular ring. In B the g's constitute a set of central orthogonal projections with LUB the unit element e. Also, a_k is an element of $g_k B$ with $a_k^2 = 0$. By Lemma 12 there exists in B an element a with $ag_k = a_k$ for all k. We have

$$e_{2k-1}ae_{2k} = e_{2k-1}ag_k e_{2k} = e_{2k-1}a_k e_{2k} = a_k .$$

Further, $e_i ae_j = 0$ for all other i, j. If j is even, say $j = 2k$, we have

$$e_i ae_j = e_i ag_k e_{2k} = e_i a_k e_{2k} = e_i e_{2k-1} a_k$$

which is 0 unless $i = 2k - 1$. If j is odd, $j = 2k - 1$:

$$e_i ae_j = e_i ag_k e_{2k-1} = e_i a_k e_{2k}e_{2k-1} = 0.$$

LEMMA 14. *Let A be a complete $*$-regular ring. Let $\{e_i\}$, $i = 1, 2, \cdots$ be a sequence of orthogonal projections in A. Let elements a_i in e_iAe_{i+1} be given ($i = 1, 2, \cdots$). Then there exists in A an element a satisfying $e_iae_{i+1} = a_i$ for all i, $e_iae_j = 0$ for $j \neq i + 1$.*

PROOF. We make two successive applications of Lemma 13. First we produce an element b with $e_{2k-1}be_{2k} = a_{2k-1}$ for all k, $e_ibe_j = 0$ otherwise. Then we apply Lemma 13 to the projections e_2, e_3, $\cdots$. Let f be their LUB. The result is an element c in fAf with $e_{2k}ce_{2k+1} = a_{2k}$, $e_ice_j = 0$ otherwise for $i, j \geq 2$. Since c is in fAf we also have $e_1c = ce_1 = 0$. By defining $a = b + c$ we satisfy the requirements of the lemma.

LEMMA 15. *Let A be a complete $*$-regular ring. Let $\{e_i\}$, $i = 1, 2, \cdots$ be a sequence of orthogonal projections in A with LUB 1. Suppose that $e_{2k-1} \sim e_{2k}(k = 1, 2, \cdots)$. Let elements a_i in e_iAe_i be given. Then there exists in A an element a with $e_iae_i = a_i$ for all i, $e_iae_j = 0$ for $j \neq i$.*

PROOF. Let x_k, y_k be elements implementing the equivalence of e_{2k-1} and e_{2k}, so that $x_k \in e_{2k-1}Ae_{2k}$, $y_k \in e_{2k}Ae_{2k-1}$, $x_ky_k = e_{2k-1}$, $y_kx_k = e_{2k}$. By four applications of Lemma 13 there exist elements b, c, u, v with $e_{2k-1}be_{2k} = a_{2k-1}x_k$, $e_{2k}ce_{2k-1} = a_{2k}y_k$, $e_{2k-1}ue_{2k} = x_k$, $e_{2k}ve_{2k-1} = y_k$, and $e_i(b, c, u$ or $v)e_j = 0$ otherwise. Define $a = bv + cu$. We must prove $e_iae_j = \delta_{ij}a_i$. We shall suppose that i is even (say $i = 2k$); the proof when i is odd is similar. We have $e_{2k}be_j = 0$ for all j; hence (Lemma 8) $e_{2k}b = 0$. Again $e_{2k}c - e_{2k}ce_{2k-1}$ left annihilates every e_j and so it is 0. Similarly $e_{2k-1}u = e_{2k-1}ue_{2k}$. Thus

$$e_{2k}a = e_{2k}ce_{2k-1}ue_{2k} = a_{2k}y_kx_k = a_{2k}.$$

Hence $e_{2k}ae_{2k} = a_{2k}$ and $e_{2k}ae_j = 0$ for $j \neq 2k$.

4. Matrix units

Throughout this section and the next we shall assume given a complete $*$-regular ring A, and an infinite sequence $\{e_i\}$ of orthogonal equivalent projections in A with LUB 1. Ultimately we shall discover this to be impossible; but provisionally we analyze in some detail the infinite matrices that arise in this way.

Select elements e_{1i}, e_{i1} that implement the equivalence of e_1 and e_i. Thus $e_{1i} \in e_1Ae_i$, $e_{i1} \in e_iAe_1$, $e_{1i}e_{i1} = e_1$, $e_{i1}e_{1i} = e_i$. Define $e_{ij} = e_{i1}e_{1j}$. If we agree to write $e_{ii} = e_i$ we find that the rule $e_{ij}e_{jk} = \delta_{jk}e_i$ for matrix multiplication is satisfied. Define D to be the subring of A commuting with all e_{ij}.

The proofs of Lemmas 11 and 13 in [4] can be repeated without change, and with Lemma 15 at hand we can repeat the proof of Lemma 12 there. We shall merely state the results.

LEMMA 16. *If an element of D annihilates any e_{ij} it is 0.*

LEMMA 17. *The mapping $x \to e_ix$ is an isomorphism of D onto e_iAe_i.*

LEMMA 18. *$e_iAe_j = De_{ij}$, and the representation in this form of an element of e_iAe_j is unique.*

It follows from Lemma 17 that we can transplant the involution from e_iAe_i to D and convert D into a complete $*$-regular ring. But this involution on D is not necessarily induced on D by the one in A (or in other words, for different i the various involutions induced on D may differ). This pitfall will however not disturb us.

Now given any element a in A we have by Lemma 18 that $e_i a e_j = \alpha_{ij} e_{ij}$ with α_{ij} a unique element of D. In this way we associate with a the matrix (α_{ij}). Addition of matrices is of course performed coordinatewise. Multiplication presents a puzzle if the row and column being multiplied both have an infinite number of non-zero entries. If however one of them has only a finite number of non-zero entries multiplication is performed as usual. Suppose for instance that the matrix (α_{ij}) for a has $\alpha_{1i} = 0$ for $i > n$. Then $e_1 a - e_1 a (e_1 + \cdots + e_n)$ left annihilates every e_j and so (Lemma 7) is 0. Let (β_{ij}) be a second matrix, representing b. Then on computing the 11-entry of ab we find

$$e_1 a b e_1 = e_1 a (e_1 + \cdots + e_n) b e_1 = \alpha_{11} \beta_{11} + \cdots + \alpha_{1n} \beta_{n1}.$$

Lemmas 14 and 15 give us the following information.

LEMMA 19. *Let two arbitrary sequences α_i, β_i in D be given. Then A contains an element whose matrix has the α's down the main diagonal, the β's down the diagonal next above the main diagonal, and zeros elsewhere.*

We conclude this section by noting the effect of the involution on the matrix units. Since $e_{ij} \in e_j * A e_i$, we have $e_{ij}* = \lambda_{ij} e_{ji}$ with λ_{ij} a uniquely determined element of D. Applying $*$ to the equation $e_{ij} e_{ji} = e_i$ we find that $\lambda_{ji} \lambda_{ij} - 1$ annihilates e_i and so (Lemma 16) is 0. Similarly $\lambda_{ij} \lambda_{ji} = 1$. Thus λ_{ij} is a regular element of D. We shall abbreviate λ_{1i} to λ_i, and then λ_{i1} becomes λ_i^{-1}.

5. Proof of finiteness

The next step in the argument will be to prove that the first row of the matrices can be arbitrarily prescribed.

LEMMA 20. *For any sequence α_i of elements in D, there exists in A an element whose matrix has the α's in its first row and 0's elsewhere.*

PROOF. It will suffice to construct an element x with 1's across its first row. For by Lemma 19 there exists a diagonal matrix y with α_1, α_2, $\cdots$ down the diagonal, and $e_1 x y$ will then furnish the matrix we are seeking.

We apply Lemma 19 to produce an element a with 1's down the main diagonal, -1's down the diagonal next above, and 0 elsewhere. Let $1 - f$ be the right projection of a, so that the right annihilator of a is fA. From the equation $af = 0$ we compute that the matrix of f has constant columns. Let ϕ_i be the element of D occupying the ith column of f. Then $e_i f e_1 = \phi_i e_{i1}$, $e_1 f e_i = \phi_i e_{1i}$. Hence

$$(2) \qquad\qquad \phi_i e_{1i} = (\phi_1 e_{i1})*.$$

The case $i = 1$ of (2) gives us $\phi_1 e_1 = e_1 \phi_1 *$. Left-multiplying this by e_{i1} we get $e_{i1} \phi_1 = e_{i1} \phi_1 *$ (recall that ϕ_1 commutes with e_1). By applying this to (2) we obtain $\phi_i e_{1i} = (e_{i1} \phi_1 *)* = \phi_1 \lambda_i^{-1} e_{1i}$. Hence

$$(3) \qquad\qquad \phi_i = \phi_1 \lambda_1^{-1}.$$

In the regular ring D the left annihilator of ϕ_1 is of the form $D\gamma$ with γ an idempotent. By (3) we have $\gamma\phi_i = 0$ for all i, whence $\gamma f = 0$. Again $\phi_1 D = (1 - \gamma)D$, so that there exists an element ζ in D with $\phi_1\zeta = 1 - \gamma$. Let c be the element in A with $\zeta, \lambda_2\zeta, \lambda_3\zeta, \cdots$ down the main diagonal and zeros elsewhere. Using (3) we compute that fc has $1 - \gamma$ across its first row.

Let t be any element of A right annihilating $a\gamma$. Then γt is in fA. But $\gamma f = 0$, from which it follows that $\gamma t = 0$. Hence γ left-annihilates the right annihilator of $a\gamma$, and $\gamma \in Aa\gamma$. Say $\gamma = ba\gamma$, $b = (\beta_{ij})$. Computing out the first row of $ba\gamma$, we obtain the equations $\beta_{11}\gamma = \gamma$, $(-\beta_{11} + \beta_{12})\gamma = 0$, $(-\beta_{12} + \beta_{13})\gamma = 0, \cdots$. It follows that $\beta_{1i}\gamma = \gamma$ for all i. The matrix for $b\gamma$ has γ across its first row. Add this to the previously acquired matrix with $1 - \gamma$ across its first row and we have a matrix with 1's across its first row.

The following reformulation of Lemma 20 will be convenient.

LEMMA 21. *Let A be complete $*$-regular ring. Let $\{e_i\}$, $i = 1, 2, \cdots$ be a sequence of equivalent orthogonal projections in A. Let elements a_i in e_1Ae_i be given. Then there exists in A an element a with $e_1ae_i = a_i$ for all i.*

The contradiction is going to be achieved by applying Lemma 21 to a different set of projections, obtained via the Gram-Schmidt procedure from the elements e_{1i} and the matrix having 1's in its first row and zero's elsewhere. We shall write w for this matrix. For future reference we observe that $w^2 = w \neq 0$, so that the left and right projections of w are not orthogonal. The defining properties of w are: $e_1 w = w$, $we_i = e_{1i}$ for every i. We note the two consequences

(4) $$we_{i1} = e_1,$$

(5) $$we_{1i*} = \lambda_i e_1.$$

We apply $*$ to the equation (4), recalling that $e_{1i*} = \lambda_i^{-1}e_{1i}$. The result is

(6) $$e_{1i}w* = \lambda_i e_1.$$

We define $w_n = w(1 - e_1 - e_2 - \cdots - e_n)$ and observe

(7) $$w_n = w - e_{11} - e_{12} - \cdots - e_{1n},$$

(8) $$w_nw_n* = w(1 - e_1 - \cdots - e_n)w* = w_nw*.$$

From the definition of w_n we derive immediately

(9) $$w_ne_i = 0 (i \leq n), \qquad w_ne_i = e_{1i} \qquad (i > n),$$

and a further result is

(10) $$w_ne_{1i}* = 0 \qquad (i \leq n).$$

For the sake of uniformity of notation it is desirable to write $w = w_0$. For $n = 0, 1, \cdots$ we have that w_nw_n* is in e_1Ae_1 and so is of the form μ_ne_1 with μ_n in D. A computation of w_nw_n*, using (7), (5), and (6), yields

(11) $$\mu_n = \mu_0 - \lambda_1 - \lambda_2 - \cdots - \lambda_n.$$

Next we claim that μ_n is a regular element of D (has a two-sided inverse). Since D is a regular ring, it suffices to prove that μ_n is neither a left nor a right divisor of 0. If $\alpha\mu_n = 0$, $\alpha \epsilon D$, then $\alpha w_n w_n * \alpha * = 0$, whence $\alpha w_n = 0$. For $i > n$ we get $0 = \alpha w_n e_i = \alpha e_{1i}$ by (9). Hence $\alpha = 0$. If $\mu_n\beta = 0$, then $\beta * w_n w_n * \beta = 0$, whence $w_n * \beta = 0$, $e_i w_n * \beta = 0$, $e_{1i} * \beta = 0$, $\lambda_i e_{i1}\beta = 0$, $\beta = 0$.

We define $u_0 = w_0 = w$, $u_1 = e_1 - \lambda_1\mu_0^{-1}w_0$, and in general

$$(12) \qquad u_n = e_{1n} - \lambda_n\mu_{n-1}^{-1}w_{n-1}.$$

From (6) and (8) we derive

$$(13) \qquad u_n w* = 0 \qquad\qquad (n > 0).$$

From (10) we get

$$(14) \qquad u_n e_{1i}* = 0 \qquad\qquad (i < n).$$

Putting together (7), (13) and (14) we obtain $u_n u_i* = 0$ for $i < n$. Since on applying $*$ we find also $u_i u_n* = 0$ we have proved

$$(15) \qquad u_i u_j* = 0 \qquad\qquad (i \neq j).$$

Using (9) and (11) we next compute that for $n > 0$

$$(16) \qquad u_n e_{n1} = (1 - \lambda_n\mu_{n-1}^{-1})e_1 = \mu_n\mu_{n-1}^{-1}e_1.$$

From (16), and the fact that $e_1 u_n = u_n$, we see that $u_n A = e_1 A$; hence e_1 is the left projection of u_n for $n > 0$. Since $u_0 e_1 = w e_1 = e_1$, this fact also holds for $n = 0$. So the elements $u_0, u_1, u_2, \cdots$ all have the same left projection. If we write g_i for the right projection of u_i we have that the g's are equivalent projections. By (15) the g's are orthogonal projections. We now apply Lemma 21 (with a harmless inversion of order) to this new set of equivalent orthogonal projections. Write v_i for the relative inverse of u_i, so that $u_i v_i = e_1$, $v_i u_i = g_i$. There exists an element a in A satisfying

$$(17) \qquad g_0 a g_0 = g_0 v_0 g_0,$$

$$(18) \qquad g_i a g_0 = -g_i v_i\lambda_i\mu_{i-1}^{-1}g_0 \qquad\qquad (i > 0).$$

Throughout the computation it should be borne in mind that $g_i v_i = v_i$, $u_i g_i = u_i$. We have by (17),

$$(19) \qquad w a g_0 = u_0 a g_0 = u_0 g_0 a g_0 = u_0 v_0 g_0 = e_1 g_0.$$

Multiply (18) on the left by u_i, noting that $u_i v_i = e_1$ and commutes with elements of D:

$$(20) \qquad u_i a g_0 = -\lambda_i\mu_{i-1}^{-1}e_1 g_0 \qquad\qquad (i > 0).$$

We claim that $e_{1r} a g_0 = 0$ for all r. Suppose that this known for $i < r$. Then by (7) and (19):

$$(21) \qquad w_{r-1} a g_0 = w a g_0 = e_1 g_0.$$

By (21) and the definition (12) of the u's:

$$(22) \qquad u_r a g_0 = e_{1r} a g_0 - \lambda_r \mu_{r-1}^{-1} e_1 g_0 \qquad (r > 0).$$

Equations (20) and (22) imply $e_{1r} a g_0 = 0$, as desired. Left-multiplying by e_{r1} we find that $e_r a g_0 = 0$ for all r. Hence (Lemma 7) $a g_0 = 0$, and then $e_1 g_0 = 0$ by (21). But e_1 and g_0 are the left and right projections of w, and we noted above that they are not orthogonal. This contradiction completes the proof of our first main result.

THEOREM 1. *A complete *-regular ring cannot have an infinite set of equivalent orthogonal projections.*

From Theorem 1 we can quickly pass to the usual formulation of finiteness.

THEOREM 2. *Let e and f be projections in a complete *-regular ring with $e \geqq f$, $e \sim f$. Then $e = f$.*

PROOF. Suppose that $e \neq f$. We shall produce a third projection g with $g \leqq f$, $g \neq f$, $g \sim f$, and $e - f \sim f - g$. Note that $e - f$ and $f - g$ are orthogonal projections. Iteration of this procedure will yield an infinite set of equivalent orthogonal projections, contradicting Theorem 1.

Let x and y be the elements implementing the equivalence of x and $y : x \in eAf$, $y \in fAe$, $xy = e$, $yx = f$. Write $g_1 = yfx$. Then $yf \cdot fx = g_1$, $fx \cdot yf = f$, $fxg_1 = fx$, $g_1 yf = yf$, whence $f \sim g_1$. Write $x_1 = x - fx$, $y_1 = y - yf$. Then $x_1 y_1 = e - f$, $y_1 x_1 = f - g_1$, $(e - f)x_1 = x_1(f - g_1) = x_1$, $(f - g_1)y_1 = y_1(e - f) = y_1$, whence $e - f \sim f - g_1$. Let g be the right projection of g_1; then g is a projection with $g \leqq f$. We have $g_1 g = g_1$, $gg_1 = g$, from which it follows that $(f - g)(f - g_1) = f - g_1$, $(f - g_1)(f - g) = f - g$. Thus $g \sim g_1$ and $f - g \sim f - g_1$. From this we derive $f \sim g$, $e - f \sim f - g$. It remains to be seen that g and f are not equal. If $g = f$, then since $Ag = Ag_1$ we have $Af = Ag_1$. That is, f is a left multiple of $g_1 = yfx$, say $f = zyfx$. Left multiply by x and right multiply by y: the result is $e = xzyf$, whence $ef = e$, $e = f$. This contradicts our assumption that e and f are unequal.

6. The continuity axioms

In this section we shall complete the proof that the projections in a complete *-regular ring form a continuous geometry. Now that we have proved finiteness, the arguments in [3] need very little change; so we shall merely present an outline, with additional detail where needed.

(1) *Annihilators.* In a *-regular ring the right annihilator of any element is a principal right ideal generated by a projection. In a complete *-regular ring the right annihilator of *any* subset has this form, the generating projection being the greatest lower bound of the individual annihilating projections.

The right annihilator of a right ideal is of the form uA with u a central projection—see [3, Theorem 2.3, Corollary 1].

(2) *Central cover.* The LUB of central projections is central (Lemma 9). Thus for any element a there is a largest central projection v annihilating a. We shall

call the projection $u = 1 - v$ the *central cover* of a; it is the smallest central projection satisfying $ua = a$.

Equivalent projections have the same central cover. Suppose e, f are equivalent via x and y. Then since $f = yx = yex$, any central projection annihilating e also annihilates f. Hence e and f have the same central cover.

If a has central cover u, the right annihilator of aA is $(1 - u)/A$. It follows that a and b have orthogonal central covers if and only if $aAb = 0$.

(3) *Finite additivity of equivalence.* Suppose $e_1, \cdots, e_n$ are orthogonal projections with sum e, $f_1, \cdots, f_n$ orthogonal projections with sum f, and $e_i \sim f_i$. Then $e \sim f$. In fact if x_i, y_i implement the equivalence of e_i, f_i then $\sum x_i$, $\sum y_i$ do the same for e and f.

(4) *Additivity of equivalence in the orthogonal case.* Let e_i be orthogonal projections with LUB e, f_i orthogonal projections with LUB f; suppose that $e_i \sim f_i$ for all i and that e and f are orthogonal. We can prove that e and f are equivalent from Lemma 12, in much the same way as the proof of Lemma 13. Write $g_i = e_i + f_i$, $g = e + f$ and suppose that x_i, y_i implement the equivalence of e_i, f_i. We drop down to the subring B of all elements of gAg commuting with each g_i. In B we find by Lemma 12 elements x, y with $xg_i = x_i$, $yg_i = y_i$ for all i. The elements x and y implement the equivalence of e and f.

(5) *Comparability of orthogonal projections.* Let e and f be projections with $eAf \neq 0$. Pick any non-zero element in eAf, and let e_1 and f_1 be its left and right projections. Then we have $e_1 \leqq e$, $f_1 \leqq f$, $e_1 \sim f_1$. Now suppose further that e and f are orthogonal. Then, by the additivity of equivalence within e and f, this procedure may be pursued by transfinite induction until projections e', f' are reached with $e' \leqq e$, $f' \leqq f$, $e' \sim f'$, $(e - e')A(f - f') = 0$. The central covers of $e - e'$ and $f - f'$ are orthogonal. In this way we arrive at the comparability (in the generalized sense) of e and f: there exists a central projection u such that ue is equivalent to a portion of uf, $(1 - u)f$ is equivalent to a portion of $(1 - u)e$.

(6) *The case of Type I.* Call a projection e *abelian* if every projection in eAe is central in eAe (it is too much to expect all of eAe to be commutative—for instance eAe might be a division ring). We say that A is of type I if it possesses an abelian projection with central cover 1; A is of type II if it has no non-zero abelian projections. Any complete ∗-regular ring is uniquely a direct sum of rings of types I and II.

We are in a position to repeat verbatim the proofs of Lemmas 4.7, 4.8, 4.10 and 4.11 in [3]; for Lemma 4.9 we can substitute the remark made above that equivalent projections have the same central cover. We arrive at the following theorem: let A be a complete ∗-regular ring of type I, and $\{e_i\}$ a set of orthogonal projections in A; then there exists a direct summand of A in which all but a finite number of the e's vanish. This suffices to prove unrestricted additivity of equivalence in the type I case. As a matter of fact, it also proves the continuity axioms, so that our business with type I has quite concluded.

(7) *Subdivision of projections.* Let A be a complete ∗-regular ring of type II,

and e a projection in A. Then $e = f + g$ with f and g equivalent orthogonal projections; the proof of [3, Lemma 4.12] can be repeated.

We need to note the uniqueness, up to equivalence, of this bisection process. At present we can do this only within orthogonal projections. Suppose then that e and e_1 are equivalent orthogonal projections and that $e = f + g$, $e_1 = f_1 + g_1$ are subdivisions into equivalent orthogonal projections. Apply generalized comparability to f and f_1. After dropping down to a direct summand we can assume say that $f_1 \sim h$ with $h \leq f$. The equivalence of f and g induces an equivalence of h with a projection $k \leq g$. Then $g_1 \sim f_1 \sim h \sim k$. So $e = f + g \sim f_1 + g_1 \sim h + k \leq e$. By finiteness $h + k = e$ which means that $h = f$, $k = g$. Hence all of f, g, f_1, g_1 are equivalent.

This puts us in a position to repeat the process of bisection and we see that in a complete *-regular ring of type II any projection can be subdivided into four equivalent orthogonal ones.

(8) *Additivity of equivalence when there is room inside the complement.* Let e_i be orthogonal projections with LUB e, f_i orthogonal projections with LUB f, and suppose that $e_i \sim f_i$ for all i. Assume further that $e \sim g$ with g orthogonal to f. Let x and y implement the equivalence of e and g. The mapping $a \to yax$ is an isomorphism of eAe onto gAg, and it sends every idempotent into an equivalent idempotent. In the present context however we have to face the following difficulty: the mapping does not necessarily send projections into projections. This obstacle can be overcome by a procedure of the Gram-Schmidt type.

It is convenient to state a preliminary lemma. Let r and s be idempotents in a *-regular ring with $rs = sr = s$. Let r_1 and s_1 be the right projections of r and s. Then $r - s \sim r_1 - s_1$. If we bear in mind the equations $r_1 s_1 = s_1 r_1 = s_1$, $rr_1 = r$, $r_1 r = r_1$, $ss_1 = s$, $s_1 s = s_1$, we can verify directly that $r - rs_1$ and $r_1 - r_1 s$ implement an equivalence of $r - s$ and $r_1 - s_1$.

Now let the index set for the projections e_i and f_i be well ordered. For each ordinal α define h_α to be the LUB of e_β with $\beta < \alpha$. Then the h's are a well ordered ascending set of projections with LUB e. Let k_α be the right projection of the idempotent $yh_\alpha x$; the k's are projections ascending to g. Write $m_\alpha = k_{\alpha+1} - k_\alpha$. We have that $e_\alpha = h_{\alpha+1} - h_\alpha$ is equivalent to $yh_{\alpha+1}x - yh_\alpha x$, and by the lemma of the preceding paragraph the latter is equivalent to m_α. Hence $m_\alpha \sim f_\alpha$. Since the m's are orthogonal projections with LUB g, and since we know that additivity of equivalence holds within the orthogonal projections f and g, we derive $f \sim g$. Hence $e \sim f$.

(9) *Unrestricted additivity of equivalence* is now available by repeating the proof of [3, Theorem 5.5].

(10) *Comparability of arbitrary projections* can now be proved, since we no longer need the restriction to orthogonal projections.

(11) *The continuity axioms.* The proofs of Lemma 6.4 and Theorem 6.5 in [3] can be repeated without change. We have proved:

THEOREM 3. *The projections in a complete ∗-regular ring form a continuous geometry.*

7. Lattice-theoretic preliminaries

Let L be a lattice with 0 and 1. We say that L is *orthocomplemented* [1, p. 123] if there exists a mapping $a \to a'$ of L onto itself which is one–one, order inverting, and satisfies $a'' = a$, $a \cap a' = 0$, $a \cup a' = 1$. We shall call a' the orthogonal complement of a, and we say that a and b are *orthogonal* if $b \leq a'$ (this relation is evidently symmetric between a and b).

Let L be a complemented modular lattice. We say that L can be coordinatized if there exists a regular ring A such that L is isomorphic to the lattice of principal right ideals in A. Von Neumann has proved [7, p. 50] that if an orthocomplemented L can be coordinatized by A, then A is ∗-regular. Putting this together with our Theorem 3 we have: *if an orthocomplemented complete modular lattice can be coordinatized it is a continuous geometry.*

Let L again be a lattice with 0 and 1. We say that two elements are *perspective* if they have a common complement. Let $S = \{x_i\}$ (i ranging over an index set I) be a subset of L; if S is infinite assume L to be complete. We say that S is independent if the following is true: whenever I is split into two disjoint subsets I_1 and I_2, and y_1, y_2 are the unions of the x_i's with i ranging over I_1, I_2, then $y_1 \cap y_2 = 0$. In terms of these concepts we can state the sufficient condition for coordinatization which is established in [7]: if a complemented modular lattice L has n independent perspective elements with union $1(n \geq 4)$, then L can be coordinatized. We are going to show that this condition is nearly always fulfilled in an orthocomplemented complete modular lattice. We shall use certain fragments from the theory of continuous geometries, supplemented by the additional arguments made possible by orthocomplementation.

We note two pieces of notation: we write $L(a, b)$ for the sublattice of all elements x satisfying $a \leq x \leq b$, and we write $\sum a_i$ for the union of the elements a_i, i ranging over an index set.

LEMMA 22. *Let a and b be elements of a complemented modular lattice L. Let c and d be any elements of L with $c \leq a \cap b$, $d \geq a \cup b$. Then $a \sim b$ in L if and only if $a \sim b$ in $L(c, d)$.*

LEMMA 23. *Let a, b be perspective elements in a complemented modular lattice; let x be a common complement of a and b. Then the mapping $c \to (c \cup x) \cap b$ is an isomorphism of the lattice $L(0, a)$ onto $L(0, b)$ and in this mapping every element is perspective to its image.*

LEMMA 24. *Let a and be be elements of a modular lattice with $a \cap b = 0$. Let x, y, z, t be elements with $x, y \leq a$ and $z, t \leq b$. Then $(x \cup z) \cap (y \cup t) = (x \cap y) \cup (z \cap t)$.*

LEMMA 25. *Let a, b, c be independent elements of a complemented modular lattice, and suppose that $a \sim b$, $b \sim c$. Then $a \sim c$.*

These four lemmas can be found in [6] as Theorems 3.1, 3.3, 1.2 and 3.5. Lemma 24 is also a special case of [1, p. 73, Theorem 7].

LEMMA 26. *Let x, y_1 be elements in an orthocomplemented lattice with x orthogonal to each y_i. Then x is orthogonal to $\sum y_i$.*

PROOF. We have $y_i \leq x'$ for every i, hence $\sum y_i \leq x'$, hence x is orthogonal to $\sum y_i$.

The next lemma is an immediate corollary of Lemma 26.

LEMMA 27. *In an orthocomplemented lattice orthogonal elements are independent.*

LEMMA 28. *Let a_i be orthogonal elements in an orthocomplemented complete modular lattice. Let x_i, y_i be elements with x_i, $y_i \leq a_i$, $x_i \cap y_i = 0$. Then*

$$\sum x_i \cap \sum y_i = 0.$$

PROOF. Write a, x, y for $\sum a_i$, $\sum x_i$, $\sum y_i$. Write b_i, t_i, u_i for $\sum a_j$, $\sum x_j$, $\sum y_j$, taken over all j with $j \neq i$. Then $a = a_i \cup b_i$, and a_i, b_i are orthogonal by Lemma 26. Also $x = x_i \cup t_i$, $y = y_i \cup u_i$, t_i, $u_i \leq b_i$. By Lemma 24, $x \cap y = (x_i \cap y_i) \cup (t_i \cap u_i) = t_i \cap u_i \leq b_i$. Hence $x \cap y$ is orthogonal to a_i. By Lemma 26, $x \cap y$ is orthogonal to $\sum a_i = a$. Since $x \cap y \leq a$, this implies $x \cap y = 0$.

We can now prove that perspectivity is additive over orthogonal elements.

LEMMA 29. *Let a_i, b_i be elements in an orthocomplemented complete modular lattice. Suppose that the a's are all orthogonal, the b's are all orthogonal, and each a_i is orthogonal to each b_j. Suppose further that $a_i \sim b_i$ for all i. Then $\sum a_i \sim \sum b_i$.*

PROOF. By Lemma 22, $a_i \sim b_i$ in $L(0, a_i \cup b_i)$. Hence we can find x_i with $a_i \cup x_i = b_i \cup x_i = a_i \cup b_i$, $a_i \cap x_i = b_i \cap x_i = 0$. For $\sum a_i$, $\sum b_i$, $\sum x_i$ write a, b, x. Evidently $a \cup x = b \cup x = a \cup b$. Since the elements $a_i \cup b_i$ are orthogonal, and a_i, $x_i \leq a_i \cup b_i$, it follows from Lemma 28 that $a \cap x = 0$. Similarly $b \cap x = 0$. Hence $a \sim b$ in $L(0, a \cup b)$. By Lemma 22, a and b are perspective in the original lattice.

LEMMA 30. *An orthocomplemented complete modular lattice cannot have an infinite number of orthogonal perspective elements.*

PROOF. Suppose that on the contrary $\{a_i\}$ constitutes an infinite set of orthogonal perspective elements, and write $a = \sum a_i$. We work in the lattice $L(0, a)$. Split the set $\{a_i\}$ into four disjoint sets of equal cardinal number, and write $b_j (j = 1, \cdots, 4)$ for the LUB's of the four subsets. By Lemma 26 the b's are orthogonal, whence by Lemma 27 they are independent. By Lemma 29 the b's are mutually perspective. Hence $L(0, a)$ fulfils the condition for coordinatization, and it is consequently isomorphic to the lattice of projections in a complete *-regular ring. The projections corresponding to the a's are orthogonal, and by Lemma 1 they are equivalent. But this contradicts Theorem 1. (Alternatively we could argue that by Theorem 3, $L(0, a)$ is a continuous geometry, and consequently [6, Theorem 3.8] it cannot have an infinite independent set of perspective elements.)

With Lemma 30 at hand it would be possible to finish the proof of our main theorem purely lattice-theoretically. But it will be shorter for us to continue to appeal to the coordinatization theorem, and make use of Theorem 3.

8. The center

The results in this section are obtained without the use of orthocomplementation, and may be useful in future studies of complemented complete modular lattices. What we have done is to change slightly the known continuous geometry proofs so as to avoid the use of the continuity axioms.

We repeat some definitions from [6, page 38]. Let L be a complemented modular lattice. We write $(a, b, c)D$ to mean that the distributive law holds for a, b and c: it turns out that any of the six versions of the distributive law implies all five others. We write $(a, b)D$ if $(a, b, c)D$ for every c in L; $(a)D$ if $(a, b)D$ for every b in L. The set of elements a with $(a)D$ coincides with the set of elements having unique complements and is called the *center* of L. If a is in the center of L and b is its complement, then L is the direct product of $L(0, a)$ and $L(0, b)$.

Following notation used on page 44 of [6] we write $(a, b)P$ if $(a, b)D$ and $a \cap b = 0$.

LEMMA 31. *If $(a, c)D$ and $(b, c)D$ in a complemented modular lattice then $(a \cup b, c)D$.*

LEMMA 32. *Let a and b be elements of a complemented modular lattice with $a \cap b = 0$. Then the following statements are equivalent: (1) $(a, b)P$ (2) there do not exist non-zero elements a_1, b_1 with $a_1 \leq a$, $b_1 \leq b$, $a_1 \sim b_1$.*

LEMMA 33. *Let a, b, c be non-zero elements of a complemented modular lattice with $a \sim b$, $b \sim c$. Then there exist non-zero elements a_1, c_1 with $a_1 \leq a$, $c_1 \leq c$, $a_1 \sim c_1$.*

Lemma 31 is a portion of Lemma 5.1 in [6]; Lemma 32 is Theorem 5.7 in [6]; Lemma 33 is Lemma 7 in [2].

The next lemma is an immediate corollary of Lemma 32.

LEMMA 34. *If in a complemented modular lattice $(a, b)P$, $c \leq a$ and $d \leq b$, then $(c, d)P$.*

LEMMA 35. *Let a, b, c, be elements of a complemented modular lattice with $a \cap c = 0$, $a \sim b$ and $(b, c)P$. Then $(a, c)P$.*

PROOF. Suppose the contrary. Then by Lemma 32 there exist non-zero elements a_1, c_1 with $a_1 \leq a$, $c_1 \leq c$, $a_1 \sim c_1$. By Lemma 23 the perspectivity between a and b induces a perspectivity between a_1 and an element b_1 with $b_1 \leq b$. Thus $b_1 \sim a_1$, $a_1 \sim c_1$, $b_1 \cap c_1 = 0$. Lemma 33 now applies to yield non-zero perspective elements b_2, c_2 with $b_2 \leq b_1$, $c_2 \leq c_1$. But by Lemma 34, $(b_1, c_1)P$. This contradicts Lemma 32.

LEMMA 36. *Let a, u_i be elements of a complete complemented modular lattice, and let $u = \sum u_i$. Suppose $(a, u_i)P$ for every i. Then: (1) $a \cap u = 0$, (2) $(a, u)P$.*

PROOF. (1) Let b be a complement of a. Since $(a, u_i)D$ we have

$$u_i = (a \cup b) \cap u_i = (a \cap u_i) \cup (b \cap u_i).$$

Since $a \cap u_i = 0$ this implies $u_i = b \cap u_i$, whence $b \geq u_i$. Hence $b \geq u$, and $a \cap u \leq a \cap b = 0$.

(2) It remains to prove $(a, u)D$. Suppose the contrary. Then by Lemma 32 there exist non-zero elements $c \leqq a$, $v \leqq u$ with $c \sim v$. We consider $v \cap u_i$. By Lemma 23 the perspectivity between v and c induces a perspectivity between $v \cap u_i$ and an element d with $d \leqq c$. Since $v \cap u_i \leqq u_i$, $d \leqq a$ and $(a, u_i)P$ it follows from Lemmas 34 and 32 that $v \cap u_i = 0$. Next we note that $v \sim c$, $(c, u_i)P$ and $v \cap u_i = 0$. Hence Lemma 35 applies to yield $(v, u_i)P$. By part (1) of the present lemma (already proved), $v \cap u = 0$. Since $v \leqq u$, this means $v = 0$, a contradiction.

THEOREM 4. *Let a be any element in a complemented complete modular lattice L. Then the center of $L(0, a)$ consists exactly of all elements of the form $z \cap a$ with z in the center of L.*

This is Theorem 1.6 in [8]. With Lemma 36 at hand we can repeat its proof without further change.

THEOREM 5. *Let L be a complemented complete modular lattice, Z its center. Then Z is again complete; more precisely, Z is a complete sublattice of L. For any set $\{z_i\}$ in Z and any a in L we have $a \cap \sum z_i = \sum (a \cap z_i)$.*

PROOF. Let $\{z_i\}$ be any subset of Z. In order to prove that Z is a complete sublattice of L, it suffices for us to prove that $z = \sum z_i$ is in Z; by duality the greatest lower bound will also be in Z. Given a in L, we have to prove $(z, a)D$. Write $b = \sum (a \cap z_i)$, and let c be a relative complement of b within a, so that $a = b \cup c$, $b \cap c = 0$. We have $a \cap z_i \leqq b$ and hence $a \cap z_i \leqq b \cap z_i$. Also $c \cap z_i \leqq a \cap z_i$. Hence $c \cap z_i \leqq b \cap c \cap z_i = 0$. Since z_i is central, $(c, z_i)P$ holds. Then by Lemma 36 we have $(c, z)P$, that is, $c \cap z = 0$ and $(c, z)D$. Since $b \leqq z$ we have $(b, z)D$ by the modular law. By Lemma 31, $(a, z)D$ holds, and we have proved that z is central. Again $a \cap z = (b \cap z) \cup (c \cap z) = b$, which is the final statement of the theorem.

With Theorem 5 at hand we can repeat the discussion occurring on pages 17–19 of [8], obtaining the following fact: if $\{z_i\}$ is an independent central set with LUB 1 in a complemented complete modular lattice L, then L is isomorphic to the complete direct product of the lattices $L(0, z_i)$. We shall make tacit use of this result several times in the arguments that follow.

We may also define the *central cover* of an element a in L as the greatest lower bound of all central elements z with $z \geqq a$. It will be convenient to call an element *faithful* if its central cover is 1.

9. The case of type II

DEFINITIONS. An element a in a lattice L with 0 is called a *D-element* if the lattice $L(0, a)$ is distributive. (In [8] the term used is "minimal"). We say that L is of type II if it has no D-elements other than 0. A complemented complete modular lattice is said to be of type I if it has a faithful D-element.

Let L be a complemented complete modular lattice and a a non-zero D-element in L with central cover z. Then a is faithful in $L(0, z)$. We then examine the complementary direct factor for D-elements. The process can be pursued by transfinite induction and we see: any complemented complete modular lattice is uniquely a direct product of lattices of types I and II.

LEMMA 37. *Any orthocomplemented modular lattice L which is not distributive contains two non-zero orthogonal perspective elements.*

PROOF. Since L is not distributive it contains an element a which is not central. Let b be the orthogonal complement of a. We cannot have $(a, b)D$ for then by [6, Theorem 5.3] a would be central. By Lemma 32 there exist non-zero perspective elements a_1, b_1 with $a_1 \leqq a$, $b_1 \leqq b$. Of course a_1 and b_1 are orthogonal.

LEMMA 38. *Let L be an orthocomplemented modular lattice of type II. Then L contains four non-zero orthogonal perspective elements.*

PROOF. It will in fact be convenient to prove by induction that L contains n non-zero orthogonal perspective elements. Assume that we have $n - 1$: a_1, $\cdots$, a_{n-1}. By Lemma 37, $L(0, a_{n-1})$ contains two non-zero orthogonal perspective elements, say b_{n-1} and b_n. By Lemma 23 the perspectivity between a_{n-1} and $a_i (i = 1, \cdots, n - 2)$ induces a perspectivity between b_{n-1} and an element $b_i \leqq a_i$. The elements $b_1, \cdots, b_n$ are all orthogonal and hence (Lemma 27) independent. Lemma 25 shows that any two of the b's are perspective.

THEOREM 6. *Any orthocomplemented complete modular lattice of type II can be coordinatized (and hence is a continuous geometry).*

PROOF. We pick, by Lemma 38, four non-zero orthogonal perspective elements. Working within the orthogonal complement of their union, we may again select four such elements. The process can be pursued by transfinite induction, after which an application of Lemma 29 yields four orthogonal perspective elements with union 1. This implies coordinatization.

10. The case of type I

For almost all purposes the type I case is much the easier of the two. But the problem of introducing coordinates is one where type I is more awkward. In addition to the "non-Desarguian" exceptional case we have to take special care to iron out ragged edges.

The next five lemmas are valid without orthocomplementation.

LEMMA 39. *In a complemented complete modular lattice of type I, 1 is expressible as a union of D-elements.*

PROOF. Let a be the union of all D-elements, and let b be a complement of a. Then b contains no D-elements. It follows from Lemma 32 that $(c, b)P$ for any D-element c (note that an element which is perspective to a D-element is itself a D-element). By Lemma 36, $(a, b)P$. But by [6, Theorem 5.3] this implies that a and b are central. Since a is faithful (it contains a faithful D-element) this is possible only if $b = 0$, $a = 1$, as desired.

LEMMA 40. *In a complemented complete modular lattice of type I any element contains a D-element with the same central cover.*

PROOF. Let x be the given element; we can suppose $x \neq 0$. It will suffice for us to prove that x contains a non-zero D-element b, for we then repeat the argument within the complement of the central cover of b, etc. If x contains no non-zero D-elements then, by Lemma 32, $(c, x)P$ for every D-element c. By Lemmas 36 and 39, $(1, x)P$ and $x = 0$.

LEMMA 41. *Let L be a complemented complete modular lattice, x a D-element in L, y an element with $y \leqq x$. Then $y = z \cap x$ where z is in the center of L.*

PROOF. This follows from Theorem 4 and the fact that everything in $L(0, x)$ is central.

LEMMA 42. *Let x and y be D-elements in a complemented modular lattice. If $(x, y)P$ then $x \cup y$ is a D-element.*

PROOF. In the lattice $L(0, x \cup y)$, x and y are complements. Hence, by [6, Theorem 5.3] x and y are central in $L(0, x \cup y)$. Thus $L(0, x \cup y)$ is the direct product of $L(0, x)$ and $L(0, y)$ and is distributive.

LEMMA 43. *In a complemented complete modular lattice of type I any two faithful D-elements are perspective.*

PROOF. Let a and b be faithful D-elements; they will be non-zero unless the whole lattice is 0. It cannot be the case that $(a, b)P$. For then by Lemma 42 $a \cup b$ is a D-element. By Lemma 41, $a = z \cap (a \cup b)$ with z central. Since a is faithful and $z \geqq a$ we must have $z = 1$, $a = a \cup b$, contradicting $a \cap b = 0$. By Lemma 32 there therefore exist non-zero perspective elements a_1, b_1 with $a_1 \leqq a$, $b_1 \leqq b$. By Lemma 41, $a_1 = u \cap a$ with u central. Thus in the direct factor $L(0, u)$ the portions $u \cap a$, $u \cap b$ of a and b are perspective. We repeat the argument within the complement of u and proceed by transfinite induction.

Let us call an orthocomplemented complete modular lattice n-*homogeneous* if 1 is expressible as the union of n orthogonal faithful D-elements. Note that by Lemma 43 the latter are necessarily perspective.

THEOREM 7. *An orthocomplemented complete modular lattice of type I is a direct product of n-homogeneous lattices $(n = 1, 2, \cdots)$.*

PROOF. Let $a_1, \cdots, a_r$ be a maximal set of orthogonal faithful D-elements; by Lemma 30 such a set must be finite. Let $b = a_1 \cup \cdots \cup a_r$. If the element b' is faithful, by Lemma 40 it contains a faithful D-element which can be used to enlarge the set of a's. Hence there exists a non-zero central element z with $z \cap b' = 0$. The lattice $L(0, z)$ is r-homogeneous. Having found one such factor, one proceeds by transfinite induction.

For $n \geqq 4$ an n-homogeneous orthocomplemented complete nodular lattice can be coordinatized and so is a continuous geometry. The case $n = 1$ is merely that of a Boolean algebra. Stubbornly holding out are the cases $n = 2$ and 3, describable roughly as "direct integrals" of projective lines and projective planes.

To handle a lattice which is 2- or 3-homogeneous we make the following remark. Suppose that L contains four non-zero orthogonal perspective D-elements (by supplementary arguments this can actually be shown to be impossible). Then we can drop down to a direct factor where they are faithful, expand to a maximal set, and (as in the proof of Theorem 7) drop further down to a factor which is n-homogeneous for some $n \geqq 4$. This last factor is a continuous geometry. We can suppose we have factored out in advance the largest direct factor which is a continuous geometry, and accordingly we are entitled to assume that L does not contain four non-zero orthogonal perspective D-elements.

LEMMA 44. *Let L be an orthocomplemented complete modular lattice of type I which does not contain four non-zero orthogonal perspective D-elements. Let $\{b_i\}$ be any orthogonal set in L. Then there exists a non-zero central element u such that all but three of the elements $u \cap b_i$ vanish.*

PROOF. Let z_i be the central cover of b_i. If z_1 is orthogonal to all the remaining z's we take $u = z_1$. Suppose then that $z_1 \cap z_2 \neq 0$. If $z_1 \cap z_2$ is orthogonal to all the rest we take $u = z_1 \cap z_2$. If we can do this twice more we will reach $v = z_1 \cap z_2 \cap z_3 \cap z_4 \neq 0$. But then the elements $v \cap b_j$ $(j = 1, \cdots 4)$ will (by Lemma 40) contain D-elements with central cover v; these D-elements will be orthogonal and, by Lemma 43, perspective.

With Lemma 44 at hand it is easy to complete the proof of the continuity axioms. By duality it suffices to examine the axiom that concerns well ordered ascending sets. Let then $\{a_\alpha\}$ be well ordered ascending; we can assume that for any limit ordinal λ, a_λ is the union of the preceding a's. We must prove that for any c

$$(23) \qquad\qquad c \cap \sum a_\alpha = \sum (c \cap a_\alpha).$$

Write $b_\alpha = a_{\alpha+1} \cap a'_\alpha$. The b's are orthogonal and we apply Lemma 44 to them. The result is that, after dropping down to a direct factor, all but three at most of the b's are 0. But this means that $\{a_\alpha\}$ is ultimately constant and makes (23) obvious. Thus (23) holds in a direct factor. Repetition of this process transfinitely shows that (23) holds in the whole lattice. We have completed the proof and we end the paper the way it began:

THEOREM. *Any orthocomplemented complete modular lattice is a continuous geometry.*

UNIVERSITY OF CHICAGO

BIBLIOGRAPHY

1. G. BIRKHOFF, Lattice Theory, New York, 1948.
2. I. HALPERIN and J. v. NEUMANN, *On the transitivity of perspective mappings*, Ann. of Math., 41 (1940), 87–93.
3. I. KAPLANSKY, *Projections in Banach algebras*, Ann. of Math., 53 (1951), 235–249.
4. I. KAPLANSKY, *Algebras of type I*, Ann. of Math., 56 (1952), 460–472.
5. G. W. MACKEY, *On infinite-dimensional linear spaces*, Trans. Amer. Math. Soc., 57 (1945), 155–207.
6. J. v. NEUMANN, Continuous Geometry, Part I, Princeton, 1936.
7. J. v. NEUMANN, Continuous Geometry, Part II, Princeton, 1937.
8. J. v. NEUMANN, Continuous Geometry, Part III, Princeton, 1937.
9. C. E. RICKART, *Banach algebras with an adjoint operation*, Ann. of Math., 47 (1946), 528–550.

Afterthought

The opening page of the paper accurately describes how I came to conjecture that the theorem in the paper's title is true. It took most of the summer of 1954 to hack out a proof.

There have been two subsequent accounts. Amemiya and Halperin [1] generalized the theorem significantly and couched their proofs in purely lattice-theoretic terms. Skornyakov [5] devoted Chapter 8 to the theorem; his book contains many additional things of interest.

When [2] appeared, I was amazed. Let R be a *-regular ring, not assumed to be complete. Ara and Menal proved that $xx^* = 1$ implies $x^*x = 1$, and the proof is short and elegant. If one could strengthen this to a proof that $xy = 1$ implies $yx = 1$, then this would supersede my paper with a stronger theorem and a (hopefully) simplified proof. (Note for the record: I am just asking, not conjecturing.)

I would like to publicize their result, and so I shall take the space to repeat the proof. I have slightly recast it, separating out two lemmas, and so forth.

LEMMA 1. *Let R be a ring with 1 and involution*. Assume that $tt^* = 0$ in R implies that $t = 0$. Suppose that $xx^* = 1$ and that e is a projection right-annihilating $1 - x$. Then e and x commute.*

Proof. It suffices to prove that e left-annihilates $1 - x$, and for this $e(1 - x) \cdot (1 - x^*)e = 0$ suffices. This yields to a small computation, using $xe = e$ (whence $ex^* = e$) and $xx^* = 1$.

LEMMA 2. *Let R be a ring with 1 and involution*. Assume that $xx^* = 1$ and that $1 - x$ is left-invertible. Then $1 - x$ and x are invertible. In particular, $xx^* = 1$.*

Proof. By taking * we have that $1 - x^*$ is right-invertible. From the equation

$$(1 - x)x^* = x^* - 1 \tag{1}$$

we deduce that $1 - x$ is right-invertible. Thus $1 - x$ is invertible on both sides. So is $1 - x^*$. From (1) again we see that x^*, and hence x, is invertible.

THEOREM. *In a *-regular ring $xx^* = 1$ implies $x^*x = 1$.*

Proof. Let e be the projection generating the right annihilator of $1 - x$. By Lemma 1, e commutes with x. We can drop down to the subring commuting with e without changing our problem. So e is now central. We may thus assume that e is either 1 or 0. If e is 1, $1 - x$ is 0, x is 1. If e is 0, the right annihilator of $1 - x$ is 0, whence $1 - x$ is left-invertible. By Lemma 2, $x^*x = 1$.

REMARK. By slightly changing the formulation and proof we can drop the unit element.

I conclude with two notes.

(a) There is a minor flaw in the paper. It concerns a side issue and does not jeopardize the main proof. The flaw is pointed out in [3, p. 76] and [3, Theorem 54, p. 82] supplies a correction.

(b) Reference [4] is more accessible than the original notes on continuous geometry; in addition, it contains valuable material added by Israel Halperin.

References

1. I. Amemiya and I. Halperin, Complemented modular lattices, *Canadian J. of Math.* 11(1959), 481–520.
2. P. Ara and P. Menal, On regular rings with involution, *Arch. Math.* 42(1984), 126–30.
3. I. Kaplansky, *Rings of Operators*, Benjamin, (New York), 1968.
4. J. von Neumann, *Continuous Geometry*, edited by I. Halperin with a foreword, a list of changes, and comments on the text, Princeton Univ. Press, Princeton, NJ, 1960.
5. L. A. Skornyakov, *Complemented Modular Lattices and Regular Rings*, Moscow, 1961. English translation, Edinburgh-London, 1964.

ANNALS OF MATHEMATICS
Vol. 68, No. 2, September, 1958
Printed in Japan

PROJECTIVE MODULES

By IRVING KAPLANSKY[1]

(Received November 6, 1957)

1. Introduction

Let R be an integral domain in which all finitely generated ideals are principal. Let M be a projective R-module. It is well known that if M is finitely generated it is free. By a familiar stepwise procedure this can be extended to the case where M is countably generated. Is it possible to drop the countability assumption? In endeavoring to answer this, I stumbled on a theorem which appears to be worth recording: *any projective module is a direct sum of countably generated modules.*

Three results on the structure of projective modules are deduced:

(1) Over a local ring any projective module is free,

(2) Over a commutative semi-hereditary ring any projective module is a direct sum of finitely generated ideals.

(3) Over a regular ring any projective module is a direct sum of principal ideals. In each case the main theorem is used to reduce the problem to the countably generated case, and then a special argument is given that takes advantage of countability.

The main theorem is proved very simply; indeed it is really a piece of " universal algebra ". We actually prove it in a more general setting: if a module is direct sum of countably generated modules, so is any direct summand. As an application, we show in § 3 how to add the finishing touch to Kulikov's solution of a problem proposed by Baer.

2. Reduction to the countably generated case

Once for all: every ring has a unit element which acts as unit operator on any module. Further: " module " means " left module ".

THEOREM 1. *Let R be a ring, M an R-module which is a direct sum of (any number of) countably generated R-modules. Then any direct summand of M is likewise a direct sum of countably generated R-modules.*

PROOF. Let M be a direct sum of modules M_i, i ranging over an index set, each M_i countably generated. Suppose that $M = P \oplus Q$. We must prove that P is a direct sum of countably generated modules.

The plan is to express M as the union of a well ordered increasing se-

---

[1] Work on this paper was done with the aid of a contract with the Office of Scientific Research of the Air Force.

372

quence of submodules $\{S_\alpha\}$ with the following properties :

(1) if α is a limit ordinal, S_α is the union of all S_β with $\beta < \alpha$,

(2) $S_{\alpha+1}/S_\alpha$ is countably generated,

(3) each S_α is the direct sum of a subset of the M_i's,

(4) $S_\alpha = P_\alpha + Q_\alpha$, where $P_\alpha = S_\alpha \cap P$, $Q_\alpha = S_\alpha \cap Q$. Let us first see how this gives the desired result. We have that P_α is a direct summand of S_α, which is a direct summand of M. Hence P_α is a direct summand of M, and it follows that it is a direct summand of $P_{\alpha+1}$. Since

$$S_{\alpha+1}/S_\alpha = P_{\alpha+1}/P_\alpha \oplus Q_{\alpha+1}/Q_\alpha ,$$

$P_{\alpha+1}/P_\alpha$ is countably generated, being a homomorphic image of a countably generated module. It is clear that, for α a limit ordinal, P_α is the union of all P_β with $\beta < \alpha$. These facts combine to show that P is a direct sum of countably generated modules.

We proceed to construct the S_α's. No problem exists at a limit ordinal. Supposing that S_α is at hand, we show how to construct $S_{\alpha+1}$. Choose some M_j which is not contained in S_α. Let a countable generation of M_j be given by $x_{11}, x_{12}, x_{13}, \cdots$ Split x_{11} into its P- and Q-components. Each of these new elements, in the expression of M as the direct sum of the M_i's, has a non-zero entry in only a finite number of the M_i's. Assemble this finite collection of M_i's, and let $x_{21}, x_{22}, x_{23} \cdots$ denote a countable set of generators for their union. Next we repeat on x_{12} the treatment just given to x_{11}. The result will be a new countable set $x_{31}, x_{32}, x_{33} \cdots$, the third row of an infinite matrix we are constructing. We proceed in this fashion, pursuing the elements along successive diagonals in the order $x_{11}, x_{12}, x_{21}, x_{13}, x_{22}, x_{31}$, etc. Finally, $S_{\alpha+1}$ is taken to be the submodule generated by S_α and all the x's. That $S_{\alpha+1}$ has all the properties we desire is plain.

3. A problem of Baer

Let G be a torsion-free abelian group which is a direct sum of groups of rank one. Let H be a direct summand of G. Is H a direct sum of groups of rank one ? This problem was studied by Baer [1], who obtained several partial results. In [4] Kulikov gave an affirmative answer, assuming only that H is countable. But we can now pass readily to the uncountable case. Any abelian group of rank one is countable. By Theorem 1, H is a direct sum of countable groups. Now apply Kulikov's theorem.

REMARK. The way I have phrased the proof does not permit instant generalization to modules over principal ideal rings (a module of rank one need not be countably generated). However a slight modification of the proof of Theorem 1 yields the following : if R is an integral domain and M a torsion-free R-module which is a direct sum of modules of countable rank, then any direct summand of M is likewise a direct sum of modules of countable rank.

4. Local rings

We take the term " local ring " in its general sense, i.e., a ring (possibly non-commutative) where the non-units form an ideal.

THEOREM 2. *Any projective module over a local ring is free.*

We decompose the proof into two lemmas, the first of which is proved by an obvious stepwise procedure ; note that the countability assumption is indispensable.

LEMMA 1. *Let R be any ring, M a countably generated R-module. Assume that any direct summand N of M has the following property : any element of N can be embedded in a free (resp. finitely generated) direct summand of N. Then M is free (resp. a direct sum of finitely generated modules).*

With Lemma 1 and Theorem 1 at hand, Lemma 2 is all that is needed to complete the proof of Theorem 2.

LEMMA 2. *Let P be a projective module over a local ring. Then any element of P can be embedded in a free summand of P.*

PROOF. Write $F = P \oplus Q$, F free. Let x be the element of P whose insertion into a free direct summand of P is required. Select a basis u_i of F such that the expression of x in terms of that basis has the smallest possible number of non-zero entries. Say $x = a_1 u_1 + \cdots + a_n u_n$ is the expression in question. We note that none of the a's can be a right linear combination of the remaining ones. For suppose that $a_n = a_1 b_1 + \cdots + a_{n-1} b_{n-1}$. If we replace u_i by $u_i + b_i u_n$ for $i = 1, \cdots, n - 1$ and leave all other basis elements unchanged we get a new basis for F. Thus

$$x = a_1(u_1 + b_1 u_n) + \cdots + a_{n-1}(u_{n-1} + b_{n-1} u_n)$$

is a shorter expression for x, a contradiction.

Let $u_i = y_i + z_i$ be the decomposition of u_i into P- and Q-components. Necessarily

$$(1) \qquad a_1 y_1 + \cdots + a_n y_n = a_1 u_1 + \cdots + a_n u_n ,$$

since each is the P-component of x. Write $y_i = \sum_{j=1}^{n} c_{ij} u_j + t_i$, where t_i

is a linear combination of the basis elements other than $u_1, \cdots, u_n$. Combine this equation with (1), and equate the coefficients of u_j. The result is

$$(2) \qquad\qquad a_1 c_{1j} + \cdots + a_n c_{nj} = a_j \, .$$

It follows from (2) that the elements c_{jj}, and c_{ij} with $i \neq j$, must be non-units; otherwise some a_i would be a right linear combination of the others, contrary to what was shown above. Hence the matrix (c_{ij}) is non-singular; for it has units down the main diagonal and non-units elsewhere, and over a local ring this suffices to make a matrix non-singular. Thus the elements $y_1, \cdots, y_n$ form a basis for F, when augmented by the u's other than $u_1, \cdots, u_n$. If we write S for the submodule spanned by $y_1, \cdots, y_n$ we have $x \in S \subset P$, S is a free direct summand of F, whence S is a free direct summand of P. The proof of Theorem 2 is complete.

In the treatment by Cartan and Eilenberg of the case where P is finitely generated, they derive simultaneously the statement that any generating set for P contains a basis. It is easy to see that this is false in the infinite case.

5. Semi-hereditary rings

A ring is said to be left semi-hereditary if every finitely generated left ideal is projective. In the commutative case one of course omits the distinction between left and right.

THEOREM 3. *Let R be a commutative semi-hereditary ring, P a projective R-module. then P is a direct sum of modules each of which is isomorphic to a finitely generated ideal in R.*

We begin by noting that Theorem 3 is known when P is finitely generated [3, Ch. I, Prop. 6.1]. Because of Theorem 1 and Lemma 1, our problem therefore reduces to proving the following lemma.

LEMMA 3. *Let P be a projective module over a commutative semi-hereditary ring R. Then any element of P can be embedded in a finitely generated direct summand of P.*

PROOF. Suppose that $F = P \oplus Q$ with F free, and let $x \in P$ be the element whose embedding in a finitely generated direct summand of P is required. Write S for the set of all elements y in F which satisfy an equation of the form $by = cx$, with $b, c \in R$, b not a divisor of 0. It is immediate that S is a submodule of F containing x. Actually, S is contained in P. For if $y = w + z$ is the decomposition of y into P- and Q-components, we have $by \in P$, $bz = 0$, $z = 0$ (since b is not a divisor of 0 and F is free). We shall show that S is the desired finitely generated direct summand of P.

Take any basis for F and let $x = a_1u_1 + \cdots + a_nu_n$ be the expression for x in terms of this basis. Write G for the (free) submodule spanned by $u_1, \cdots, u_n$. We note that S is contained in G. For if $y \in S$, $by = cx = c(a_1u_1 + \cdots + a_nu_n)$ with b not a divisor of 0. It follows that the unique expression of y in terms of the u's can only involve $u_1, \cdots, u_n$.

Since R is semi-hereditary, the principal ideal a_1R is projective. Thus the mapping of R into a_1R given by $r \to a_1r$ is onto a projective module, and its kernel is a direct summand of R. In other words: the annihilator of a_1 in R is a direct summand of R. Since we may study our problem separately in the two summands, we are entitled to assume that a_1 is not a divisor of 0. We note the following consequence: any non-zero element of S carries a non-zero coefficient of u_1.

We propose to show that S is a direct summand of G by proving that $G^* = G/S$ is projective. Let v_i denote the image of $u_i (i = 1, \cdots, n)$ in the natural mapping of G onto G^*. The elements $v_2, \cdots, v_n$ span a submodule H^* of G^*. We claim that H^* is free. For suppose $d_2v_2 + \cdots + d_nv_n = 0$ $(d_i \in R)$. Then $d_2u_2 + \cdots + d_nu_n \in S$. But we saw that any non-zero element of S has a non-zero term in u_1. Hence the d's are 0.

It follows readily from the definition of S that S is closed under "division" by a_1: if $t \in F$, $a_1t \in S$, then $t \in S$. Hence multiplication by a_1 induces an isomorphism of G^* into itself. Since $a_1u_1 + \cdots + a_nu_n \in S$, we have $-a_1v_1 = a_2v_2 + \cdots + a_nv_n$, $a_1v_1 \in H^*$, whence $a_1G^* \subset H^*$. Thus multiplication by a_1 transforms G^* into a submodule of H^*. Since H^* is free and G^* is finitely generated, G^* is projective [3, Ch. I, Prop. 6.1], as desired. We now know that S is a direct summand of G, which yields the information that S is finitely generated. Furthermore, since G is a direct summand of M, we find that S is a direct summand of M and hence also of P. This concludes the proof of Lemma 3 and Theorem 3.

COROLLARY. *Let R be an integral domain in which all finitely generated ideals are principal. Then any projective R-module is free.*

I do not know whether the hypothesis of commutativity can be dropped in Theorem 3. There is, however, a special class of non-commutative semi-hereditary rings for which a different technique works. We recall that a ring is *regular* if for any a there exists an x such that $axa = a$.

THEOREM 4. *Any projective module over a regular ring is a direct sum of modules isomorphic to principal left ideals.*

With the aid of Theorem 1, we see that the following lemma is adequate.

LEMMA 4. *If P is a projective module over a regular ring R, then any* *finitely generated submodule of P is a direct summand of P.*

PROOF. We may assume that P is free on a finite number (say n) of basis elements. The proof is most briefly given by appealing to the regularity of R_n, the n by n matrix ring over R [2]. Let S be the given submodule of P, and let I be the set of all matrices in R_n whose rows all belong to S. Then I is a finitely generated left ideal in R_n, and so has a complementary left ideal J. If we let T denote the submodule of P consisting of all rows that appear in J, we have $P = S \oplus T$.

UNIVERSITY OF CHICAGO AND PRINCETON UNIVERSITY

BIBLIOGRAPHY

1. R. Baer, *Abelian groups without elements of finite order*, Duke Math. J. 3(1937), 68–122.
2. B. Brown and N. H. McCoy, *The maximal regular ideal of a ring*, Proc. Amer. Math. Soc. 1(1950), 165–171.
3. H. Cartan and S. Eilenberg, Homological Algebra, Princeton, 1956.
4. L. Kulikov, *Direct decompositions of groups*, Ukrain. Mat. Z. 4(1952), 230–275 and 347–372; Amer. Math. Soc. Translations, Series 2, vol. 2, 23–87.

Afterthought

I can only repeat from the opening paragraph of the paper how it came about that I hoped that projective modules might be direct sums of countably generated modules and found that it was true. In retrospect I am amazed that it was not done long ago.

What did I mean in saying that the main theorem is really a piece of "universal algebra"? Certainly I did not mean it in the technical sense, for universal algebra is a field I have yet to master.

On reviewing the subject years later, I did recast the main theorem as a piece of lattice theory. But I find this formulation fussy and boring, and so I will not subject it to public scrutiny. Let me just give a sample of the territory covered: The main theorem is valid for nonabelian groups. That is, if a group G is a direct product of countable groups, so is any direct factor of G.

This exploration, however, called my attention to another context where a theorem of this kind was waiting to be picked up, a context which I find interesting. Let V be a vector space carrying a symmetric inner product $(,)$. A basis x_i for V is diagonal if $(x_i, x_j) = 0$ for $i \neq j$. If V has countable dimension, and the characteristic is not 2, a diagonal basis always exists, but in the uncountable case a diagonal basis need not exist.

THEOREM. *If V has a diagonal basis, W is an orthogonal direct summand of V, and the characteristic is not 2, then W has a diagonal basis.*

My search of the literature did not find this theorem. The proof is a slight variant of the proof of Theorem 1 in the paper. This theorem does not fit a lattice-theoretic framework; the relevant partially ordered set is the set of all orthogonal direct summands of V and in general is not a lattice. I have not tried for a still-higher level of generality by postulating a partially ordered set with appropriate axioms; presumably the result would be even more fussy and boring.

There is one specific update: On page 376 I asked whether commutativity could be dropped in Theorem 3. The answer is yes, and this was promptly proved by Albrecht [2]. Papers [1], [3], and [4] develop the theory further.

References

1. Takeo Akasaki, Projective R-modules with chain conditions on R/J, *Proc. Japan Acad.* 46(1970), 94–97.
2. Felix Albrecht, On projective modules over semi-hereditary rings, *Proc. Amer. Math. Soc.* 12(1961), 638–39.
3. Hyman Bass, Projective modules over free groups are free, *J. of Alg.* 1(1964), 367–73.
4. István Beck, Projective and free modules, *Math. Zeit.* 129(1972), 231–34.

THE HOMOLOGICAL DIMENSION
OF A QUOTIENT FIELD

IRVING KAPLANSKY

In memory of Professor TADASI NAKAYAMA

Let R be an integral domain, Q its quotient field. For any R-module A we write $d(A)$ for its homological dimension (minimal length of a projective resolution), or $d_R(A)$ if it is necessary to call attention to R.

The number $d(Q)$ is pertinent to several questions concerning R-modules [3], [4]. We summarize what is known about it. It is an easy exercise that $d(Q) \leqq 1$ if Q is a countably generated R-module. Also [3, Lemma 3.2], $d(Q) = 1$ if R is Noetherian of Krull dimension one and this result is generalized in [4]. In 1957 I noted a result in the converse direction (mentioned on page 386 of [3]): if R is a valuation ring and $d(Q) = 1$, then Q is countably generated.

In the present note I prove a more general converse: if $d(Q) = 1$ and R is quasi-local (i.e. has exactly one maximal ideal) then Q is a countably generated R-module. The basic device is similar to the artifice used in [1] and is simpler than my 1957 treatment of the valuation ring case.

We begin the discussion by letting R be any integral domain, Q its quotient field. Let S denote the set of all elements a^{-1}, a ranging over the elements of R that are not zero and not units. Of course S spans Q as an R-module, and we choose to resolve Q extravagantly by mapping a free module F onto Q, F having a basis $\{u_a\}$ indexed by the elements of R not zero or units, the mapping $f : F \to Q$ being given by $u_a \to a^{-1}$. Let the kernel of f be G. We assume that G is free (this will be the case if R is quasi-local and $d(Q) \leqq 1$, since all projectives over R are then free [1]). Let $\{v_a\}$ be a basis of G.

Let S_0 be a countable semigroup in S (i.e. a countable subset closed under multiplication). We describe a procedure that will be iterated ad infinitum. Let F_0 be the submodule of F spanned by the basis elements corresponding to S_0. Let $G_0 = G \cap F_0$ (i.e. the kernel of f when restricted to F_0). It is easily

Received May 4, 1965.

139

seen that G_0 is countably generated. (One way to do this: number the elements of S_0, let b_n be the product of the first n, and let u_n be the basis element with $f(u_n) = b_n$; then the elements $u_n - (b_n/b_{n+1})u_{n+1}$ span G_0). Pick a countable generation of G_0, write each generator as a linear combination of v's, write each resulting v as a linear combination of u's, and assemble the elements $f(u)$ for these u's. Together with S_0, these $f(u)'s$ will generate a new countable semi-group S_1.

Iteration of the above procedure yields an increasing sequence of countable semigroups $S_0, S_1, S_2, \ldots$. Let S^* denote their union. Then S^* is again a countable sub-semi-group of S. When we (as done above for S_0) write F^* for the submodule of F spanned by the basis elements corresponding to S^*, and $G^* = F^* \cap G$, we find that G^* is a free direct summand of G, being spanned by the totality of v's that arose in the construction.

Let A be the R-submodule of Q spanned by S^*. Let B be any R-module lying between A and Q that is spanned by reciprocals of elements of R. Denote by C the submodule of F spanned by the elements u_a corresponding to all elements a^{-1} lying in B. Let $D = G \cap C$. The exact sequence

$$0 \to D \to C \to B \to 0$$

furnishes a short free resolution of B. In the application shortly to be made, the module B will be non-projective; hence we will have

(1) $$d(D) = d(B) - 1.$$

Next we note the induced free resolution of B/A:

$$0 \to D/G^* \to C/F^* \to B/A \to 0.$$

In the forthcoming application it will also be the case that B/A is non-projective. But even if B/A is projective we always have the inequality

(2) $$d(D/G^*) \geq d(B/A) - 1.$$

Since G^* is a direct summand of G, it is also a direct summand of the intermediate D. Thus D is the direct sum of D/G^* and the free module G^*. Hence

(3) $$d(D) = d(D/G^*).$$

On putting together (1), (2), and (3) we find

$$(4) \qquad\qquad d(B/A) \leq d(B).$$

The analysis up to this point may be useful in future generalizations, but we turn specifically now to the proof of our main result.

THEOREM 1. *Let R be a quasi-local integral domain with quotient field Q, and assume $d(Q) \leq 1$. Then Q is a countably generated R-module.*

Proof. We construct S^* and A as above. (An intial S_0 is needed; it can be taken to consist of the powers of any element of S). We shall prove $A = Q$ by deriving a contradiction from the contrary assumption. If $A \neq Q$ we pick $q \in R$ (not 0 or a unit) such that q^{-1} is not in A. Our choice for B is $q^{-1}A$. We have that B properly contains A, and also that B is isomorphic to A. The module A is spanned by a sequence of elements each of which is a proper multiple of the succeeding one. From this we readily deduce $d(A) = 1$, and so we likewise have $d(B) = 1$. In particular, B is not projective; the discussion above is thus applicable and we have that (4) holds. Hence $d(B/A) \leq 1$. But we now argue that $d(B/A) = 2$.

The argument is based on two lemmas on homological dimension which are perhaps most easily quoted from [2]. Let T denote the ring $R/(q)$. We note that $B/A = B/qB$. Then [2, Th. 1.7] $d_T(B/A) \leq d_R(B) = 1$. The possibility that $d_T(B/A) = 0$ is readily excluded. For suppose that B/A is free as a T-module. We have a sequence u_i of generators of B/A with the property $u_i = \lambda_i u_{i+1}$, where λ_i is a non-unit in T. For large enough i it must be the case that u_i has (in its expression in terms of a T-basis of B/A) a unit for one of its coefficients. But then the equation $u_i = \lambda_i u_{i+1}$ is impossible at that coordinate.

Hence $d_T(B/A) = 1$. The hypotheses of [2, Th. 1.3] are fulfilled and we deduce $d(B/A) = 2$. With this contradiction the proof of Theorem 1 is complete.

Theorem 1 cannot be extended automatically to the global case, as is shown by the case of Noetherian domains of Krull dimension one. Perhaps in some sense this is the only exception. We can at any rate handle the case of polynomial rings over a field.

THEOREM 2. *Let K be an uncountable field and R a polynomial ring over K in n indeterminates, $n \geq 2$. Let Q be the quotient field of R. Then $d(Q) \geq 2$. If $n = 2$, $d(Q) = 2$.*

Proof. Let M denote the maximal ideal of R at the origin. We have that Q is also the quotient field of the localization R_M, and that $Q = Q_M$. Thus $dR_M(Q) \leq d_R(Q)$. Now R_M is a unique factorization domain which has uncountably many primes (if x and y are two of the indeterminates and α ranges over K, the elements $x + \alpha y$ are distinct primes in R_M). This means that the quotient field of R_M is not a countably generated R_M-module. By Theorem 1, $dR_M(Q) \geq 2$, and so $d_R(Q) \geq 2$. The final statement of the theorem is evident since for $n = 2$ the global dimension of R is 2.

I am indebted to H. Bass for a spirited discussion that resulted in several significant improvements.

BIBLIOGRAPHY

[1] I. Kaplansky, *Projective modules*, Annals of Math. **68** (1958), 372-7.
[2] I. Kaplansky, *Homological dimension of rings and modules*, Univ. of Chicago mimeographed notes, 1959.
[3] E. Matlis, *Divisible modules*, Proc. Amer Math. Soc. **11** (1960), 385-391.
[4] E. Matlis, *Cotorsion modules*, Memoirs Amer. Math. Soc. no. **49**, 1964.

University of Chicago

Afterthought

I first heard about projective dimension in a talk given by Sammy Eilenberg at the University of Chicago. The talk preceded by several years the publication of [1].

I was instantly fascinated and began experimenting to see for which modules I could compute this new dimension. Early in the game I looked at the quotient field Q of an integral domain R and noticed that the projective dimension of Q is 1 if Q is a countably generated R-module (and $Q \neq R$). The proof is evident as soon as one observes that Q can be generated by a sequence a_i^{-1} with each a_i dividing a_{i+1}.

It has happened to me several times and I always have the same reaction: When a theorem is seen to be true in the countable case and looks mysterious otherwise, I am aroused. In the present case my thoughts returned to this problem repeatedly. As the introduction to the paper notes, at an early stage the case where R is a valuation domain yielded (by a messy transfinite induction) the following result: Countable generation of Q was necessary as well as sufficient. It was gratifying later to see a method that was simpler and furthermore extended the result to any quasi-local domain.

Ross Hamsher proved a splendid generalization in [2, Cor. 2.4]: If $d(Q) = 1$ then any countable subset of Q/R lies in a countably generated module direct summand of Q/R. This is a generalization since Q/R is an indecomposable module when R is quasi-local [4, Cor. 4.2, reproved in 2, Lemma 3.1]. In addition, he was able to complete a circle of ideas about divisible modules initiated by Eben Matlis in [3].

Barbara Osofsky, in a series of papers that began with [5], investigated the area in depth, covering the complications brought on by larger cardinal numbers, commutative rings that are not quasi-local, and noncommutativity. The paper [6] is an excellent report on the state of the art as of 1974. Papers [7] and [8] present further developments.

References

1. Henri Cartan and Samuel Eilenberg, *Homological Algebra*, Princeton Univ. Press, Princeton, NJ, 1956.
2. Ross Hamsher, On the structure of a one-dimensional quotient field, *J. of Algebra* 19(1971), 416–25.
3. Eben Matlis, Divisible modules, *Proc. Amer. Math. Soc.* 11(1960), 385–91.
4. _____, Decomposable modules, *Trans. Amer. Math. Soc.* 125(1966), 147–79.
5. Barbara L. Osofsky, Global dimension of valuation rings, *Trans. Amer. Math. Soc.* 127(1967), 136–49.
6. _____, The subscript of $\aleph_n$, projective dimension and the vanishing of $\varprojlim^{(n)}$, *Bull. Amer. Math. Soc.* 80(1974), 8–26.
7. _____, Projective dimension of "nice" directed unions, *J. of Pure and Applied Alg.*, 13(1978), 179–219.
8. _____, Remarks on the projective dimension of $\aleph$-unions, pp. 223–34 in *Ring Theory*, Waterloo 1978, Springer Lecture Notes 734, 1979.

STUDIA MATHEMATICA, T. XXXI. (1968)

Composition of binary quadratic forms*

by

IRVING KAPLANSKY (Chicago, Ill.)

1. Introduction. Gauss's complete discussion of the composition of binary quadratic forms over the integers ([6], sections 235 - 244 and several later sections) was a tour de force that makes remarkable reading to this day.

Several of the great mathematicians of the nineteenth and early twentieth centuries took up the theme and gave fresh accounts of the work. This material takes up twenty condensed pages in Dickson's history ([3], p. 60 - 79).

The idea of giving still another account of this venerable subject arose when I attempted to extend the theory to Bézout domains (integral domains where every finitely generated ideal is principal). Now the modern view of composition is that it is really just multiplication of suitable modules. (This idea is attributed by Dickson to Dedekind, quoting the eleventh supplement in [5]. A recent exposition is [1], p. 212 - 5.) But when one proceeds to a detailed execution, there are difficulties. The correspondence between quadratic forms and modules needs touching up. There is some trouble disentangling a module from its conjugate, overcome by "orienting" the module; there is also a need to use "strict" equivalence of modules, meaning multiplication by elements of *positive* norm. Both of these points seem to require an ordered integral domain, and on closer inspection one sees further obstacles if the base ring has units other than ± 1.

I might have concluded that ordering was indispensable for composition, had it not been for the existence of still another method, the technique of "united forms", also attributed by Dickson to Dedekind (tenth supplement in [5]; as late as 1929 Dickson [4], Ch. IX, thought this to be the best method to put in his book). It is a fact that this discussion is valid verbatim for any principal ideal domain of characteristic $\neq 2$. But I could not get it to work for Bézout domains (the difficulty comes

---

* Work on this paper was supported in part by the National Science Foundation.

up in the preliminary lemma asserting that forms can represent elements prime to any given element).

In due course a workable idea presented itself: the technique of *pairs* consisting of a module and an element. The use of pairs shrinks the need for orientation down to the picking of a square root of the discriminant; it builds strict equivalence harmlessly into the definitions; and it accommodates units other than ± 1. Even over the ring of integers the present discussion may have expository merit.

2. Modules. Let K be a field, L a separable quadratic extension of K. We shall in general use early letters of the alphabet for elements of K, late letters for L.

We write $*$ for the automorphism of L over K, T for the trace, N for the norm. N equips L with the structure of a quadratic form over K. The mapping $*$ preserves N, and multiplication by an element x of L multiplies all norms by Nx. We shall need the well-known converse

LEMMA 1. *Let f be a one-one linear transformation of L onto itself which multiplies all norms by a fixed factor. Then f is multiplication by a non-zero element of L, or such a multiplication followed by $*$.*

Let R be an integral domain with quotient field K. We study R-submodules of L. The extra structure on L endows these modules with additional structure. For instance, if A is an R-submodule of L we write A^* for the set of all x^* with $x \in A$; A^* is again an R-module. The elements Nx, x ranging over A, generate a (possibly fractional) ideal in R which we call NA, the norm of A. When A and B are R-modules, so is their product AB (this meaning as usual the set of sums of terms xy, $x \in A$, $y \in B$).

When A is free of dimension 2, we define DA, the *discriminant* of A as follows: take a basis x, y and set $DA = (xy^* - x^*y)^2$. Note that $DA \in K$. (More accurately, DA should be called the *discriminant relative to the chosen basis*; a change of basis will multiply DA by the square of a unit in R.)

The expression $xy^* - x^*y$ will also play a role. We note that if $u = ax + by$, $v = cx + dy$, then $uv^* - u^*v = (ad - bc)(xy^* - x^*y)$.

We finally note a natural equivalence relation: A and B are *equivalent* f $B = xA$ with x a non-zero element of L.

3. Pairs. A pair $[A, a]$ consists of an R-submodule A of L and a non-zero element a in K. We extend to pairs the various concepts introduced in Section 2:

$$[A, a]^* = [A^*, a], \quad N[A, a] = NA/a,$$

$$[A, a][B, b] = [AB, ab], \quad D[A, a] = DA/a^2,$$

the last being defined when A is free 2-dimensional.

Equivalence of pairs is defined as follows: $[A, a] \sim [B, b]$ if there exists a non-zero element x in L with $B = xA$, $b = (Nx)a$. It is an easy exercise to see that conjugation, norm, product, and discriminant are all well defined on equivalence classes of pairs.

4. Quadratic forms. A "concrete" binary quadratic form over a commutative ring with unit R is an expression $ax^2 + bxy + cy^2$, $a, b, c \in R$. Abstractly, the structure is that of a quadratic form on a free 2-dimensional R-module together with a distinguished basis. When the basis is changed, we pass to an equivalent form; if the change of basis has determinant 1, we speak of *proper* equivalence.

The discriminant is $b^2 - 4ac$. Under equivalence it gets multiplied by the square of a unit, and under proper equivalence it is invariant.

In the setup of Section 2, take a free module A. The norm puts a quadratic form on A. When we take a basis of A, we get a concrete form (with coefficients in K), and it is easily checked that the two discriminants introduced coincide.

With a pair $[A, a]$ we associate the quadratic form on A given by the norm divided by a. Again the two discriminants coincide.

5. The correspondence. In this section we have to assume characteristic $\neq 2$.

With an equivalence class of pairs we wish to associate a proper equivalence class of binary quadratic forms. Our aim does not extend beyond doing this for a fixed discriminant, say Δ (on pairs this is meaningful only up to the square of a unit, of course, but for the concrete forms we mean discriminant exactly Δ).

Δ is an element of K having a square root in L. Arbitrarily fix a square root δ. Let a pair $[A, a]$ of discriminant Δ be given. We say that the basis x, y of A is *admissible* if $(xy^* - x^*y)/a = \delta$. Admissible bases exist: with any choice of basis x, y we have $(xy^* - x^*y)^2/a^2 = u^2\Delta$, u a unit in R, so that $(xy^* - x^*y)/a = \pm u\delta$. We need only replace x by $\pm u^{-1}x$.

On A relative to an admissible basis we take the form N/a; the result is a concrete quadratic form f whose proper equivalence class we take as the image of the equivalence class of $[A, a]$.

Suppose we pass to a different admissible basis of A. Then since the change of basis must have determinant 1, the proper equivalence class of f is unaffected.

Let the pair $[B, b]$ be equivalent to $[A, a]$ *via* the element z, so that $B = zA$, $b = (Nz)a$. If the basis x, y is admissible for A, we see at once that zx, zy is admissible for B. The concrete form thus obtained for $[B, b]$ is identical with the one for $[A, a]$.

We have now shown that we have a well defined map from equivalence classes of pairs to proper equivalence classes of forms. We proceed to show that it is one-to-one and onto.

As regards onto, we explicitly exhibit the inverse image. For $f = ax^2 + bxy + cy^2$ of discriminant Δ we invent the pair $[A, a]$, where A is the module spanned by a and $(b - \delta)/2$. These elements are in fact an admissible basis for A and we find the image of $[A, a]$ to be f.

Suppose finally that $[A, a]$ and $[B, b]$ both have discriminant Δ and lead to properly equivalent forms. Now this means that after a change of basis of determinant 1, say on the first form, the two forms become identical. We can suppose that this change of admissible basis has already been done for A. We thus have admissible bases, say x, y for A and u, v for B, giving rise to identical concrete forms. This says that the mapping (say F) from A to B given by sending x into u and y into v multiplies norms by b/a. We can extend F to a mapping of L into L and then apply Lemma 1 to conclude that F is either multiplication by an element z (necessarily of norm b/a) or such a multiplication followed by $*$. We can check which it is by looking at determinants. Multiplication by z has determinant Nz; $*$ has determinant -1; since $(xy^* - x^*y)/a = (uv^* - v^*u)/b$, the determinant of F is b/a (see Section 2). Hence $*$ does not appear, and we have proved $[A, a]$ and $[B, b]$ to be equivalent, as required. We summarize:

THEOREM 1. *Let K be a field of characteristic $\neq 2$, L a quadratic extension of K. Let R be an integral domain with quotient field K. Fix a discriminant Δ and a square root of Δ. For a pair $[A, a]$ of discriminant Δ, A a free 2-dimensional R-submodule of L, pick an admissible basis as above, thus getting a binary quadratic form. This implements a one-to-one correspondence between all equivalence classes of pairs with discriminant Δ and all proper equivalence classes of binary quadratic forms with discriminant Δ.*

6. Composition. Let us suppose that to the concrete forms f, g of discriminant Δ we have associated the pairs $[A, a]$ and $[B, b]$. The obvious way to get a product for f and g is to look to the product pair $[AB, ab]$. But two difficulties arise. For a general integral domain R, AB need not be a free module. This difficulty disappears if R is a Bézout domain, so we assume this henceforth. Secondly, $[AB, ab]$ need not have discriminant Δ, and we have no procedure for meshing different discriminants. We shall not give the exact conditions for $[AB, ab]$ again to have discriminant Δ, but pass at once to the best behaved case: primitive forms. We say that a pair is *primitive* if its norm is R; a concrete form is *primitive* if its coefficients lie in R and generate R. One easily sees that the two notions correspond when we pass from pairs to forms as above.

Then the crucial fact is that *the primitive pairs of a fixed discriminant form a group under multiplication.* This is of course well known and goes back to Gauss. For the reader's convenience we state the relevant facts in a theorem, and sketch the proof.

One definition is needed: in our context an *order* is a free 2-dimensional module which is a ring containing 1.

THEOREM 2. *Let R be a Bézout domain with quotient field K, L a separable quadratic extension of K. Then*

(a) *two orders are identical if and only if they have the same discriminant (up to the square of a unit in R),*

(b) *a (free 2-dimensional) module A is an invertible ideal over the unique order P having the same discriminant as the pair $[A, NA]$, and $AA^* = N(A)P$,*

(c) *for any modules A and B, $N(AB) = N(A)\,N(B)$.*

Sketch of proof. (a) An order has a basis 1, r with r integral over R. Its discriminant is $(r-r^*)^2$. Given a second order with basis 1, t, suppose its discriminant $(t-t^*)^2 = u^2(r-r^*)^2$, u a unit in R. Then $t-t^* = \pm u(r-r^*)$, $t \pm ur$ is invariant under $*$, hence lies in K, hence in R (any Bézout domain is integrally closed). So the two orders coincide.

(b) We perform the brief computation of [2], Prop. 1.4.1. We pick a basis for A of the form a, z with $a \in K$, $Tz = b$, $Nz = c$. Then $NA = (a^2, ab, c) = (e)$, say, and $DA = a^2(b^2-4c)$. Let $t = az/e$. We find that the module P spanned by 1 and t is an order whose discriminant $a^2(b^2-4c)/e^2$ coincides with the discriminant of the pair $[A, e]$. We find $PA = A$ and $AA^* = eP$, showing that A is an invertible ideal over P. A cannot be an invertible ideal over a different order because quite generally an object cannot be an invertible ideal over two different integral domains.

(c) We have $AA^* = eP$, $BB^* = fQ$, where $(f) = NB$, and Q is the order attached to B. One easily sees that PQ is again an order. Then the equation $(AB)(AB)^* = ef\,PQ$ identifies PQ as the order attached to AB, showing that $(ef) = N(AB)$.

Consider now the primitive pairs with a given discriminant $\varDelta$, and let P be the order with discriminant $\varDelta$. It is immediate from Theorem 2 that these pairs form a group, with $[P, 1]$ as the unit. The pairs equivalent to $[P, 1]$ form a subgroup, and so the equivalence classes of primitive pairs also form a group, which we call the *extended class group* of P, say $H(P)$.

By the class group $G(P)$ of P we mean as usual the invertible ideals of P modulo principal ideals. The map $[A, a] \to A$ induces a homomorphism of $H(P)$ onto $G(P)$, the kernel being isomorphic to units of R modulo norms of units of P.

We summarize the situation. Let R be a Bézout domain of characteristic $\neq 2$, and let the setup be as above. Fix a discriminant Δ, a square root δ, and let P be the order with discriminant Δ. The equivalence classes of primitive pairs of discriminant Δ are in one-one correspondence with the proper equivalence classes of binary quadratic forms of discriminant Δ. On the former we have the group structure given by the extended class group $H(P)$. We transfer the group structure to the forms and we have defined composition.

What happens if we replace δ by the other square root $-\delta$? The correspondence changes (in a harmless way — each pair is replaced by its conjugate). However the group structure on the forms is unchanged; it is entirely intrinsic.

7. Connection with united forms. We verify that Dedekind's method of united forms, whenever applicable, gives the same composition as that obtained above.

In brief, the setup is this: we are given a, b, c, d in R with the first three generating R. We wish to see that $ax^2+bxy+cdy^2$ and $cx^2+bxy+ady^2$ compose to yield $acx^2+bxy+dy^2$. All three forms have the discriminant $\Delta = b^2-4acd$. Pick a square root δ of Δ. Then suitable corresponding pairs are $[A, a]$, $[B, c]$ and $[C, ac]$ where A, B, C are spanned by a, c, ac respectively and $z = (b-\delta)/2$. We have $z^2-bz+acd = 0$, so AB is spanned by $ac, az, cz, bz - acd$. The term acd can be deleted since it is a multiple of ac. The terms az, bz, cz combine to z. Hence $AB = C$ and the pair $[C, ac]$ is the product of the pairs $[A, a]$ and $[B, c]$.

8. The ordered case. Let R be an ordered integral domain. Suppose as in the discussion above that a fixed square root δ has been picked for one discriminant Δ. The other discriminants that are pertinent (i.e. that go with forms "embeddable" in our fixed field L) have the form $k^2\Delta$, k non-zero in K. For any such we have a natural choice for a square root: $k\delta$ with $k > 0$.

What we can get out of this is best described by going backwards from forms to pairs: we get a coherently defined map on all proper equivalence classes of binary quadratic forms to all equivalence classes of pairs. But we are not yet ready for composition, for the mapping is not necessarily one-one. Indeed, it is one-one if and only if ± 1 are the only units in R.

Suppose finally that R is an ordered integral domain, Bézout, and that its only units are ± 1. Then we get composition defined on all binary quadratic forms with discriminants having ratio a square, just as Gauss did for the ring of integers. The composition is quite intrinsic, at least granted the ordering of R. If R has a unique ordering, it is entirely intrinsic.

We briefly discuss other aspects of the ordered case. It is natural to distinguish two cases, according to the sign of Δ.

(1) $\Delta < 0$. All norms are positive. The extended class group divides into two cosets according to the sign of a in the pair $[A, a]$. Nothing essential is lost by insisting that a be positive, i.e. discarding the negative definite forms.

(2) $\Delta > 0$. Since there exists elements with negative norm, any pair is equivalent to one with a positive, and this normalization may be made if one prefers.

Things become simpler still when the only units are ± 1. If $\Delta < 0$, the positive part of the extended class group coincides with the class group; the pairs are superfluous. If $\Delta > 0$ and -1 is the norm of a unit in P, the pairs are again superfluous. If $\Delta > 0$ and -1 is not the norm of a unit in P, the use of pairs amounts to the same thing as to the customary notion of strict equivalence: $B = zA$ with $Nz > 0$. Even for the ring of integers, the pairs do have the merit of treating the various cases in a unified way.

9. Final remarks. (1) All the results in this paper carry over to the case where L is the direct sum of two copies of K and the involution is the mapping interchanging the two summands (the corresponding binary quadratic forms have discriminants which are perfect squares). It was solely for expository reasons that this case was not incorporated in the body of the paper.

(2) Over a general Bézout domain (i.e. with no ordering or with units other than ± 1) can composition be defined without the restriction to a fixed discriminant? I see no natural way to do this. Perhaps impossibility could be proved rigorously by putting the matter in a functorial setting.

(3) If one is willing to make enough arbitrary choices, a product can be defined. For instance, this was done by Smith for the ring of Gaussian integers ([7], p. 423-427 in the pagination of his collected works).

(4) For characteristic 2 it is at present not clear whether composition is definable under any reasonable conditions.

(5) There is a different point of view on the whole subject, which has certain advantages, but represents a radical departure from the Gauss tradition. Allow equivalence of binary quadratic forms to mean that the determinant of the transformation can be any unit. Modify equivalence of pairs by identifying each pair $[A, a]$ with its conjugate $[A^*, a]$. Then: there is a one-one correspondence between equivalence classes of pairs and equivalence classes of binary quadratic forms. We can proceed forthwith to define composition, with no worries about discri-

minants and no special concern for characteristic 2; the only trouble is that the "product" is in general two-valued. On primitive pairs with a fixed discriminant the structure obtained is that of an abelian group in which every element has been identified with its inverse.

References

[1] H. Cohn, *A second course in number theory*, New York 1962.

[2] E. C. Dade, O. Taussky and H. Zassenhaus, *On the theory of orders*, Math. Ann. 148 (1962), p. 31-64.

[3] L. E. Dickson, *History of the theory of numbers*, Vol. III, New York 1934.

[4] — *Introduction to the theory of numbers*, Chicago 1929.

[5] G. R. Dirichlet and R. Dedekind, *Vorlesungen über Zahlentheorie*.

[6] C. F. Gauss, *Disquisitiones arithmeticae*.

[7] H. J. S. Smith, *On complex binary quadratic forms*, Proc. Roy. Soc. 13 (1864), p. 278-298 = *Collected works*, Vol. 1, p. 418-442.

Reçu par la Rédaction le 14. 12. 1967

Afterthought

The idea of extending Gaussian composition of binary quadratic forms to wide classes of commutative rings independently occurred to several other people. I have tried to assemble a complete bibliography; it also includes other items.

There is a lot to digest here. The thought of covering every aspect in a definitive account is tempting but also daunting. I am not undertaking it here.

The technique of using pairs has not met with great success. In [7] I followed it up in the four-dimensional case, and Susanna Epp, in [4] and [5], went on to the eight-dimensional case. There is one additional instance where someone followed up the use of pairs: paper [11] by Leonard and Williams.

There is one question I feel I ought to raise in retrospect. Why didn't I try to extend Gauss' original method to commutative rings? The answer is psychological. I pictured the great masters of the nineteenth century examining the Gauss account and throwing up their hands, saying "there must be a better way." And I was convinced that the better way was to move from forms to modules and then just multiply. Now I'm not quite so sure.

References

1. Hubert S. Butts and Bill J. Dullin, Composition of binary quadratic forms over integral domains, *Acta Arithm.* 20(1972), 223–51.
2. Hubert S. Butts and Dennis Estes, Modules and binary quadratic forms over integral domains, *Lin. Alg. and Appl.* 1(1968), 153–80.
3. Hubert S. Butts and Gordon Pall, Modules and binary quadratic forms, *Acta Arithm.* 15(1968), 23–44.
4. Susanna Samuels Epp, Submodules of Cayley algebras, *J. of Alg.* 24(1973), 104–26.
5. _____, The Brandt condition in Cayley algebras, *J. of Algebra* 38(1976), 213–24.
6. Burton W. Jones, The composition of quadratic binary forms, *Amer. Math. Monthly* 56(1949), 380–91.
7. Irving Kaplansky, Submodules of quaternion algebras, *Proc. London Math. Soc.* 19(1969), 219–32.
8. Martin Kneser, Komposition binärer quadratischer Formen, *Abh. Braunschweig. Wiss. Ges.* 33(1982), 41–42; *Math. Rev.* 84f, no. 10028.
9. _____, Composition of binary quadratic forms, *J. of Number Theory* 15(1982), 406–13.
10. _____, Komposition quadratischer Formen, pp. 161–73 in *Algebra-Tagung Halle*, Halle, 1987; *Math. Rev.* 89c, no. 11060.
11. Philip A. Leonard and Kenneth S. Williams, Representability of binary quadratic forms over a Bézout domain, *Duke Math. J.* 40(1973), 533–39.
12. S. Lubelski, Unpublished results on number theory II, Composition theory of binary quadratic forms, *Acta Arithm.* 7(1961), 9–17.
13. Erwin Mrowka, Ambige Klassen und Kompositionen binärer Quadratischer Formen, *Wiss. Z. Hochsch. Verkehrwes. Dresden* 5(1957), no. 2, 293–300; *Math. Rev.* 23A, no. 1606.

14. Gordon Pall, Composition of binary quadratic forms, *Bull. Amer. Math. Soc.* 54(1948), 1171–75.
15. ——————, Some aspects of Gaussian composition, *Acta Arithm.* 24(1973), 401–9.
16. ——————, Pythagorean triples, Gaussian composition, and spinor genera, *Adv. in Math.* 19(1976), 1–5.
17. Bart Rice, Quaternions and binary quadratic forms, *Proc. Amer. Math. Soc.* 27(1971), 1–7.
18. Olga Taussky, Composition of binary integral quadratic forms via integral 2 by 2 matrices and composition of matrix classes, *Lin. and Mult. Algebra* 10(1981), 309–18.
19. Jacob Towber, Composition of oriented binary quadratic form-classes over commutative rings, *Adv. in Math.* 36(1980), 1–107.
20. André Weil, Gauss et la composition des formes quadratiques binaires, pp. 895–912 in *Aspects of Mathematics and Its Applications*, North-Holland, Amsterdam, 1986.

Reprinted from JOURNAL OF ALGEBRA   Vol. 20, No. 1, January 1972
All Rights Reserved by Academic Press, New York and London   *Printed in Belgium*

Adjacent Prime Ideals

IRVING KAPLANSKY

Department of Mathematics, University of Chicago, Chicago, Illinois 60637

Received January 15, 1971

TO PROFESSOR RICHARD BRAUER, TO COMMEMORATE HIS SEVENTIETH BIRTHDAY,
FEBRUARY 10, 1971

In an important paper [3] in which Krull proved his basic theorems concerning the behavior of prime ideals under integral extensions, he raised a further question [3, p. 755] which seems to have remained unanswered.

Let R and T be integral domains with T integral over R and R integrally closed. (For the problem about to be stated it makes no difference to assume that T is also integrally closed.) Let Q, Q_0 be prime ideals in T with Q properly containing Q_0 and such that no prime ideal lies properly between them. Set $P = Q \cap R$, $P_0 = Q_0 \cap R$. Is it necessarily the case that there is no prime ideal properly between P and P_0? In other words, do adjacent prime ideals contract to adjacent ones if the big domain is integral over the little one and the little one is integrally closed?

The answer is "no". I hasten to add that in my example neither ring is Noetherian, and there are a few remarks about this at the end of the note.

Before giving a specific example, I present two propositions which set the stage for families of examples.

PROPOSITION 1. *Let B be an integrally closed integral domain possessing exactly two nonzero prime ideals M and N, both of them maximal. Assume that B_M is not a valuation domain. Let $T = B[x]$ be the polynomial ring in one variable over B. Then there exist prime ideals M_1, N_1 in T such that M_1 is contained in MT but not NT, N_1 contains NT but not MT, $M_1 \subset N_1$ (all inclusions proper), and M_1 and N_1 are adjacent (i.e., there is no prime ideal of T properly between M_1 and N_1).*

Proof. Pick u in the quotient field K of B so that neither u nor u^{-1} lies in B_M. Pick v in N but not in M. Since B_N is one dimensional, we have that $uv^n = w$ is a non-unit in B_N for sufficiently large n. Since v is a unit in B_M, we have that neither w or w^{-1} lies in B_M. We define M_1 to be the kernel of the homomorphism of $B[x] = T$ into K that sends x into w. After proving that $M_1 + NT \neq T$ we take N_1 to be any prime ideal of T containing $M_1 + NT$. We have a number of things to verify.

94

(1) $0 \subset M_1 \subset MT$ (proper inclusions). If $w = a/b$ with $a, b \in B$ then $bx - a \in M_1$ shows that $M_1 \neq 0$. Any nonzero element of M belongs to MT but not to M_1. Finally, suppose that $M_1 \not\subset MT$. Then w satisfies an equation with coefficients in B (hence in B_M) with at least one coefficient a unit in B_M. Since B_M is quasilocal and integrally closed, it follows from Ref. [7, p. 19] (and also Ref. [2, Theorem 67]) that w or $w^{-1} \in B_M$, a contradiction.

(2) $M_1 \not\subset NT$. Since $w \in B_N$ we have $w = c/s$, with $c, s \in B$ and $s \notin N$. Then $sx - c \in M_1$ shows that $M_1 \not\subset NT$.

(3) $M_1 + NT \neq T$. If $1 \in M_1 + NT$, then w satisfies a polynomial equation with coefficients in B and constant term not in N. Since B_N is integrally closed, this implies that $w^{-1} \in B_N$, a contradiction since w is a non-unit in B_N.

(4) With N_1 selected to be any prime ideal in T containing $M_1 + NT$, we have to prove $N_1 \not\supset MT$. But if $N_1 \supset MT$, then $N_1 \supset M + N = B$, a contradiction.

(5) M_1, N_1 are adjacent. Suppose that the prime ideal Q of T lies properly between M_1 and N_1. The chain $Q \supset M_1 \supset 0$ of three prime ideals cannot all have the same contraction in B [2, Theorem 37]. (I am quoting from Ref. [2] several results which can be found in all standard references.) Hence $Q \cap B$ must be M or N. From $Q \cap B = M$ we get the contradiction $N_1 \supset Q \supset MT$. Hence $Q \cap B = N$. If $Q \neq NT$, we get the chain $NT \subset Q \subset N_1$ of prime ideals all contracting to N in B. Hence $Q = NT$, and $M_1 \subset Q = NT$ is a contradiction.

PROPOSITION 2. *Let A be a one-dimensional quasilocal integrally closed domain that is not a valuation domain. Let B be a domain integral over A, and with quotient field finite-dimensional over the quotient field of A. Assume that B has exactly two prime ideals lying over the maximal ideal of A. Let x be an indeterminate over B. Then there exist in $B[x]$ adjacent prime ideals whose contractions to $A[x]$ are not adjacent.*

Proof. Write M, N for the maximal ideals of B. To apply Proposition 1 we need to know that B_M is not a valuation domain. But A is the intersection of B_M and the quotient field of A [1, Lemma 1.29 on p. 18]; so if B_M were a valuation domain, the same would be true for A.

Now we can see that the ideals M_1, N_1 furnished by Proposition 1 do what is required. Write $R = A[x]$, $T = B[x]$, and let P be the unique maximal ideal of A. Since MT and NT both contract to PR in R, we have the inclusions

$$M_1 \cap R \subset PR \subset N_1 \cap R,$$

and these inclusions are proper since T is integral over R, so that distinct comparable prime ideals in T have distinct contractions in R [2, Theorem 44]. Thus $M_1 \cap R$, $N_1 \cap R$ are not adjacent, whereas M_1, N_1 are adjacent by Proposition 1.

It remains to invent integral domains A and B that meet the requirements of Proposition 2. Let K be any field of characteristic $\neq 2$, and let L be a proper superfield of K in which K is algebraically closed (for instance, adjoin an indeterminate to K). Let C be the ring of all polynomials in a variable t over L, subject to the restriction that the constant term must be in K. The polynomials with constant term 0 form a maximal ideal J in C. We take $A = C_J$, and we let B be the result of adjoining $\sqrt{1 + t}$ to A. We leave to the reader the routine verification that the hypotheses of Proposition 2 have been fulfilled.

The standard Noetherian example of an integral extension where adjacent primes fail to contract to adjacent primes is furnished by taking R to be Nagata's celebrated counter-example to the saturated chain condition and T its integral closure [4, pp. 203–5; 7, pp. 327–9]. (By the saturated chain condition I mean the statement that all maximal chains of prime ideals between two given prime ideals have the same length.) Thus there is a connection between the problem of adjacent primes and the saturated chain condition; indeed the present note was inspired by the observation that, with A as in Proposition 2, $A[x]$ furnishes a simpler integrally closed domain violating the saturated chain condition than the one given by Mrs. Sally [6].

To pinpoint the connection, I submit the following remark. With domains $R \subset T$ and T integral over R, assume the saturated chain condition in T. Let us stick to prime ideals of finite rank. Then if rank is preserved in contracting from T to R (as it is by the going down theorem if R is integrally closed, or if R is merely integral over an integrally closed domain) we see at once that adjacent primes contract to adjacent primes.

The study of adjacent primes in the Noetherian case should probably be postponed, pending the answer to the following question. Does every integrally closed Noetherian domain satisfy the saturated chain condition? Is this true more generally for domains integral over an integrally closed Noetherian domain? In Ref. [5], Ratliff discusses a family of questions of this type, with references to earlier literature.

References

1. S. Abhyankar, "Ramification Theoretic Methods in Algebraic Geometry," Princeton Univ. Press, Princeton, N. J., 1959.
2. I. Kaplansky, "Commutative Rings," Allyn and Bacon, 1970.

3. W. KRULL, Beiträge zur Arithmetik kommutativer Integritätsbereiche. III Zum Dimensionsbegriff der Idealtheorie, *Math. Zeit.* **42** (1937), 745–766.
4. M. NAGATA, "Local Rings," Interscience, New York, 1962.
5. L. RATLIFF, Characterizations of catenary rings, preprint, 67 pp.
6. J. SALLY, Failure of the saturated chain condition in an integrally closed domain, *Notices Amer. Math. Soc.* **17** (1970), 560.
7. O. ZARISKI AND P. SAMUEL, "Commutative Algebra," Vol. II, Van Nostrand, Princeton, N. J., 1960.

Printed in Belgium by the St. Catherine Press Ltd., Tempelhof 37, Bruges

Afterthought

This note has for the most part been superseded. However, it retains historical interest as the first example answering Krull's question in the negative, and it established going between as a fourth playground for commutative ring-theorists, joining lying over, going up, and going down.

Of course, a non-Noetherian example could not be considered satisfactory. The final words in the note were prophetic. Ogoma's exhibition [1] of a noncatenary Noetherian integrally closed domain was soon followed by Ratliff's exhibition [2, Theorem 2.3] of the failure of going between in the Noetherian case.

In conclusion I note an error and rectify it. (I am indebted to Thomas Shores for pointing out the error in a letter dated May 23, 1972.) In Proposition 2 an additional hypothesis is needed: that B is integrally closed. Then, in the discussion that follows the proof of Proposition 2, one needs to add that the ring B constructed there is indeed integrally closed. This is an easy and standard bit of reasoning; however, since an error is being corrected, I will take the space to record a proposition that covers what is needed.

PROPOSITION. *Let R be an integrally closed domain containing $1/2$, K its quotient field, L the extension field of K obtained by adjoining a square root v of a unit u in R. Then the integral closure of R in L is $R + Rv$.*

Proof. Given $a + bv$ $(a, b \in K)$ integral over R, we have to prove that a and b are in R. Now $a - bv$ is also integral over R. Hence so is their sum $2a$. Since $1/2 \in R$, we have $a \in R$. Then bv is integral over R, so is its square $b^2 u$, which therefore lies in R, $b^2 \in R$ since u is a unit, and finally $b \in R$ since R is integrally closed.

References

1. T. Ogoma, Noncatenary pseudogeometric normal rings, *Jap. J. Math.* 6(1980), 147–63.
2. L. J. Ratliff, Jr., Four notes on *GB*-rings, *J. Pure Appl. Algebra* 23(1982), 197–207.

PACIFIC JOURNAL OF MATHEMATICS
Vol. 86, No. 1, 1980

SUPERALGEBRAS

Irving Kaplansky

Dedicated to Gerhard Hochschild on the occasion of his 65th birthday

1. **Introduction.** The theory of graded Lie algebras, now more widely called Lie superalgebras, underwent a very rapid development starting about 1973, inspired by the interest expressed in the subject by physicists. I was active in the field for about a year, during 1975 and 1976. Thus far I have published only the announcement [16] (jointly with Peter Freund of Chicago's Physics Department, to whom I am enormously indebted); in addition, the summary [29] is to appear.

The present mature state of the field, and the fact that Hochschild (partly in collaboration with Djoković) made several important contributions, make this an appropriate occasion to publish some further details. Although Victor Kac has brilliantly solved the main problems, there remains the possibility that the different methods I used retain some independent interest.

The large bibliography is intended to be complete on mathematical references not contained in [9]; there is also a selection of physics papers. I hope this bibliography will be useful to some readers.

This article is written so as to keep the overlap with [29] to a minimum.

2. **Invariant forms.** When I began studying Lie superalgebras I imitated [46] and selected as an initial goal the classification of those simple Lie superalgebras (over an algebraically closed field of characteristic 0) that admit a suitable invariant form.

For basic definitions and facts about Lie superalgebras, I refer to [25]. I shall just recall that if ϕ is a superrepresentation of the Lie superalgebra L then

$$(x, y) = \mathrm{STr}\,(\phi(x)\phi(y))$$

is an invariant form on L, where STr denotes the supertrace. This can be extended to "projective representation", following the model of [28, p. 66], but since the setup will shortly be axiomatic anyway, I shall not pursue the details here.

Assume now that the form ψ on L induced by ϕ is nondegenerate. Write $L = L_0 + L_1$, with L_0 and L_1 the even and odd parts of L. We have that ψ is symmetric on L_0, skew on L_1, and that L_0 and L_1 are orthogonal relative to ψ. It follows that ψ remains nondegenerate when restricted to L_0. Hence L_0 is the direct sum of a semisimple algebra and an abelian algebra.

Since the assumption that the form comes from a representation plays no further role in the investigation it is feasible to weaken the hypothesis by assuming outright that L admits an invariant form and that L_0 is semisimple $\oplus$ abelian. We assume that L is simple.

3. **Cartan decomposition.** The role of a Cartan subalgebra of L is satisfactorily played by a Cartan subalgebra H of L_0. The decomposition of L_0 relative to H is fully known, for the abelian part of L_0 creates minimal interference. So the even roots and root spaces have standard properties.

The decomposition of L_1 relative to H creates odd roots and root spaces with properties not quite so standard. Odd roots may be isotropic. Also, two-dimensional root spaces are possible; but this happens only in one algebra: the 14-dimensional projective special linear algebra of 4×4 matrices. In this algebra there moreover exist odd roots λ, μ with $(\lambda, \mu) \neq 0$ and $\lambda + \mu$, $\lambda - \mu$ both (even) roots. This is again a unique exception and will be ruled out in the axioms about to be given.

4. **Axioms for roots.** The system of roots that has arisen can now be treated axiomatically. We postulate a finite-dimensional vector space V over a field of characteristic 0. V is equipped with a nondegenerate symmetric form (,). In V a finite set Γ of non-zero vectors is given; we call the members of Γ "roots". Γ is a disjoint set-theoretic union of two subsets whose members we call "even" and "odd". There are seven axioms.

1. Γ spans V.

2. Along with any vector Γ contains its negative. A root and its negative have the same parity.

3. The even roots in Γ constitute the system of roots of an (ordinary) semisimple Lie algebra. (The form on each simple component is a scalar multiple of the Killing form, the scalar varying with the component.)

4. For any two non-orthogonal odd roots the sum or the difference is a root, but not both.

REMARK. It is probably feasible to classify the larger class of root systems that arise if the phrase "but not both" is deleted; I have not tried, since no application is in sight.

In the final three axioms α is an even root and λ is an odd isotropic root.

5. $2(\alpha, \lambda)/(\alpha, \alpha) = 0,\ \pm 1,\ \text{or}\ \pm 2.$

6. If $2(\alpha, \lambda)/(\alpha, \alpha) = -1$ then $\lambda + \alpha$ is a root.

7. If $2(\alpha, \lambda)/(\alpha, \alpha) = -2$ then $\lambda + \alpha$ and $\lambda + 2\alpha$ are roots. $\lambda + \alpha$ is odd.

The roots in the Lie superalgebras of §2 (i.e., with an invariant form, and an even part which is semisimple $\oplus$ abelian) satisfy these axioms, with the solitary 14-dimensional exception mentioned in §3.

5. **The structure theorem.** Indecomposable systems satisfying these axioms were classified in a piece of work I completed in August 1975. The result can today be stated briefly. One gets precisely the systems attached to the following simple Lie superalgebras: special linear, orthosymplectic, and the exceptional algebras of dimensions 17, 31, and 40. The proof was elementary but long.

It is a routine matter to exhibit these root systems, so two samples will suffice.

Special linear. Take an orthogonal direct sum $X \oplus Y$ where X has an orthonormal basis $e_1, \cdots, e_m$ and Y has a negative orthonormal basis $f_1, \cdots, f_n$ (this means that the f's are orthogonal and each $(f_j, f_j) = -1$). The even roots consist of all $e_i - e_r$ and $f_j - f_s$ $(i \neq r, j \neq s)$. The odd roots are the $2mn$ vectors $\pm(e_i + f_j)$.

$G(3)$, *the 31-dimensional algebra.* Let p, q, r be vectors satisfying $(p, p) = (q, q) = (r, r) = -2$, $(q, r) = (r, p) = (p, q) = 1$. Let f be a vector perpendicular to p, q, r satisfying $(f, f) = 2$. The roots are as follows (the negatives are to be inserted as well).

Even: $p, q, r, q - r, r - p, p - q, 2f.$

Odd isotropic: $f \pm p, f \pm q, f \pm r.$

Odd non-isotropic: $f.$

6. **A model of** $G(3)$. I present a model of $G(3)$ which may be useful for some purposes. Take the even part L_0 to be $G_2 \oplus A_1$ and the odd part L_1 as $C \otimes V$, where C denotes the 7-dimensional space of elements of trace 0 in a Cayley matrix algebra and V is a 2-dimensional space carrying a nonsingular alternate form $(,)$. Let G_2 act on C in the standard way and A_1 on V as linear transformations skew relative to $(,)$. It remains to define the multiplication $L_1 \times L_1 \to L_0$. This is done via two auxiliary maps ϕ and ψ.

$\phi \colon C \times C \to G_2$. This is the map which appears on page 143 of [21]:

$$\phi(c, d) = [L_c L_d] + [L_c R_d] + [R_c R_d],$$

where L and R denote left and right multiplication.

$\psi \colon V \times V \to A_1$. For v, w in V define $\psi(v, w)$ to be the linear transformation on V that sends x into $(x, v)w + (x, w)v$. The product from $L_1 \times L_1$ to L_0 is now defined by

$$(c \otimes v)(d \otimes w) = (v, w)\phi(c, d) + 4 \operatorname{tr}(c, d)\psi(v, w)$$

where tr denotes the trace on the Cayley matrix algebra, normalized so that $\operatorname{tr}(1) = 1$. One must of course verify the Jacobi identity.

7. **Jordan superalgebras.** For the basic facts on Jordan superalgebras, I refer to [27]. In my version of the theory, completed in June, 1976, I used the classical method of idempotents and Peirce decompositions, rather than Kac's Lie method. The key hurdle that had to be surmounted was to exclude the case where the even part is unit element plus radical (called the "nodal" case in the literature on nonassociative algebras). Here is the proof.

PROPOSITION. *Let $J = J_0 + J_1$ be a Jordan superalgebra over a field of characteristic 0. Let N be the radical of J_0. Assume that J has a unit element 1 and that every element of J_0 is an element of N plus a scalar multiple of 1. Then $N + NJ_1$ is an ideal in J.*

Proof. It is easy to see that $NJ_1 \cdot J_1 \subset N$ is the only nontrivial inclusion that needs verification. Thus, for $a, b \in J_1$ and $n \in N$ we need to show that $z = na \cdot b$ lies in N. Assume not. Let c be another element in J_1. We have that $R_b R_c + R_c R_b$ is a derivation of J (this is a special case of a general principle for converting algebra identities into superalgebra identities). Likewise, R_b^2 is a derivation. These derivations restrict to derivations on J_0, and by ordinary Jordan theory carry N into N (this is where characteristic 0 is used). Thus $zb = (na \cdot b)b \in NJ_1$. It follows that $b \in NJ_1$ and then that $(na \cdot c)b \in NJ_1$. Next

$$(na \cdot b)c + (na \cdot c)b \in NJ_1 .$$

Hence $zc \in NJ_1$. c is arbitrary in J_1 and so $zJ_1 \subset NJ_1$, $J_1 = NJ_1$, and $J_1 = 0$ by a Nakayama lemma argument. Everything is trivial if $J_1 = 0$. The proof is complete.

Added in proof (May 28, 1980). I missed some references, and many additional ones have now appeared. I have compiled a supplementary bibliography.

REFERENCES

1. B. L. Aneva, S. G. Mihov, and D. C. Stojanov, *On some properties of representations of conformal superalgebra*, Theor. Mat. Fiz., **31** (1977), 179–189 (Russian with

English summary; reviewed in Zbl. 366, no. 17013).

2. R. Arnowitt and P. Nath, *Spontaneous symmetry breaking of gauge supersymmetry*, Phys. Rev. Letters, **36** (1976), 1526-1529.

3. Nigel Backhouse, *Some aspects of graded Lie algebras*, pp. 249-254 in *Group Theoretical Methods in Physics*, Academic Press, New York, 1977.

4. Nigel Backhouse, *The Killing form for graded Lie algebras*, J. Math. Physics, **18** (1977), 239-244.

5. ————, *A classification of four-dimensional Lie superalgebras*, J. Math. Physics. **19** (1978), 2400-2402.

6. F. A. Berezin, *Representations of the supergroup $U(p, q)$*, Functional Anal. Appl., **10** (1976), no. 3, 70-71; translation 221-223.

7. F. A. Berezin and D. A. Leîtes, *Supermanifolds*, Doklady Akad. Nauk SSSR, **224** (1975), 505-508; translation **16** (1975), 1218-1222.

8. F. A. Berezin and V. S. Retah, *The structure of Lie superalgebras with semisimple even part*, Functional Anal. Appl., **12** (1978), no. 1, 64-65; translation 48-49.

9. L. Corwin, Y. Ne'eman, and S. Sternberg, *Graded Lie algebras in mathematics and physics* (Bose-Fermi symmetry), Reviews of Modern Physics, **47** (1975), 573-603.

10. Geoffrey Dixon, *Fermi-Bose and internal symmetries with universal Clifford algebras*, J. Math. Physics, **18** (1977), 2204-2206.

11. D. Ž. Djoković, *Classification of some 2-graded Lie algebras*, J. Pure Applied Alg.. **7** (1976), 217-230.

12. ————, *Representation theory for symplectic 2-graded Lie algebras*, J. Pure Appl. Algebra, **9** (1976-7), 25-38.

13. ————, *Isomorphism of some simple 2-graded Lie algebras*, Canad. J. Math.. **29** (1977), 289-294.

14. D. Ž. Djoković and G. Hochschild, *Semisimplicity of 2-graded Lie algebras II*, Illinois J. Math., **20** (1976), 134-143.

15. Mohamed El-Agawany and Artibano Micali, *Le théorème de Poincaré-Birkhoff-Witt pour les algèbres de Lie graduées*, R. C. Acad. Sci. Paris, **285A** (1977), 165-168.

16. Peter G. O. Freund and I. Kaplansky, *Simple supersymmetries*, J. Mathematical Phys.. **17** (1976), 228-231.

17. C. Fronsdal, *Differential geometry in Grassman algebras*, Letters in Math. Physics, **1** (1976), 165-170.

18. F. Gürsey and L. Marchildon, *The graded Lie groups $SU(2, 2/1)$ and $OSp(1/4)$*, J. Mathematical Phys., **19** (1978), 942-951.

19. J. Hietarinta, *Supersymmetry generators of arbitrary spin*, Physical Reviews D, **13** (1976), 838-850.

20. G. Hochschild, *Semisimplicity of 2-graded Lie algebras*, Illinois J. Math., **20** (1976), 107-123.

21. N. Jacobson, *Lie Algebras*, Interscience, 1962.

22. V. G. Kac, *Classification of simple Lie superalgebras*, Functional Anal. Appl., **9** (1975), 3, 91-92; translation 263-265.

23. ————, *Letter to the editor*, Functional Anal. Appl., **10** (1976), no. 2 93; translation 163.

24. ————, *A sketch of Lie superalgebra theory*, Comm. Math. Phys., **53** (1977), 31-64.

25. ————, *Lie superalgebras*, Advances in Math., **26** (1977), no. 1, 8-96.

26. ————, *Characters of typical representations of classical Lie superalgebras*, Comm. in Algebra, **5** (1977), 889-897.

27. ————, *Classification of simple Z-graded Lie superalgebras and simple Jordan superalgebras*, Comm. in Algebra, **5** (1977), 1375-1400.

28. I. Kaplansky, *Lie Algebras and Locally Compact Groups*, Chicago Lectures in Mathematics, Univ. of Chicago Press, 1971.

29. ————, *Lie and Jordan superalgebras*, to appear in Proc. of Charlottesville Conf.

on nonassociative algebras in physics held March, 1977.

30. B. G. Konopel'čenko, *Extensions of the Poincaré algebra by spinor generators*. JETP Letters, **21** (1975),612-614; translation 287-288.

31. B. Kostant, *Graded manifolds, graded Lie theory, and prequantization*, Diff. Geom. Meth. Math. Phys., Proc. Symp. Bonn 1975, Springer Lecture Notes **570** (1977), 177-306.

32. D. A. Leites, *Cohomology of Lie superalgebras*, Functional Anal. and its Appl., **9** (1975), no. 4, 75-76; translation 340-341.

33. D. J. R. Lloyd-Evans, *Geometric aspects of supergauge theory*, J. of Math. Physics **18** (1977), 1923-1927.

34. F. Mansouri, *A new class of superalgebras and local gauge groups in superspace*. J. of Math. Physics, **18** (1977), 2395-2396.

35. J. P. May, *The cohomology of restricted Lie algebras and of Hopf algebras*. Bull. Amer. Math. Soc., **71** (1965), 372-377.

36. ———, Same title, J. of Alg., **3** (1966), 123-146.

37. J. W. Milnor and J. C. Moore, *On the structure of Hopf algebras*, Ann. of Math., **81** (1965), 211-264.

38. W. Nahm, V. Rittenberg, and M. Scheunert. *The classification of graded Lie algebras*, Phys. Letters, **61B** (1976), 383-384.

39. ———, *Graded Lie algebras: Generalization of Hermitian representations*. J. of Math. Physics, **18** (1977), 146-154.

40. ———, *Irreducible representations of the osp(2,1) and spl(2,1) graded Lie algbras*, ibid., 155-162.

41. A. Pais and V. Rittenberg. *Semi-simple graded Lie algebras*, J. Math. Physics, **16** (1975), 2062-2073; erratum ibid., **17** (1976), 598.

42. N. T. Petrov and R. P. Zaikov, *Superalgebras*, C. R. Acad. Bulgare Sci., **29** (1976), 1241-1243 (reviewed in Zbl. 366, no. 17014).

43. V. S. Retah, *Massey operations in Lie superalgebras and deformations of complex-analytic algebras*, Functional Anal. Appl., **11** (1977), no. 4, 88-89; translation 319-321.

44. L. E. Ross, *Representations of graded Lie algebras*, Trans. Amer. Math. Soc., **120** (1965), 17-23.

45. M. Scheunert, W. Nahm, and V. Rittenberg, *Classification of all simple graded Lie algebras whose Lie algebra is reductive. I*, J. Math. Physics, **17** (1976), 1626-1639, II, ibid., 1640-1644.

46. G. Seligman, *On Lie algebras of prime characteristic*, Memoirs Amer. Math. Soc., no. 19, 1956.

47. S. Sternberg, *Some recent results on supersymmetry*, Diff. Geom. Math. Phys., Symp. Bonn 1975, Springer Lecture Notes, **570** (1977), 145-176.

48. S. Sternberg and J. A. Wolf, *Hermitian Lie algebras and metaplectic representations*, Trans. Amer. Math. Soc., **238** (1978), 1-43.

49. H. Tilgner, *Graded generalizations of Weyl and Clifford algebras*, J. Pure Appl. Algebra, **10** (1977), 163-168.

50. ———, *A graded generalization of Lie triples*, J. Algebra, **47** (1977), 190-196.

51. ———, *Extensions of Lie-graded algebras*, J. Math. Physics, **18** (1977), 1987-1991.

Received December 10, 1978.

UNIVERSITY OF CHICAGO
CHICAGO, IL 60637

Afterthought

The year 1975 was an interesting one in the history of Lie superalgebras. Three people independently went to work on classifying the simple ones: Gerhard Hochschild of Berkeley (later joined by Dragomir Djoković), Victor Kac of Moscow, and myself of Chicago. For both Hochschild and me the project was suggested by a physicist: Murray Gell-Mann in his case and Peter Freund in mine.

Naturally enough, each of us pursued an analytic continuation of earlier work. Hochschild believed that the only good algebras are those for which all representations are completely reducible; I was enthusiastic about classifying root systems; and Kac had a strong background from his work on infinite-dimensional Lie algebras and Lie algebras of characteristic p, so strong that he was able to do the whole job.

The story of how we became aware of each other's work is worth telling. During the summer of 1975, as a guest of UCLA, I completed the classification of the relevant root systems, as reported in this paper. I sent our preprints. One went to Shoshichi Kobayashi, who had done similar work on the nonsuper case. He promptly wrote me that he believed that his colleague Gerhard Hochschild was studying the super case. My preprint went to Hochschild and he reciprocated. In September 1975 the late Boris Weisfeiler briefly visited Chicago. I told him about what I was doing. He mentioned that Victor Kac was at work in the area and gave me Kac's home address. Out went my preprint and back came a clarifying letter. From this point on everybody pretty well knew what everybody else was doing.

As a final note I report that paper [29] in the bibliography of the paper never appeared. Little is lost. Everything in this paper is superseded, except for one question that I raised. Let A be a finite-dimensional, semisimple Jordan superalgebra over an algebraically closed field of characteristic 0. In studying the structure of A we may assume that A is indecomposable (i.e., not a direct sum). In addition to simple algebras there is just one further known example: the result of adjoining a unit element to a direct sum of copies of the three-dimensional simple algebra without unit. Are there any others?

SIAM J. MATRIX ANAL. APPL.
Vol. 11, No. 2, pp. 213–217, April 1990

ALGEBRAIC POLAR DECOMPOSITION*

IRVING KAPLANSKY†

Abstract. Choudhury and Horn made a conjecture concerning conditions for a complex matrix to admit a decomposition as a product of an orthogonal matrix and a symmetric matrix. This conjecture, in a stronger form, is confirmed.

Key words. complex orthogonal, complex symmetric, polar decomposition

AMS(MOS) subject classifications. 15A23, 15A57

1. Introduction. Let A be a complex square matrix. It is classical that A can be written as the product of a unitary matrix and a positive semidefinite Hermitian matrix; the Hermitian part is always unique and the unitary part is unique if A is invertible. In [1] Choudhury and Horn studied an algebraic variant. In this variant one seeks to write $A = QS$, with Q complex orthogonal (rather than unitary) and S complex symmetric (rather than Hermitian). There appears to be no reasonable way to restrict Q or S or both so as to make the decomposition unique, and so we forget about uniqueness. However, existence of the decomposition merits scrutiny. In fact, the decomposition is not always possible, as the matrix

$$\begin{pmatrix} 1 & i \\ 0 & 0 \end{pmatrix}$$

shows. So it is natural to impose conditions. If $A = QS$, then $A'A = SQ'.QS = S^2$ and $AA' = QS.SQ' = QS^2Q^{-1}$. Thus two necessary conditions are visible: similarity of $A'A$ and AA' and the possession by $A'A$ of a square root. On page 225 of [1] it is conjectured that these two conditions are sufficient. The conjecture is true, and, moreover, the square root condition is redundant.

In view of the algebraic nature of the investigation, it is to be expected that any algebraically closed field of characteristic $\neq 2$ is acceptable. The case of characteristic 2 is indeed different and will not be examined in this paper.

There is one more note before the formal statement of the theorem. The hypothesis that $A'A$ is similar to AA' will be weakened to the hypothesis that $(A'A)^m$ and $(AA')^m$ have the same rank for all m. This is not being done for the sake of generalization, but rather because the rank condition is trivially inherited by the direct summands that we shall encounter below. In any event the generalization is nominal, for in the nilpotent case (the only case that matters) one can see a priori that the rank condition implies similarity.

THEOREM. *Let A be a square matrix over an algebraically closed field of characteristic $\neq 2$. Then A can be written QS, with Q orthogonal and S symmetric, if and only if $(A'A)^m$ and $(AA')^m$ have the same rank for every m.*

2. Three lemmas. The lemmas in this section will facilitate the proof of the theorem.

As far as possible, we shall operate in a basis-free fashion. This calls for the following setup. We assume given an algebraically closed field k of characteristic $\neq 2$ and two linear spaces V and W of the same dimension over k, each carrying a nonsingular symmetric

---

* Received by the editors November 14, 1988; accepted for publication (in revised form) March 15, 1989.

† Mathematical Sciences Research Institute, 1000 Centennial Drive, Berkeley, California 94720. This work was supported by National Science Foundation grant DMS 8505550.

bilinear form. For both V and W we use the notation $(,)$ for the form. We are given a linear transformation A from V to W. It has a transpose A' that maps W into V. We have $(Av, w) = (v, A'w)$ for all v in V, w in W. Our objective is to write $A = QS$, where S is symmetric on V and Q maps V orthogonally into W, that is, $(Qv_1, Qv_2) = (v_1, v_2)$ for all v_1, v_2 in V. When this is so, we have $A' = SQ'$, $A'A = S^2$; furthermore, the equation $A = QS$ implies that A and S have the same null space. So a necessary condition for $A = QS$ to be achievable is that $A'A$ has a symmetric square root with the same null space as A. This condition is also sufficient. This essentially appears on p. 220 of [1], but for completeness a proof is included.

LEMMA 1. *Let A, V, and W be as above. Suppose that $A'A$ has a symmetric square root S with the same null space as A. Then there exists an orthogonal linear transformation Q from V to W satisfying $A = QS$.*

Proof. We define Q first from the range of S to the range of A by setting $QSx = Ax$. From the equality of the null spaces of S and A we first see that this is a valid definition and then that the mapping is one-to-one. We have

$$(Sx, Sx) = (S^2x, x) = (A'Ax, x) = (Ax, Ax).$$

So Q preserves the bilinear form as far as Q is thus far defined. By Witt's theorem, Q can be extended to an orthogonal mapping from all of V onto all of W.   □

LEMMA 2. *Over an algebraically closed field of characteristic $\neq 2$, let V be a $(2m + 1)$-dimensional linear space with a nonsingular symmetric bilinear form. Let T be a nilpotent symmetric linear transformation on V with elementary divisors of degrees m and $m + 1$. Then T has a symmetric square root with a one-dimensional null space equal to T^mV.*

Proof. It is evident that T has a square root. By [1, Thm. 4] T has a symmetric square root S (while in [1] the ground field is the field of complex numbers, the proof there is valid in this more general context). Necessarily, S has a single elementary divisor of degree $2m + 1$ and so it has a one-dimensional null space equal to $S^{2m}V = T^mV$.   □

LEMMA 3. *Over an algebraically closed field of characteristic $\neq 2$ let V be a $2m$-dimensional linear space with a nonsingular symmetric bilinear form. Let T be a nilpotent symmetric linear transformation on V with elementary divisors of degrees m and m. Let u be a vector in V with $T^{m-1}u \neq 0$, $(T^{m-1}u, u) = 0$. Then T has a symmetric square root with a one-dimensional null space spanned by $T^{m-1}u$.*

Proof. Take a basis $x_1, \cdots, x_m, y_1, \cdots, y_m$ of V with each $(x_i, y_i) = 1$ and the form vanishing on all other pairs of basis elements. Let U be the linear transformation given by

$$Ux_i = x_{i+1}(i = 1, \cdots, m-1), \qquad Ux_m = 0,$$

$$Uy_i = y_{i-1}(i = 2, \cdots, m), \qquad Uy_1 = 0.$$

Note that U is nilpotent and that it has the same elementary divisors as T: thus U and T are similar. One readily checks that U is symmetric. From this, one knows that U and T are orthogonally similar. So V admits a basis of the same kind relative to T, and we shall use the same notation x_i, y_i for this basis. $T^{m-1}V$ is two-dimensional, spanned by x_m and y_1. So $T^{m-1}u$ has the form $ax_m + by_1$. In computing $(T^{m-1}u, u)$, the only portions of u that make a contribution are the terms in x_1 and y_m, and this part of u must have the form $ax_1 + by_m$. Since $(ax_m + by_1, ax_1 + by_m) = 2ab$, we must have $a = 0$ or $b = 0$ in order for $(T^{m-1}u, u)$ to vanish. Thus $T^{m-1}u$ is a scalar multiple of x_m

or y_1. By symmetry we can suppose that $T^{m-1}u$ is a scalar multiple of x_m. We now define S as having the single Jordan block

$$y_m, x_1, y_{m-1}, x_2, \cdots, y_2, x_{m-1}, y_1, x_m,$$

that is, S sends each of these vectors to the next and sends x_m to 0. It is routine to check that S is symmetric. We have $S^2 = T$, and the null space of S is spanned by x_m, and hence by $T^{m-1}u$. □

3. Reduction to the case where $A'A$ is nilpotent. We return to A, V, and W as in § 2. In this section we shall reduce the problem of achieving an algebraic polar decomposition of A to the case where $A'A$ is nilpotent.

It is standard that V has a unique direct sum decomposition $V = V_1 + V_2$ with each V_i invariant under $A'A$, $A'A$ invertible on V_1, and $A'A$ nilpotent on V_2. Write $W = W_1 + W_2$ for the analogous decomposition of W relative to AA'. There are three things to be checked:

(a) A maps V_2 into W_2 and A' maps W_2 into V_2.

(b) A maps V_1 one-to-one onto W_1 and A' maps W_1 one-to-one onto V_1.

(c) The two decompositions are orthogonal relative to the bilinear forms.

By symmetry only half of each of these statements needs proof. The following characterization of the summands V_1 and V_2 will be used: V_2 consists of the elements annihilated by some power of $A'A$, and V_1 consists of the elements lying in the range of large powers of $A'A$ (see p. 113 of [2] for details concerning this).

(a) If $x \in V_2$, then $(A'A)^n x = 0$ for some n, $(AA')^n Ax = 0$, $Ax \in W_2$.

(b) Suppose that $x \in V_1$ and $Ax = 0$. Then $A'Ax = 0$ and $x = 0$, since $A'A$ is invertible on V_1. Thus A induces a one-to-one map of V_1 into W_1. In particular, $\dim W_1 \geqq \dim V_1$. By symmetry the opposite inequality also holds. Thus $\dim V_1 = \dim W_1$ and it follows that the map A of V_1 into W_1 is onto as well as one-to-one.

(c) To establish the orthogonality of v_1 and v_2 for $v_i \in V_i$, we take n large enough so that $(A'A)^n v_2 = 0$ and v_1 is in the range of $(A'A)^n$, say $v_1 = (A'A)^n v'$. Then $(v_1, v_2) = ((A'A)^n v', v_2) = (v', (A'A)^n v_2) = 0$.

We begin the proof of the theorem. For an invertible linear transformation the QS decomposition is known. Hence our business is finished regarding V_1 and W_1. It remains to treat the restrictions of A and A' to V_2 and W_2, but first we have to observe that the hypothesis of the theorem is inherited. Since rank is additive on direct sums, we see that, for every m, $(A'A)^m$ on V_2 and $(AA')^m$ on W_2 have the same rank. We change notation, replacing V_2 and W_2 by V and W. Henceforth $A'A$ and AA' are nilpotent.

4. The case where $A'A$ is nilpotent. The procedure will be to detach a well-behaved direct summand. The idea is not new; for instance, it appears in essence in [3]. In the present context appropriate modifications are needed to cope with two vector spaces.

We form the longest product

$$\cdots A'AA'A$$

that is nonzero and call it B. Let r be the number of terms in this product. The parity of r makes a difference.

r even. $B = (A'A) \cdots (A'A)$. It cannot be the case that (Bx, x) vanishes for all x in V, for then $B = 0$ by polarization (this uses characteristic $\neq 2$). Choose x in V with $(Bx, x) \neq 0$. If $C = (AA')^{r/2}$, then B and C have the same rank by hypothesis and in particular $C \neq 0$. Thus we have y in W with $(Cy, y) \neq 0$. We line up the following $r + 1$ elements of V:

$$x_0 = x, \qquad x_1 = A'y, \qquad x_2 = A'Ax, \qquad x_3 = A'AA'y, \cdots$$

$$x_{r-1} = A'(AA')^{\frac{r}{2}-1} y, \qquad x_r = (A'A)^{\frac{r}{2}} x = Bx.$$

For $i + j = r$ we have $(x_i, x_j) = (Bx, x)$ for i even and (Cy, y) for i odd. Thus (x_i, x_j) is nonzero for $i + j = r$. Because of the maximal property of r, (x_i, x_j) vanishes for $i + j > r$. So the matrix of elements (x_i, x_j) has nonzero elements on the antidiagonal (the diagonal running from the upper right corner to the lower left corner) and 0's to the right; it is an invertible matrix. As is well known, this implies that the elements $x_0, \cdots, x_r$ are linearly independent. Furthermore, the bilinear form is nonsingular on the subspace X, which they span [4, Thm. 1 on p. 4], and X is an orthogonal direct summand of V [4, Thm. 2 on p. 6]. The construction is now repeated on the other side of the ledger, producing elements $y_0 = y$, $y_1 = Ax$, $y_2 = AA'y$, $\cdots$, $y_r = Cy$, which form a basis of an orthogonal direct summand Y of W. A sends x_i into y_{i+1} for $i = 0, \cdots, r - 1$ and annihilates x_r. The null space of A is one-dimensional, spanned by x_r. $A'A$ is a symmetric linear transformation on X admitting the two Jordan blocks $x_0, x_2, x_4, \cdots, x_r$ and $x_1, x_3, x_5, \cdots, x_{r-1}$. The range of $(A'A)^{r/2}$ on X is spanned by x_r. Lemma 2 applies with $T = A'A$ and $m = r/2$. The symmetric square root of $A'A$ thus obtained has $(A'A)^m X$ as its null space, that is, its null space is spanned by x_r and thus coincides with the null space of A. This gets us ready to apply Lemma 1 to A and A', restricted to X and Y, so that the QS decomposition is achieved there. Let X' and Y' be the orthogonal complements of X and Y in V and W. Then A sends X' into Y', A' sends Y' into X', and A and A' remain transposes when restricted to X' and Y'. The proof of the theorem is finished by induction as soon as we observe that the hypothesis of the theorem is inherited; this is seen by additivity of the rank just as at the end of the preceding section.

r odd. $B = A(A'A) \cdots (A'A)$. The procedure is similar but has to be changed a little, since B now sends V into W. We select x in V and y in W simultaneously to satisfy $(Bx, y) \neq 0$. The elements x_i and y_i are picked in essentially the same way as before:

$$x_0 = x, \qquad x_1 = A'y, \qquad x_2 = A'Ax, \cdots, \qquad x_{r-1} = (A'A)^{(r-1)/2} x,$$

$$x_r = A'(AA')^{\frac{(r-1)}{2}} y = By,$$

$$y_0 = y, \qquad y_1 = Ax, \qquad y_2 = AA'y, \cdots, \qquad y_{r-1} = (AA')^{(r-1)/2} y,$$

$$y_r = A(A'A)^{\frac{(r-1)}{2}} x = B'x.$$

The passages to the subspaces X and Y go as before. In achieving the polar decomposition of A restricted to X, we use Lemma 3 in place of Lemma 2. The details are as follows. The null space of A is spanned by x_r. $T = A'A$ admits the two Jordan blocks $x_0, x_2, \cdots, x_{r-1}$ and $x_1, x_3, \cdots, x_r$. With the choices $u = x_1$ and $m = (r + 1)/2$ the hypotheses of Lemma 3 are in place. This makes Lemma 1 applicable and achieves the polar decomposition of A restricted to X.

Summary. The desired polar decomposition of the original linear transformation A has been achieved in a succession of steps. First, the portion where $A'A$ is invertible was detached in § 3. Then the summand where $A'A$ is nilpotent was decomposed into a number of pieces, and on each of these, special circumstances made $A = QS$ possible, with the aid of suitable lemmas from § 2. In this way the proof of the theorem is complete.

REFERENCES

[1] D. CHOUDHURY AND R. A. HORN, *A complex orthogonal-symmetric analog of the polar decomposition*, SIAM J. Algebraic Discrete Methods, 8 (1987), pp. 218–225.
[2] P. R. HALMOS, *Finite-Dimensional Vector Spaces*, Springer-Verlag, Berlin, New York, 1974.
[3] I. KAPLANSKY, *Orthogonal similarity in infinite-dimensional spaces*, Proc. Amer. Math. Soc., 3 (1952), pp. 16–25.
[4] ———, *Linear Algebra and Geometry—A Second Course*, Chelsea, New York, 1974.

Afterthought

This paper pretty well speaks for itself. But as an afterthought I offer three salutes and three remarks. The salutes:

To Dipa Choudhury and Roger Horn for proposing an intriguing problem.

To Lajos László who in his review (MR 88b: 15011) devoted a final paragraph to stating the problem. As a matter of routine I browse *MR* fairly carefully, but it is virtually impossible to browse all journals with similar care. Without this review, I might never have noticed the problem.

To *Mathematical Reviews* for the important service if provides to all mathematicians.

The remarks:

I filled half a notebook with computations in characteristic 2, but nothing emerged meriting public scrutiny.

With something approaching the speed of light the theorem (or at least a statement of the theorem) has already appeared in the book [1]. See pages 476 and 489.

In [2] the theorem is proved in matrix style and many related results are added.

References

1. Roger Horn and Charles Johnson, *Topics in Matrix Analysis*, Cambridge Univ. Press, Cambridge, 1991.
2. Roger Horn and Dennis Merino, Contragredient equivalence: A canonical form and some applications; to appear in *Lin. Alg. and Appl.*

Reprinted from JOURNAL OF ALGEBRA
All Rights Reserved by Academic Press, New York and London

Vol. 133, No. 2, September 1990
Printed in Belgium

Nilpotent Elements in Lie Algebras

IRVING KAPLANSKY

*Math Sciences Research Institute, 1000 Centennial Drive,
Berkeley, California 94720*

Communicated by Susan Montgomery

Received August 25, 1988

DEDICATED TO THE MEMORY OF YITZ HERSTEIN

1. INTRODUCTION

Let k be the field of q elements. Then the number of nilpotent n by n matrices with entries in k is a power of q, and the power in question is $n^2 - n$. This very pretty result was proved by Yitz in collaboration with Nat Fine [1]. The proof involved a sum taken over all partitions of n. This rather complicated computation was averted in proofs promptly furnished by Reiner [4] and by Gerstenhaber [2]. Still another proof will be sketched at the end of this paper.

Any nilpotent matrix has trace 0. So there is no harm in restricting consideration to the n by n matrices of trace 0. Now we have a simple Lie algebra, at least if n is not divisible by the characteristic of k. The exponent $n^2 - n$ is identifiable as the dimension of this Lie algebra minus its rank, the dimension being $n^2 - 1$ and the rank $n - 1$. We are ready to ask whether a similar result holds for any simple Lie algebra.

For the $A-G$ algebras, i.e., the analogues of the simple Lie algebras in characteristic 0, this has been investigated. The results are affirmative, except that certain "bad" primes have not yet been mastered. First, Steinberg [6] proved the analogous result for the unipotent elements of the corresponding algebraic groups. Then Springer [5] transferred the results to the Lie algebras. (I am indebted to Robert Steinberg for this reference.)

That leaves the additional simple Lie algebras that one has in characteristic p, beginning with the Witt algebra.

467

0021-8693/90 $3.00

2. THE WITT ALGEBRA

I take the Witt algebra in its realization as the Lie algebra of derivations of $k[x]$, $x^p = 0$. We then have the expected result. (The dimension is p and the rank 1, so that the dimension minus the rank is $p - 1$.)

THEOREM.   *Let k be the field of q elements, where q is a power of p. Let A be the (commutative associative) algebra given by $k[x]$ with $x^p = 0$. Then A has q^{p-1} nilpotent derivations.*

Proof.   The proof proceeds in two steps, the first of which is to prove that the characteristic polynomial of a derivation D has the form $D^p + rD$. This follows from a brief argument if the characteristic roots of D are distinct, as we shall see below. We get around this obstacle in the time honored way by looking first at a generic derivation.

Extend the base field to $k[y_1, ..., y_p]$, where the y_i's are independent indeterminates, and then to L, the algebraic closure of this field. Let B be the algebra obtained from A by extending the base field from k to L. A derivation on B is free to take any value on x. Thus

$$Ex = y_1 1 + y_2 x + \cdots + y_p x^{p-1}$$

defines a derivation of B. If E has a repeated characteristic root then, by specialization, the same is true for any derivation of A. However, the derivation of A sending x into x has characteristic roots $0, 1, ..., p-1$. Therefore the characteristic roots of E are distinct. It follows that the characteristic vectors of E span B. There must be at least one characteristic vector carrying a nonzero coefficient of x. Write z for such a characteristic vector. Then z can replace x as a generator of B; it will not necessarily be the case that $z^p = 0$ but z^p will be a scalar. Write $Ez = sz$. Then

$$Ez^i = iz^{i-1} \cdot Ez = siz^i \qquad (i = 1, ..., p-1).$$

Next $E^p z^i = s^p i^p z^i = s^p i z^i$. Thus $E^p - s^{p-1} E = 0$. This says that the coefficients of the characteristic polynomial of E all vanish except for the linear term. This is a statement that certain polynomials in the y's vanish, an assertion that is preserved under specialization. Hence any derivation D of the algebra A has a characteristic polynomial of the form $D^p + rD$, as asserted above.

We turn to the task of counting the nilpotent derivations of A. For this we shall use the customary basis of the Witt algebra (in the version that fits the choice $x^p = 0$). Let D_i be the derivation sending x into x^{i+1} ($i = -1$,

$0, ..., p-2$). Thus D_{-1} sends x into 1, D_0 sends x into x, ..., D_{p-2} sends x into x^{p-1}. The general derivation of A has the form

$$F = a_{-1}D_{-1} + a_0 D_0 + \cdots + a_{p-2}D_{p-2}.$$

Relative to the basis $1, x, ..., x^{p-1}$, the matrix of F is

$$\begin{pmatrix} 0 & 0 & 0 & \cdots & 0 & 0 \\ a_{-1} & a_0 & a_1 & \cdots & a_{p-3} & a_{p-2} \\ 0 & 2a_{-1} & 2a_0 & \cdots & 2a_{p-4} & 2a_{p-3} \\ & & & \cdots & & \\ 0 & 0 & 0 & \cdots & (p-1)a_{-1} & (p-1)a_0 \end{pmatrix}.$$

For visual clarity I exhibit this matrix for $p = 5$, replacing $a_{-1}, ..., a_3$ by a, b, c, d, e:

$$F = \begin{pmatrix} 0 & 0 & 0 & 0 & 0 \\ a & b & c & d & e \\ 0 & 2a & 2b & 2c & 2d \\ 0 & 0 & 3a & 3b & 3c \\ 0 & 0 & 0 & 4a & 4b \end{pmatrix}.$$

Of course, the proof that follows is general. We showed above that the characteristic polynomial of F has the form $F^5 + rF$, and evidently r is the lower right 4 by 4 determinant

$$r = \begin{vmatrix} b & c & d & e \\ 2a & 2b & 2c & 2d \\ 0 & 3a & 3b & 3c \\ 0 & 0 & 4a & 4b \end{vmatrix}.$$

The condition for F to be nilpotent is $r = 0$. We distinguish two cases.

Case I. $a = 0$. Then $b = 0$. The elements c, d, and e are at liberty and there are q choices for each, for a total of q^3 (q^{p-2} for general p).

Case II. $a \neq 0$. There is just one term involving e and it carries the non-zero coefficient $2a \cdot 3a \cdot 4a$. In making r vanish there are q^3 choices for b, c, and d, $q - 1$ choices for a, and then e is determined. The contribution is thus $q^3(q-1)$ (or $q^{p-2}(q-1)$ for general p).

The grand total is $q^3 + q^3(q-1) = q^4$. For general p one gets $q^{p-2} + q^{p-2}(q-1) = q^{p-1}$. With this the proof of the theorem has concluded.

3. JORDAN ALGEBRAS

Usually, when something works for Lie algebras something very similar is valid for Jordan algebras. However, I gave up when I found that the number of symmetric nilpotent 3 by 3 matrices over the field of q elements is $q^3 + q^2 - q$. Maybe some reader will see where to go from there.

4. ANOTHER PROOF OF THE FINE–HERSTEIN THEOREM

To the four proofs already in the literature I add a fifth, but I shall present only a sketch.

I set out to give an inductive proof in as brutally direct a fashion as I could conceive. Let T be a nilpotent linear transformation on an n-dimensional vector space. Pick a vector annihilated by T for the last basis vector. Then in the matrix representing T the last row is 0. The upper left $(n-1)$ by $(n-1)$ corner is nilpotent and a count of these is known by induction. The $n-1$ remaining entries in the last column are at liberty. The difficulty is that if the null space of T has dimension two or more we will have counted T many times. This calls for the familiar inclusion–exclusion process and I was led to the identity

$$\sum_{r=0}^{n} (-1)^r q^{r(r-1)/2} F(n, r) = 0,$$

where $F(n, r)$ is the number of r-dimensional subspaces of an n-dimensional vector space over the field of q elements. Later I found this identity in the literature, as a lemma in the solution to a *Monthly* problem [3]. With this identity in hand, the details are quite routine.

5. EPILOGUE

This is how the paper stood till I got the referee's report, which contained a better proof of the theorem. I am letting my proof stand since the special form of the characteristic polynomials of the derivations may have some separate interest.

Here is the referee's proof. There are q^{p-2} nilpotent derivations with Dx nilpotent; this is the case $a = 0$ above. Let D be a nilpotent derivation with Dx non-nilpotent. Take $D^{p-2}x^{p-1}$ and let y be the result of deleting its constant term and normalizing its coefficient of x to be 1. Since $D^2y = 0$ and D has rank $p - 1$, one sees that Dy is a constant, necessarily nonzero.

There is a one-to-one correspondence between these derivations and pairs y, Dy. We have q^{p-2} choices for y and $q-1$ for Dy, resulting in $q^{p-2}(q-1)$ choices for the pair. Just as above, we add q^{p-2} and $q^{p-2}(q-1)$ to get q^{p-1}.

REFERENCES

1. N. J. FINE AND I. N. HERSTEIN, The probability that a matrix be nilpotent, *Illinois J. Math.* 2 (1958), 499–504.
2. M. GERSTENHABER, On the number of nilpotent matrices with coefficients in a finite field, *Illinois J. Math.* 5 (1961), 330–333.
3. M. HOFFMAN, Solution of problem 6407, *Amer. Math. Monthly* 91 (1984), 315–316.
4. I. REINER, On the number of matrices with given characteristic polynomial, *Illinois J. Math.* 5 (1961), 324–329.
5. T. A. SPRINGER, The unipotent variety of a semi-simple group, *in* "Algebraic Geometry, Internat. Colloq., Tata Inst., Bombay, 1968," pp. 373–391, Oxford Univ. Press, London, 1969.
6. R. STEINBERG, Endomorphisms of linear algebraic groups, *Mem. Amer. Math. Soc.* 80 (1968).

Printed by Catherine Press, Ltd., Tempelhof 41, B-8000 Brugge, Belgium

Afterthought

I have three afterthoughts.

1. I missed the relevant reference [7].

2. When Tonni Springer visited MSRI in the fall of 1990, I learned that he too had looked at the Jordan case. He went much further than I did and found that the number of nilpotent elements is an appropriate power of q except in the case of symmetric matrices (the very case I gave up on). This work is unpublished.

3. On page 455 of [6] I noted the reference to [1]; this is a paper carrying out the enumeration of nilpotent matrices of fixed rank. In the bibliography of [1] I in turn found a reference to [4]; this is devoted to enumerating matrices satisfying any given polynomial—a more general problem. (However, it should be noted that Hodges' answer involves a product over partitions which needs to be disentangled to recover the Fine-Herstein result.) The Hodges paper was received July 1, 1957, just 11 days before the Fine-Herstein paper.

I found it surprising that I had not previously encountered [4]. This inspired me to try a little search via the Citation Index, to see what might show up. There were a number of papers whose connection with the main problem was peripheral. I shall not list them here. One pertinent paper [5] did surface. I shall state Kovacs's result in the Fine-Herstein probability style; this does make the formula concise. Let $A_1, A_2, \ldots, A_k$ be $n \times n$ matrices over the field of q elements; then the probability that $A_1 A_2 \ldots A_k$ is nilpotent is $1 - (1 - q^{-n})^k$. By putting $k = 1$, one recovers the Fine-Herstein theorem.

The paper [2] is an expository article concerning problems of this kind. It has a large bibliography.

Harald Niederreiter called my attention to the related paper [3]. It and [4] can both be found in the large bibliography of [6].

References

1. D. Bollman and H. Ramírez, On the number of nilpotent matrices over Z_n, *J. reine angew. Math.* 238(1969), 85–88.

2. _____,On the enumeration of matrices over finite commutative rings, *Amer. Math. Monthly* 76(1969), 1019–23.

3. J. V. Brawley and G. L. Mullen, A note of equivalence classes of matrices over a finite field, *Internatl. J. of Math. and Math. Sci.* 4(1981), 279–87.

4. J. H. Hodges, Scalar polynomials for matrices over a finite field, *Duke Math. J.* 25(1958), 291–96.

5. A. Kovacs, Some enumeration problems for matrices over a finite field, *Lin. Alg. Appl.* 94(1987), 223–26.

6. R. Lidl and H. Niederreiter, Finite fields, *Encyclopedia of Mathematics and Its Applications*, vol. 20, Addison-Wesley, Reading, MA, 1983.

7. T. A. Springer, A formula for the characteristic function of the unipotent set of a finite Chevalley group, *J. of Algebra* 62(1980), 393–99.

Other Writings

The Euclidean Algorithm

In this note I treat the Euclidean algorithm "locally," that is, one element or one ideal at a time. Throughout, R is a commutative domain with unit.

PROPOSITION 1. *Let $x \in R$ be nonzero and a nonunit. Then the following statements are equivalent*: (a) *every ideal containing x is principal, and* (b) (x) *is a product of principal maximal ideals.*

The easy proof is left to the reader.

Following Motzkin and Samuel, I attach "levels" to some elements of R. 0 gets level -1. Units get level 0. x gets level 1 if it is not 0 or a unit and for every $a \in R$ there exist q and r with $a = qx + r$ and $r = 0$ or a unit. Let α be an ordinal and assume that levels have been assigned for ordinals less than α. Then x has level α if it does not have level less than α, and for any $a \in R$ there exist q and r where $a = qx + r$ and r has a level less than α.

PROPOSITION 2. *If an ideal contains a nonzero element with a level, it is principal.*

The proof is the obvious one. Let I be the ideal. Among all nonzero elements in I with a level, pick x with smallest level. The claim is that $I = (x)$. If not, pick $a \in I$, $a \notin (x)$, and write $a = qx + r$ as above. Then $r \neq 0$, $r \in I$, and r has smaller level than x, a contradiction.

On putting together Propositions 1 and 2, we get the following:

PROPOSITION 3. *If x is not 0 or a unit and has a level, then (x) is a product of principal maximal ideals.*

For elements of finite level we can make a quantitative sharpening.

PROPOSITION 4. *Suppose that x is not 0 or a unit and that x has finite level n. Then (x) is a product of n or fewer principal maximal ideals.*

Proof. By Proposition 3 we know that (x) is a product of principal maximal ideals. Suppose that the number of factors exceeds n. Then we can write $x = p_1 \dots p_n t$ with $(p_1), \dots, (p_n)$ maximal and t a nonunit. Write $p_1 \dots p_n = qx + r$ with level $(r) \leq n - 1$. Then either r is a unit or, by induction, (r) is a product of $n - 1$ or fewer principal maximal ideals. In either case we contradict the fact that r is divisible by $p_1 \dots p_n$.

Factorial Monoids

In [3] Nagata proved a useful theorem stating that if one obtains a factorial domain on inverting some of the principal primes in a Noetherian domain R, then R itself is factorial. See [2, pp. 35–36] for more on the subject.

I note here that the argument is essentially multiplicative, and thus extends to a cancellation monoid (a monoid is a semigroup with unit). I might let it go with that, but I will add some details.

REMARK. Everything in this note is commutative.

Let M be a cancellation monoid. A nonunit p in M is a principal prime if $p|ab$ implies $p|a$ or $p|b$. M is factorial if every nonunit is a product of principal primes; it follows that the decomposition is unique up to order and insertion of units.

We need some sort of chain condition. The weakest that suffices can be described one principal prime at a time: p is Archimedean if no element of M is divisible by all p^n (alternatively, $\bigcap p^n M$ is empty).

THEOREM. *Let $\{p_i\}$ be a set of Archimedean principal primes in the cancellation monoid M. Let N be the result of adjoining to M the inverses of the p_is. Assume that N is factorial. Then M is factorial.*

The proof needs no new thought. In brief: Take the principal primes of N and cleanse them of (positive or negative) powers of the p_is. These, together with the p_is, form a set of principal primes that work for M.

In the spirit of an unpublished portion of [1], I continue with an analysis of an arbitrary cancellation monoid M (no Archimedean assumptions). Adjoin inverses of all principal primes in M. New principal primes may appear. Continue, possibly ad transfinitum. Ultimately we reach a monoid with no principal primes.

With suitable additional assumptions one can acquire structure theorems. Here is a sample. Assume that M has no units other than 1, and that it has exactly one principal prime p, which is not Archimedean. Assume further that $M[p^{-1}]$ is factorial with exactly one principal prime. Then M can be described as follows: It consists of 1 and elements $p^i q^j$ where $j \geq 0$ and i is any integer, except that when $j = 0$ we have $i \geq 0$. A discerning reader will recognize in M the multiplicative monoid modulo units of a rank-2 discrete valuation domain.

References

1. E. G. Evans, *Studies on Commutative Rings*, Ph.d. thesis, University of Chicago, 1969.
2. R. M. Fossum, *The Divisor Class Group of a Krull Domain*, Erg. der Math. 74, Springer, 1973.
3. M. Nagata, A remark on the unique factorization theorem, *J. Math. Soc. Japan* 9(1957), 143–45.

Partally Ordered Sets and the
Burali-Forti Paradox

When I read the paper by Zwicker (reference below) I was entertained, but I put it aside as just another entry in the ever-growing list of self-referencing paradoxes. Suddenly it occurred to me that implicit in the paper was a version of Burali-Forti for partially ordered sets. I find this version to be closer to mainstream mathematics and worth recording.

In the interest of uniform terminology I am pushing the use of the terms Artinian and Noetherian for partially ordered sets, standing for the descending (resp. ascending) chain condition. In this language a well-ordered set is an Artinian chain.

Now for the paradox. I start with the brazenly illegal act of forming B: the partially ordered set obtained by placing all Artinian partially ordered sets side by side. Add a top element to B, getting A. It is evident that B and A are Artinian. Therefore A contains a carbon copy of itself, with a top element that is properly smaller. The process can be iterated indefinitely, yielding a forbidden infinite descending sequence.

The argument can be legalized by sticking to Artinian partially ordered sets with cardinal less than a given cardinal. This is a novel way to create a larger cardinal.

Reference

William S. Zwicker, Playing games with games, *Amer. Math. Monthly* 94(1987), 507–14.

Comments on Prime Ideals in Nonassociative Rings

It is standard that in an associative ring with no nilpotent ideals the intersection of the prime ideals is 0. This does not need associativity. Some care is needed in formulating the theorem and its proof without associativity. It is a nuisance to work with products of three or more elements or ideals; this can be obviated by taking as hypothesis the absence of ideals with square 0. Also, one must note that the product of two ideals need not be an ideal.

THEOREM. *Let R be a ring (there is no assumption of associativity or any substitute for it). Assume that R does not contain a nonzero ideal with square 0. Then the intersection of the prime ideals in R is 0.*

Proof. Given $x \neq 0$ in R, we must construct a prime ideal excluding x. We shall build a certain descending chain of finitely generated nonzero ideals. Take I_1 to be the ideal generated by x. With I_n at hand, take I_{n+1} to be a nonzero finitely generated ideal inside the ideal generated by the square of I_n. Let P be an ideal maximal with respect to not containing any of the I_ns; note that this is a legal Zornification. We proceed to prove that P is prime. Suppose on the contrary that we have ideals J, K not contained in P such that $JK \subset P$. Since J and K can be replaced by $J + P$ and $K + P$, we can assume that J and K properly contain P. It follows from the maximality of P that J and K each contain one of the I_ns. Say $J \supset I_r$, $K \supset I_s$, $r \leq s$. Then J and K both contain I_s. It follows that $P \supset JK \supset I_s^2$ and then that P contains the ideal generated by I_s^2, which in turn contains I_{s+1}. This is a contradiction.

In groups we can repeat the proof virtually verbatim, with normal subgroups playing the role of ideals and commutator as product. This is noted in effect in [1] and [2].

Whenever we have two parallel streams of thought, sooner or later someone is going to try to unify them. I propose to perform this shotgun wedding right now, using multiplicative lattices.

There is a substantial literature on multiplicative lattices, usually assuming both commutativity and associativity. Let us drop both. The object to be studied is a complete lattice L with an auxiliary product. We write 0 for the bottom element of L. We impose three axioms, the first two of which are basic:

245

$$xy \leq x \cap y,$$

$$(x \cup y)z = xz \cup yz, \quad z(x \cup y) = zx \cup zy.$$

In the third axiom we abstract the notion of "finitely generated" and get ready for Zorn's lemma. We say that x is finitely generated if whenever there is a chain $\{y_i\}$ with $\bigcup y_i \geq x$ then some $y_i \geq x$. The third axiom asserts that for any nonzero element u there exists a finitely generated nonzero element x with $x \leq u$.

Define p to be prime if $xy \leq p$ implies $x \leq p$ or $y \leq p$. We are now ready for the theorem: If L satisfies the three axioms, and $x^2 = 0$ in L implies $x = 0$, then 0 is an intersection of prime elements. The proof is the same.

On can go on to further topics, for instance the existence of minimal primes, conditions implying that there are only finitely many minimal primes, and so on. But I will stop at this point.

References

1. David Murdoch and Oystein Ore, On generalized rings, *Amer. J. of Math.* 63(1941), 73–86.
2. Eugene Schenkman, The similarity between the properties of ideals in commutative rings and the properties of normal subgroups of groups, *Proc. Amer. Math. Soc.* 9(1958), 375–81.

Nilpotent and Unipotent Elements in Rickart Rings

In Lemma 1 on page 216 of the reference of this paper's end, I encountered the following interesting fact: If A, B, and $A + B$ are nilpotent matrices over a field, then $\text{Trace}(AB) = 0$. The proof is computational, making use of the Jordan canonical form.

It occurred to me that the statement makes sense in a factor of Type II_1, and I soon noted that it is in fact true by the following less computational proof: A^2, B^2, and $(A + B)^2 = A^2 + B^2 + AB + BA$ are nilpotent and therefore have trace 0. So $\text{Trace}(AB + BA) = 2\,\text{Trace}(AB) = 0$. (Remark to any reader who is fond of characteristic 2: I am aware that I have shortchanged characteristic 2, but that is another story.)

I have of course used the well-known fact that a nilpotent element in a factor of Type II_1 has trace 0 (a good deal more is known). But now I wished to see this in as algebraic a way as possible. Here is the result.

DEFINITION. A left Rickart ring is a ring in which the left annihilator of any element can be generated by an idempotent.

THEOREM 1. *Every nilpotent element in a left Rickart ring is a sum of commutators.*

Proof. The idea underlying the proof is that a nilpotent element in a left Rickart ring is triangular with respect to a set of orthogonal idempotents, and so is expressible as a sum of elements with square 0. I shall not spell this out in detail, but I simply give an inductive proof of the theorem. Assume $a^n = 0$ and let the left annihilator of a^{n-1} be generated by the idempotent e. Then $ea^{n-1} = 0$. Also $aa^{n-1} = 0$, whence $a \in Re$, where R is the ring, and $ae = a$ follows. We make an induction on n. The case where $n = 2$ is disposed of by noting $ea = 0$, $a = ae - ea$. In the general case we write $a = (a - ea) + ea$. Here $(a - ea)^2 = 0$ since $a = ae$, and $(ea)^{n-1} = ea^{n-1} = 0$. Thus $a - ea$ is a commutator, ea is a sum of commutators, and it follows that a is a sum of commuators.

There is a multiplicative analogue. In formulating it something additional must be assumed, as is shown by the example of 2×2 matrices over the field of two elements. My assumption will be the existence of 1/2, but this is not the last word on the subject.

In a ring with 1 an element x is unipotent if $x - 1$ is nilpotent.

THEOREM 2. *Let R be a left Rickart ring with 1 and assume that 2 is invertible in R. Then any unipotent element in R is a product of multiplicative commuators.*

Proof. We assume $a^n = 0$ and $x - 1 = a$, define e as above, and again make an induction on n. The case $n = 2$ again gets first attention. Write $a = -2b$. We know that $1 - 2e$ and $1 + b$ are invertible, with inverses $1 - 2e$ and $1 - b$, respectively. The equation

$$x = (1 - 2e)(1 + b)(1 - 2e)(1 - b)$$

expresses x as a commutator. In verifying this note that $b^2 = 0$, $eb = 0$, and $be = b$. We turn to general n. We have $x = (1 + ea)(1 + a - ea)$. Since $(a - ea)^2$ and $(ea)^{n-1}$ are both 0, the proof is completed by induction.

References

B. Mathes, M. Omladič, and H. Radjavi, Linear spaces of nilpotent matrices, *Linear Alg. and Its Appl.* 149(1991), 215–25.

A Theorem on Graded Algebras

The following theorem arose in clarifying a portion of the classification of four-dimensional nilpotent algebras that appears on pages 99–108 of the book *Nilpotent Rings* by Kruse and Price (Gordon and Breach, 1969). It is possible that there is a phenomenon here worthy of further exploration.

THEOREM. *Let $A = A_1 + A_2 + \cdots$ be a graded algebra generated by A_1. Assume that A_2 is one-dimensional and that A_3 is nonzero. Then A is the direct sum of a trivial algebra (all products 0) in degree 1 and an algebra generated by an element of degree 1.*

Proof. Let J be the two-sided annihilator of A_1 within A_1. Let I be a vector space complement of J within A_1. Then $B = I + A_2 + A_3 + \cdots$ is an ideal and A is an algebra direct sum of J and B. We can assume $J = 0$, and then our task is to prove that A_1 is one-dimensional. We proceed in three steps.

1. It cannot be the case that $x^2 = 0$ for every $x \in A_1$. Suppose this is so. Note that $xy = -yx$ for all $x, y \in A_1$. Take $z \in A_1, z \neq 0$. Since we have normalized the annihilator J to be 0, there exists $w \in A_1$ with $zw \neq 0$. Since zw spans A_2 and $z^2 = 0$, we get $zA_2 = 0$. Thus $A_1A_2 = 0$, which contradicts $A_3 \neq 0$.
2. It cannot be the case that $x^3 = 0$ for every $x \in A_1$. Suppose this is so. Take $y \in A_1$ with $y^2 \neq 0$, so that y^2 spans A_2. Let z be any element of A_1. We have $yz = ay^2$ for a suitable scalar a. Then $y^2z = ay^3 = 0$, $A_2z = 0$. Thus $A_2A_1 = 0$, which contradicts $A_3 \neq 0$.
3. Pick $u \in A_1$ with $u^3 \neq 0$. The claim is that u spans A_1. If v is any element of A_1, then $uv = bu^2$ for a scalar b. Write $w = v - bu$. We have $uw = 0$, $uwA_1 = 0$. It follows that $wA_1 = 0$ since otherwise wA_1 would span A_2, contradicting $u^3 \neq 0$. Next $A_1wA_1 = 0$, yielding $A_1w = 0$ in a similar way. Thus w annihilates A_1 on both sides. Since we have normalized the annihilator J to be 0, we deduce $w = 0$, and $v = bu$, as desired.

The Number of Solutions of $x^3 + y^3 = 1$ in the Integers mod p

This note is entirely expository. I am writing it for the benefit of any people out there who are as ignorant of the facts about to be stated as I was until recently.

On rereading Weil's classical paper [4], I stopped to think about the initial statement that Gauss "obtained the number of solutions for all congruences $ax^3 - by^3 \equiv 1 \pmod{p}$." In the preface to [3] Schmidt says virtually the same thing.

Did Gauss find a formula? Well, not exactly. What he did was to relate the problem in question to another one that looks quite different. We are rapidly approaching the two-hundredth anniversary of the Disquitiones. I blush at the memory of my ignorance of the beautiful results in article 358.

I shall take $a = 1, b = -1$. The problem is of interest only for $p \equiv 1 \pmod 3$. It is convenient to ignore solutions with x or y equal to 0. Write $f(p)$ for the number of solutions, thus diminished, of $x^3 + y^3 = 1$ in the integers mod p.

We know that p is uniquely represented by $u^2 + 3v^2$. It follows that $4p$ is represented by $u^2 + 3v^2$ (not uniquely). But we can do better: $4p$ is uniquely represented by $u^2 + 27v^2$. The uniqueness is up to sign. Let us use the freedom to change the sign of u so as to arrange $u \equiv 1 \pmod 3$. Then: $f(p) = p + u - 8$.

The following is an easy corollary: 2 is a cubic residue of p if and only if p itself is represented by $u^2 + 27v^2$. This famous result is regarded as the first of the higher reciprocity laws.

EXAMPLES. $p = 13$, $52 = 25 + 27$, $u = -5$, $f(p) = 13 - 5 - 8 = 0$. $p = 19$, $76 = 49 + 27$, $u = 7$, $f(p) = 19 + 7 - 8 = 18$. $p = 31 = 4 + 27$, $124 = 16 + 108$, $u = 4$, $f(p) = 31 + 4 - 8 = 27$. 2 is a cubic residue of 31, $4^3 \equiv 2 \pmod{31}$.

Jacobi [2] followed up Gauss and obtained a formula, of a kind. Write $n = (p - 1)/3$. Then

$$\binom{2n}{n} \equiv -u \pmod p,$$

Since $|u| < p/2$, this determines u.

250

The Number of Solutions of $x^3 + y^3 = 1$ in the Integers mod p

EXAMPLES. $p = 13$, $n = 4$,

$$\binom{8}{4} = 70 \equiv 5 \pmod{13},$$

$p = 19$, $n = 6$,

$$\binom{12}{6} \equiv -7 \pmod{19}.$$

I conclude with two remarks.

1. I have exhibited only the tip of a very large iceberg. I wish good luck to any reader who decides to explore more deeply.

2. There is an indefinite sibling for $4p = u^2 + 27v^2$. Let p be a prime $\equiv 1 \pmod{12}$; we allow p to be negative. Then $-2p = u^2 - 27v^2$ uniquely (the uniqueness is up to the insertion of solutions of the Pell equation $x^2 - 27y^2 = 1$). The proof is routine. I did not find this remark in the literature. Also, I am unaware of any connection with other topics.

References

1. L. E. Dickson, *History of the Theory of Numbers*, vol. III, Chelsea reprint, 1952.
2. C. G. J. Jacobi, De residuis cubicis commentation numerosa, *J. für reine und angew. Math.* 2(1827), 66–69. (I found this reference in [1, p. 55].)
3. Wolfgang M. Schmidt, Equations over finite fields: An elementary approach, *Springer Lecture Notes 536*, 1976.
4. André Weil, Numbers of solutions of equations in finite fields, *Bull. Amer. Math. Soc.* 55(1949), 497–505.

Commutativity Revisited

In this note I shall give a historical review and then present two commutativity theorems that I believe are worth recording.

When Stone launched his study of Boolean rings in 1936, he noted the following: After one assumes that every element is idempotent, commutativity is free of charge [9, Theorem 1, p. 39]. What a progeny this has spawned!

Nothing happened until 1945. Then three people, Neal McCoy, Nathan Jacobson, and I, independently took a crack at the obvious generalization where one assumes that $a^n = a$ for all a. McCoy and I made some progress but were outdistanced by Jacobson [3], who went the whole way. In his collected works [5, vol. I, pp. 281–82] he gives a brief historical account.

The abstract [6] shows how far I got. There is a little more to tell, and I shall tell it for a reason. I wrote up what I had and submitted it to the Duke Journal; it was accepted. Then I learned about Jacobson's work, and he sent me a preprint of [3]. I tossed sleeplessly for a night thinking about what to do. The next morning I decided to withdraw my paper. I have never regretted the decision. My reason for relating this story is that perhaps some young mathematician who faces a similar decision may read it. I hope that he or she will similarly decide to withdraw a superseded paper. In the words of Gauss (more or less): Publish just a little, but make it good.

REMARK. Jacobson's paper proved the stronger result that commutativity follows from $a^{n(a)} = a$; a generous footnote credits me with an assist.

The next step in the story, at least as I experienced it, was the following thought. All this is very well, but we are discussing a very special class of rings—roughly speaking, continuous direct sums of finite fields. This could be remedied by assuming that each $a^{n(a)} - a$ is central rather than 0; then all commutative rings would be eligible and we would in fact be characterizing commutativity.

The late Yitz Herstein now takes the stage. When I met him for the first time at a meeting of the American Mathematical Society, I told him about the proposed generalization. The first thing he did was to put me to shame—I had missed how easy the first case is. Assume that every $a^2 - a$ is central. Then in a few lines one derives commutativity; this is set forth at the opening of [1]. It is a splendid exercise for students beginning ring theory. But the very next case ($a^3 - a$ central)

252

is a totally different matter. To this day I know no way of proving commutativity other than invoking all of Herstein's devices, including a reduction to the subdirectly irreducible case. (He once told me that this reduction was an act of desperation.) On this matter of the use of subdirect irreducibility, I am reminded of Wigner's paper on the unreasonable effectiveness of mathematics in the physical sciences. I feel that subdirect irreducibility is unreasonably effective in proving commutativity theorems.

Through four successive stages Herstein climbed to a pinnacle [2]: He proved that commutativity follows if one assumes that for every a there is a polynomial p with integral coefficients such that $a^2 p(a) - a$ is central. Then Jerry Martindale [7] went still further by weakening the assumption to the existence of an element x such that $a^2 x - a$ is central (so-called ξ-rings). This does not imply commutativity (any division ring is eligible) but a lot can be proved. I shall return to ξ-rings shortly.

A polished account of Herstein's penultimate theorem ($a^{n(a)} - a$ central) appears in [4, pp. 220–21].

In Small's collection of reviews in ring theory, commutativity theorems occupy five of the six subsections of section 28. In the 1940–1979 volume this occupies pages 827–842 and comprises 117 papers. In the 1980–1984 volume the MR classification 16A70, which is titled Commutativity Theorems, occupies pages 455–468 and comprises 102 papers. There is an astonishing variety in the conditions that imply commutativity.

REMARK. Quite a few authors make it a point to avoid Zorn's lemma. I am not impressed. One can always drop down to a subring generated by two elements. This is a countable ring, and I feel that there is little to be gained by fussing about the countable axiom of choice.

It would seem that the human race now has enough commutativity theorems, at least for a while. Nevertheless I am about to present two more. Why? There are two reasons. (a) Martindale's results are impressive but, to some extent, they leave ξ-rings twisting slowly in the wind; in Theorem 1 I am able to get closer to something definitive. (b) In doing this I need Theorem 2, which has the merit of using hypotheses that are more in the mainstream of ring theory than things like $a^2 x - a$ being central.

Let R be a subdirectly irreducible ξ-ring. It was initially hoped to prove that R must be a division ring or commutative, but this was defeated by the example on page 720 of [7], an example that comes from [8]. In the opening paragraph of [7] a weaker objective is set: If R is neither a division ring nor commutative, prove that R contains a central ideal P such that R/P is a field. This question remains open. I propose to return to the original objective and achieve it by making the stronger assumption that every subring of R is a ξ-ring. This stronger assumption is still implied by Herstein's $a^2 p(a) - a$ central hypothesis; in that sense it is acceptable.

THEOREM 1. *Let R be a subdirectly irreducible ring. Assume that every subring of R is a ξ-ring. Then R is either a division ring or commutative (of course it can be both).*

REMARK. Things are still not satisfactory. For all I know it may be the case that in Theorem 1 one can prove commutativity. I have not given this a real college try and hope that some reader will take it up. Let me just indicate one type of argument that may help. Take the division ring of ordinary quaternions. Drop to the subring D of integral linear combinations of i, j, k (the Dickson quaternions). Then D is not a ξ-ring: if we reduce D modulo an odd prime we get 2×2 matrices.

Theorem 1 is going to be deduced from the following theorem. (I borrow a term from group theory by referring to a homomorphic image of a subring as a section.) Three conditions play a role in stating Theorem 2:

(i) All commutators are central.
(ii) All nilpotent elements are central.
(iii) All one-sided ideals are two-sided.

THEOREM 2. *If every section of a ring R satisfies* (i), (ii), *and* (iii), *then R is commutative.*

REMARK. For brevity I have put demands on every section, but of course I am aware that (ii) is inherited by subrings, (iii) is inherited by homomorphic images, and (i) is inherited by both.

Proof of Theorem 2. In part the proof follows the path that Herstein made standard, but in part a new idea is needed. We begin by reducing to the case where R is generated by two elements and is subdirectly irreducible. Let S be its minimal ideal. If $S^2 \neq 0$, then S is a simple ring. Assumption (iii) then makes S a division ring, necessarily all of R. It is known that a division ring satisfying (i) is commutative. Thus we may assume that $S^2 = 0$. From assumption (ii) it follows that S is central. Let A be the annihilator of S. Herstein's arguments show that R/A is 0 or a finite field. It remains to prove that A is central, which one does by proving that it is nil. It is here that I have to depart from Herstein. I find it charming that the new idea rests on Lemmas 1 and 2, noncommutative versions of the two crown jewels of commutative ring theory: Emmy Noether's primary decomposition and the Hilbert basis theorem.

LEMMA 1. *Let R be a left Noetherian ring in which the intersection of any two nonzero left ideals is nonzero. Then any right zero-divisor in R is nilpotent.*

Proof. The proof repeats Noether's famous proof that irreducible ideals are primary. Let x be a ring zero-divisor; this means that its left annihilator is nonzero. Let J_n be the left annihilator of x^n. The ideals J_n ascend; say they become stable at r. We claim that $Rx^r \cap J_r = 0$. If y is in this intersection, then $y = zx^r$, $0 = yx^r = zx^{2r}$, $z \in J_{2r} = J_r$, $zx^r = 0$, $y = 0$. Thus the claim is sustained. We know that J_r is nonzero. Hence $Rx^r = 0$ and x is nilpotent.

LEMMA 2. *Let R be a ring generated by two elements and assume* (i). *Then R satisfies the ascending chain condition on two-sided ideals.*

My original proof of Lemma 2 imitated the standard proof of the Hilbert basis theorem, with appropriate variants. Later it was pointed out to me that one can deduce Lemma 2 from the known result that the enveloping algebra of a finite-dimensional Lie algebra over the integers is right and left Noetherian; the Lie algebra used is the Heisenberg algebra. I am therefore omitting my proof, but I have carefully filed it away.

I now resume the thread of the proof of Theorem 2. Lemmas 1 and 2 combine to show that zero-divisors in R are nilpotent. Hence A is nil, and therefore it is central and the proof of Theorem 2 is complete.

Proof of Theorem 1. We assume that R is not a division ring. Let S be the minimal ideal of R. If $S^2 \neq 0$, then S is a simple ring. By [7, Theorem 2] S is a division ring and must equal R, a contradiction. Hence $S^2 = 0$. Then [7, Theorem 3] gives us that R satisfies property (i). Lemma 2 of [7] implies that R satisfies property (ii). It remains to verify that R satisfies property (iii). Let J be a nonzero one-sided ideal in R. It suffices to prove that J contains a nonzero two-sided ideal I, for we pass to R/I and repeat the argument. We shall prove more: that J contains a nonzero central element. Take $a \neq 0$ in J. The central element $a^2x - a$ lies in J. If it is nonzero we are done. So we assume $a^2x - a = 0$. Then ax is idempotent, for by Theorem 1 of [7] a and x commute. In a ring satisfying (ii) it is known that idempotents are central. We are done unless $ax = 0$. But that implies $a = 0$, a contradiction. With this the proof of Theorem 1 is complete.

I have a final remark. There is a sufficient condition for a ring to be a ζ-ring, which I have not succeeded in fitting into the context: If a ring R has a central ideal I such that R/I is strongly regular, then R is a ζ-ring. A similar statement holds for rings where powers are central: If a ring R has a central ideal I such that R/I is nil, then for every x in R there is a power $x^{n(x)}$ which is central.

References

1. I. N. Herstein, A generalization of a theorem of Jacobson, *Amer. J. of Math.* 73(1951), 756–62.
2. ——————, The structure of a certain class of rings, *Amer. J. of Math.* 75(1953), 864–71.
3. Nathan Jacobson, Structure theory for algebraic algebras of bounded degree, *Ann. of Math.* 46(1945), 695–707.
4. ——————, Structure of rings, *Amer. Math. Soc. Coll. Publ.* vol. 37, revised edition, 1964.
5. ——————, *Collected Mathematical Papers*, 3 volumes, Birkhäuser, Boston, 1989.
6. Irving Kaplansky, The commutativity of generalized Boolean rings, abstract in *Bull. Amer. Math. Soc.* 51(1945), 60.
7. W. S. Martindale, The structure of a special class of rings, *Proc. Amer. Math. Soc.* 9(1958), 714–21.
8. J. E. McLaughlin and A. Rosenberg, Zero divisors and commutativity of rings, *Proc. Amer. Math. Soc.* 4(1953), 203–11.
9. M. H. Stone, The theory of representations for Boolean algebras, *Trans. Amer. Math. Soc.* 39(1936), 37–111.

Permissions

Springer-Verlag is grateful to the AMS and to the Annals of Mathematics for granting permission to reprint the following articles:

American Mathematical Society:

[5] *Solution of the "Problème Des Ménages"*, Bulletin of the American Mathematical Society, Vol. 49, No. 10, pp. 784–785, October, 1943.

[22] *Lattices of Continuous Functions*, Bulletin of the American Mathematical Society, Vol. 53, No. 6, pp. 617–623, June, 1947.

[28] *Rings with a Polynomial Identity*, Bulletin of the American Mathematical Society, Vol. 54, No. 6, pp. 575–580, June, 1948.

[39] *The Weierstrass Theorem in Fields with Valuations*, Proceedings of the American Mathematical Society, Vol. 1, No. 3, pp. 356–357, June, 1950.

[47] *The Structure of Certain Operator Algebras*, Transactions of the American Mathematical Society, Vol. 70, No. 2, pp. 219–255, March 1951.

[56] *Modules over Dedekind Rings and Valuation Rings*, Transactions of the American Mathematical Society, Vol. 72, No. 2, pp. 327–340, March, 1952.

Annals of Mathematics:

[8] *A Contribution to von Neumann's Theory of Games*, Annals of Mathematics, Vol. 46, No. 3, July, 1945.

[42] *Projections in Banach Algebras*, Annals of Mathematics, Vol. 53, No. 2, March, 1951.

[65] *Any Orthocomplemented Complete Modular Lattice is a Continuous Geometry*, Annals of Mathematics, Vol. 61, No. 3, May, 1955.

[68] *Projective Modules*, Annals of Mathematics, Vol. 68, No. 2, September, 1958.

Springer-Verlag would also like to extend my thanks to various other publishers for granting permission to reprint the following articles:

[2]   *Maximal Fields With Valuations*, Duke Mathematical Journal, Vol. 9, No. 2, pp. 303–21, June, 1942 and [58] *Products of Normal Operators*, Duke Mathematical Journal, Vol. 20, No. 2, pp. 257–60, June 1953 are reprinted with permission from Duke University Press.

[25]   *Locally Compact Rings*, American Journal of Mathematics, Vol. LXX, No. 2, April, 1948, is reprinted with permission from The John Hopkins University Press.

[43]   *A Theorem on Division Rings*, Canadian Journal of Mathematics, Vol. III, No. 3, 1951, is reprinted with permission from the Canadian Mathematical Society.

[44]   *A Theorem on Ring Operators*, Pacific Journal of Mathematics, Vol. 1, No. 2, June 1951 pp. 227–232, and [110] *Superalgebras*, Pacific Journal of Mathematics, Vol. 86, No. 1, 1980 pp. 93–98 are reprinted with permission from the Pacific Journal of Mathematics.

[49]   *A Generalization of Ulm's Theorem*, Summa Brasiliensis Mathematicae 2, 1951, pp. 195–202, is reprinted with permission from the Instituto de Mathematica Pura e Aplicada.

[79]   *The Homological Dimension of a Quotient Field*, Nagoya Mathematical Journal, Vol. 27, February, 1966, pp. 139–142 is reprinted with permission of Nagoya University, Japan.

[80]   *Composition of binary quadratic forms*, Studia Mathematica, Vol. 31, 1968, pp. 523–530, is reprinted with permission of the Instytut Matematyczny, Polskiej Akademii Nauk, Warsaw, Poland.

[93]   *Adjacent Prime Ideals*, Journal of Algebra, Vol. 20, No. 1, January 1972, pp. 94–97 and [120] *Nilpotent Elements in Lie Algebras*, Journal of Algebra, Vol. 133, No. 2, September 1990, pp. 467–471, are reprinted with permission of Academic Press Inc.

[119]   *Algebraic Polar Decomposition*, SIAM Journal on Matrix Analysis and Applications, Vol. 11, No. 2, April 1990, pp. 213–217, is reprinted with permission from the SIAM Journal on Matrix Analysis and Applications, Copyright 1990 by the Society for Industrial and Applied Mathematics, Philadelphia, Pennsylavania, with all rights reserved.